Lecture Notes in Computer Science　5538

Commenced Publication in 1973
Founding and Former Series Editors:
Gerhard Goos, Juris Hartmanis, and Jan van Leeuwen

Lecture Notes in Computer Science

5579

Hideyuki Tokuda   Michael Beigl
Adrian Friday   A.J. Bernheim Brush
Yoshito Tobe (Eds.)

Pervasive Computing

7th International Conference, Pervasive 2009
Nara, Japan, May 11-14, 2009
Proceedings

 Springer

Volume Editors

Hideyuki Tokuda
Keio University, Faculty of Environment and Information Studies
5322, Endo, Fujisawa City, Kanagawa 252-8520, Japan
E-mail: hxt@sfc.keio.ac.jp

Michael Beigl
Technische Universität Carolo-Wilhelmina zu Braunschweig
Mühlenpfordtstrasse 23, 38106, Braunschweig, Germany
E-mail: beigl@ibr.cs.tu-bs.de

Adrian Friday
Lancaster University, Computing Department, InfoLab 2
South Drive, Lancaster LA1 4WA, UK
E-mail: adrian@comp.lancs.ac.uk

A.J. Bernheim Brush
Microsoft Research
One Microsoft Way, Redmond, WA 98052-6399, USA
E-mail: ajbrush@microsoft.com

Yoshito Tobe
Tokyo Denki University
Department of Information Systems and Multimedia Design
2–2 Kanda-Nishiki-cho, Chiyoda-ku, Tokyo 101–8457, Japan
E-mail: yoshito_tobe@osoite.jp

Library of Congress Control Number: Applied for

CR Subject Classification (1998): C.2.4, C.3, C.5.3, D.4, H.3-5, K.4, K.6.5, J.7

LNCS Sublibrary: SL 3 – Information Systems and Application, incl. Internet/Web and HCI

ISSN   0302-9743
ISBN-10   3-642-01515-8 Springer Berlin Heidelberg New York
ISBN-13   978-3-642-01515-1 Springer Berlin Heidelberg New York

springer.com

© Springer-Verlag Berlin Heidelberg 2009
Printed in Germany

Typesetting: Camera-ready by author, data conversion by Scientific Publishing Services, Chennai, India
Printed on acid-free paper   SPIN: 12661053   06/3180   5 4 3 2 1 0

Preface

The 7th International Conference on Pervasive Computing (Pervasive 2009) was the first time the conference was held in Asia. Before reaching Asia, the conference made a long journey around the globe: starting in Zürich, traveling to Linz and Munich, then passing through Dublin before leaving Europe for the first time to be held in Toronto, followed by Sydney last year and now Nara, Japan. Over the last few years, Asia has contributed to the topics of Pervasive Computing with research ideas, engineering and many innovative products, so it was wonderful to host the conference in Asia.

When the Pervasive Computing conference series started in 2002, the integration of computing systems into everyday products was just beginning. Seven years later we now see – especially in many parts of Asia – the widespread use of computing technology embedded into our daily lives. Pervasive 2009 focused on the presentation and discussion of novel aspects of architecture, design, implementation, application and evaluation of Pervasive Computing, thus enabling a closer and more frequent use of computing systems. As the field of pervasive computing matures, the Pervasive Conference gains significance worldwide, not only among researchers, but also in industry and general society.

Pervasive 2009 attracted 147 high-quality submissions, from which we accepted 27 papers (a healthy 18% acceptance rate). This year we chose to accept both full papers (up to 18 pages) and shorter "notes" (up to 8 pages). We are pleased to say we received 113 full papers and 34 notes, accepting 20 (18%) and 7 (21%), respectively. Pervasive is a truly international and interdisciplinary conference attracting work from over 450 individual authors from 31 countries, and from a wide array of disciplines, academic and industrial organizations.

A major conference such as Pervasive requires a rigorous and objective process for selecting papers. This starts with the selection of a high-quality Program Committee: we were fortunate to be able to draw on the wisdom and experience of our 28 Program Committee members, from the most prestigious universities and research labs in Europe, North America, and Asia. This committee was aided by the input of no less than 208 external reviewers chosen by the committee on the basis of their domain knowledge and relevance to Pervasive Computing.

We ran a three-phase double-blind review process in which all papers were reviewed by two Program Committee members and two or more external reviewers. This resulted in 559 total reviews (mean 4.2 reviews per paper, exceptionally no less than 3, and as many as 7!). All papers were discussed electronically prior to a face-to-face Program Committee meeting. Additional tertiary Program Committee members were selected to ensure that every paper we discussed had three representatives at the meeting. Over 1.5 days we discussed 60 papers selected on the basis of review score, controversy, or at the wish of any discussant or Program Committee member. Additional readers were solicited during the meeting to assist with the most controversial cases. We thank Microsoft Research and the Computer Lab in Cambridge, UK, for their generous support and assistance in hosting the TPC meeting.

This event would not be possible without the help of many people: Local Arrangements Chair Tsutomu Terada, Demonstrations Chairs Sadao Obana and Guido Stromberg, Video Chairs Takeshi Okadome and Akos Vetek, Workshop Chairs Yasuyuki Sumi and Elaine M. Huang, Doctoral Colloquium Chairs Jakob E. Bardram and Tatsuo Nakajima, Late-Breaking Results Chairs Jin Nakazawa and Susanna Pirttikangas, Publications Chairs Christian Decker, Keiichi Yasumoto and Koji Kamei, Publicity Chairs Alex Varshavsky and Yoshihiro Kawahara, Finance Chairs Kazunori Takashio, Masayoshi Ohashi, and Susumu Ishihara, Tutorials Chair Soko Aoki, Student Volunteers Chair Akimitsu Kanzaki, and Local Arrangements Yasue Kishino and Tomoya Kitani.

Our final thank you also goes to all for your participation and hard work. A conference such as Pervasive 2009 is only possible through the sheer dedication of the community. Without the authors, Program Committee members, and of course, the many domain experts who acted as external reviewers, we would not be able to assemble such a high-quality technical program. It is only with the participation of you, the community, that we can have a fruitful and interesting conference, making this entirely volunteer effort worthwhile.

May 2009

Hideyuki Tokuda
Michael Beigl
A.J. Brush
Adrian Friday
Yoshito Tobe

Organization

Conference Committee

General Chairs	Hideyuki Tokuda (Keio University, Japan) Michael Beigl (Technische Universität Carolo-Wilhelmina zu Braunschweig, Germany)
Program Chairs	A.J. Brush (Microsoft Research, USA) Adrian Friday (Lancaster University, UK) Yoshito Tobe (Tokyo Denki University, Japan)
Local Arrangements Chair	Tsutomu Terada (Kobe University, Japan)
Demonstrations Chairs	Sadao Obana (ATR, Japan) Guido Stromberg (Infineon Technologies, Germany)
Videos Chairs	Takeshi Okadome (NTT, Japan) Akos Vetek (Nokia, Finland)
Workshops Chairs	Yasuyuki Sumi (Kyoto University, Japan) Elaine M. Huang (Motorola Labs, USA)
Doctoral Colloquium Chairs	Jakob E. Bardram (IT University of Copenhagen, Denmark) Tatsuo Nakajima (Waseda University, Japan)
Late-Breaking Results Chairs	Jin Nakazawa (Keio University, Japan) Susanna Pirttikangas (University of Oulu, Finland)
Publications Chairs	Christian Decker (University of Karlsruhe, Germany) Keiichi Yasumoto (NAIST, Japan) Koji Kamei (NTT, Japan)
Publicity Chairs	Alex Varshavsky (AT&T Research Lab, USA) Yoshihiro Kawahara (University of Tokyo, Japan)
Finance Chairs	Kazunori Takashio (Keio University, Japan) Masayoshi Ohashi (ATR, Japan) Susumu Ishihara (Shizuoka University, Japan)

Tutorials Chair	Soko Aoki (Keio University, Japan)
Student Volunteers Chair	Akimitsu Kanzaki (Osaka University, Japan)
Local Arrangements	Yasue Kishino (NTT, Japan)
	Tomoya Kitani (Shizuoka University, Japan)
Web	Hiroki Ishizuka (Tokyo Denki University, Japan)
	Birgit Schmidt (Germany)

Program Committee

Jakob Bardram	IT University of Copenhagen, Denmark
Louise Barkhuus	University of California, San Diego, USA
Alastair Beresford	University of Cambridge, UK
Barry Brown	University of California, San Diego, USA
Roy Campbell	University of Illinois at Urbana-Champaign, USA
Tanzeem Choudhury	Dartmouth College, USA
Sunny Consolvo	Intel Research, USA
Alois Ferscha	University of Linz, Austria
Patrik Floréen	University of Helsinki, Finland
Hans Gellersen	Lancaster University, UK
Gillian Hayes	University of California, Irvine, USA
Jeffrey Hightower	Intel Research, USA
Elaine Huang	Motorola Labs, USA
Antonio Krüger	University of Münster, Germany
Paul Lukowicz	University of Passau, Germany
Masateru Minami	University of Tokyo, Japan
Jin Nakazawa	Keio University, Japan
Shwetak Patel	University of Washington, USA
Matthai Philipose	Intel Research, USA
Aaron Quigley	University College Dublin, Ireland
Bernt Schiele	TU Darmstadt, Germany
Albrecht Schmidt	University of Duisburg-Essen, Germany
James Scott	Microsoft Research, UK
Abigail Sellen	Microsoft Research, UK
Yasuyuki Sumi	Kyoto University, Japan
Tsutomu Terada	Kobe University, Japan
Kristof Van Laerhoven	TU Darmstadt, Germany
Woontack Woo	GIST, Korea

Steering Committee

Hans Gellersen   Lancaster University, UK
Anthony LaMarca   Intel Research, USA
Marc Langheinrich   ETH Zurich, Switzerland
Aaron Quigley   University College Dublin, Ireland
Bernt Schiele   TU Darmstadt, Germany
Albrecht Schmidt   University of Duisburg-Essen, Germany
Khai Truong   University of Toronto, Canada

Reviewers

Saleema Amershi
Morgan Ames
Oliver Amft
Fredrik Andersson
Daniel Avrahami
Rafael Ballagas
David Bannach
Jörg Baus
Christian Becker
Richard Beckwith
Michael Beigl
Marek Bell
Frank Bentley
Carlos Bento
Ethan Berke
Jacob Biehl
Ulf Blanke
Susanne Boll
Philipp Bolliger
Stephen Brewster
Andreas Bulling
Jonathan Bunde-Pedersen
Andreas Butz
Yunan Chen
Keith Cheverst
Konstantinos Chorianopoulos
Luigina Ciolfi
Adrian Clear
Vlad Coroama
David Cottingham
Lorcan Coyle
Florian Daiber
George Danezis

Alexander De Luca
Anind Dey
Joan DiMicco
Simon Dobson
Afsaneh Doryab
Keith Edwards
Christos Efstratiou
Brynn Evans
David Evans
David Eyers
Mbou Eyole-Monono
Reza Farivar
Jesus Favela
James Fogarty
David Frohlich
Georg Gartner
Elizabeth Goodman
Philip Gray
Sidhant Gupta
Daniel Halperin
Thomas Riisgaard
Robert Harle
Muhammad Haroon
Simon Hay
Mike Hazas
Jennifer Healey
Urs Hengartner
Juan David
Paul Holleis
David Holman
Danny Hughes
Karin Hummel
Dieter Hutter

Tam Huynh
Jein Hwang
Jadwiga Indulska
Sozo Inoue
Stephen Intille
Umer Iqbal
Lilly Irani
Susumu Ishihara
Shahram Izadi
Mattias Jacobsson
Seiie Jang
Christian D. Jensen
Won Jeon
Oskar Juhlin
Kimmo Kalliola
Shaun Kane
Apu Kapadia
Nao Kawanishi
Fahim Kawsar
Joseph Kaye
Dagmar Kern
Julie Kientz
Sehwan Kim
David Kirk
Predrag Klasnja
Lisa Kleinman
Tadayoshi Kohno
Shinichi Konomi
Gerd Kortuem
Matthias Kranz
Christian Kray
Steinar Kristoffersen
Mareike Kritzler
Kai Kunze
Axel Küpper
Anthony LaMarca
Brian Landry
Koen Langendoen
Marc Langheinrich
Johnny Lee
Jonathan Lester
Clayton Lewis
Lin Liao
Siân Lindley
Jay Lundell

Johan Lundin
Jani Mäntyjärvi
Kenji Mase
Tara Matthews
Oscar Mayora
Rene Mayrhofer
Joseph McCarthy
Gregor McEwan
Robert Mcgrath
Sean McNee
Alexander Meschtscherjakov
Robin Message
Florian Michahelles
Takumi Miyoshi
Martin Mogensen
Mirko Montanari
Hideyuki Nakanishi
Yasuto Nakanishi
Futoshi Naya
Joseph Newman
Mark W. Newman
William Newman
David Nguyen
Daniela Nicklas
Takuichi Nishimura
Saeko Nomura
Petteri Nurmi
Georg Ogris
Hiroyuki Ohnuma
Patrick Olivier
Antti Oulasvirta
Jeffrey Pang
Kurt Partridge
Donald Patterson
Mark Perry
Erika Poole
Ivan Poupyrev
Zachary Pousman
Sören Preibusch
Cliff Randell
Anand Ranganathan
Umar Rashid
Tye Rattenbury
Josephine Reid
Jun Rekimoto

Till Riedel
Andreas Riener
Daniel Roggen
Michael Rohs
T. Scott
Ichiro Satoh
Adin Scannell
Michael Schellenbach
Hedda Schmidtke
Johannes Schöning
Tim Schwartz
John Senders
Chetan Shankar
Ross Shannon
Scott Sherwood
Anmol Sheth
Shu Shi
Chia-Yen Shih
Itiro Siio
Timothy Sohn
Tomas Sokoler
Matthew Stabeler
Ulrich Steinhoff
Maja Stikic
Oliver Storz

Martin Strohbach
Erich Stuntebeck
Sriram Subramanian
Jukka Suomela
Anthony Tang
Michael Terry
Niwat Thepvilojanapong
Andreas Timm-Giel
Khai Truong
Koji Tsukada
Elise van den Hoven
Steven Wall
Jamie Ward
Evan Welbourne
John Williamson
Andy Wilson
Christopher Wren
Danny Wyatt
Yutaka Yanagisawa
Tomoki Yoshihisa
Jaeseok Yun
Doris Zachhuber
Andreas Zinnen

Table of Contents

Digital Displays

Display Blindness: The Effect of Expectations on Attention towards
Digital Signage 1
 Jörg Müller, Dennis Wilmsmann, Juliane Exeler, Markus Buzeck,
 Albrecht Schmidt, Tim Jay, and Antonio Krüger

Users' View on Context-Sensitive Car Advertisements 9
 Florian Alt, Christoph Evers, and Albrecht Schmidt

ReflectiveSigns: Digital Signs That Adapt to Audience Attention....... 17
 Jörg Müller, Juliane Exeler, Markus Buzeck, and Antonio Krüger

Navigation

Realistic Driving Trips for Location Privacy 25
 John Krumm

Enhancing Navigation Information with Tactile Output Embedded into
the Steering Wheel... 42
 Dagmar Kern, Paul Marshall, Eva Hornecker, Yvonne Rogers, and
 Albrecht Schmidt

Landmark-Based Pedestrian Navigation with Enhanced Spatial
Reasoning... 59
 Harlan Hile, Radek Grzeszczuk, Alan Liu, Ramakrishna Vedantham,
 Jana Košecka, and Gaetano Borriello

At Home with Pervasive Applications

The Acceptance of Domestic Ambient Intelligence Appliances by
Prospective Users 77
 Somaya Ben Allouch, Jan A.G.M. van Dijk, and Oscar Peters

Adding GPS-Control to Traditional Thermostats: An Exploration of
Potential Energy Savings and Design Challenges 95
 Manu Gupta, Stephen S. Intille, and Kent Larson

KidCam: Toward an Effective Technology for the Capture of Children's
Moments of Interest 115
 Julie A. Kientz and Gregory D. Abowd

Sensors, Sensors, Everywhere

Mobile Device Interaction with Force Sensing 133
    James Scott, Lorna M. Brown, and Mike Molloy

Inferring Identity Using Accelerometers in Television Remote
Controls 151
    Keng-hao Chang, Jeffrey Hightower, and Branislav Kveton

The Effectiveness of Haptic Cues as an Assistive Technology for Human
Memory 168
    Stacey Kuznetsov, Anind K. Dey, and Scott E. Hudson

Exploring Privacy Concerns about Personal Sensing 176
    *Predrag Klasnja, Sunny Consolvo, Tanzeem Choudhury,
    Richard Beckwith, and Jeffrey Hightower*

Working Together

Enabling Pervasive Collaboration with Platform Composition 184
    *Trevor Pering, Roy Want, Barbara Rosario, Shivani Sud, and
    Kent Lyons*

Askus: Amplifying Mobile Actions 202
    *Shin'ichi Konomi, Niwat Thepvilojanapong, Ryohei Suzuki,
    Susanna Pirttikangas, Kaoru Sezaki, and Yoshito Tobe*

Boxed Pervasive Games: An Experience with User-Created Pervasive
Games 220
    *Richard Wetzel, Annika Waern, Staffan Jonsson, Irma Lindt,
    Peter Ljungstrand, and Karl-Petter Åkesson*

Tagging and Tracking

RF-Based Initialisation for Inertial Pedestrian Tracking 238
    Oliver Woodman and Robert Harle

PL-Tags: Detecting Batteryless Tags through the Power Lines in a
Building 256
    Shwetak N. Patel, Erich P. Stuntebeck, and Thomas Robertson

Geo-fencing: Confining Wi-Fi Coverage to Physical Boundaries 274
    Anmol Sheth, Srinivasan Seshan, and David Wetherall

Securing RFID Systems by Detecting Tag Cloning 291
    *Mikko Lehtonen, Daniel Ostojic, Alexander Ilic, and
    Florian Michahelles*

Methods and Tools

Towards Ontology-Based Formal Verification Methods for Context
Aware Systems 309
    Hedda R. Schmidtke and Woontack Woo

Situvis: A Visual Tool for Modeling a User's Behaviour Patterns in a
Pervasive Environment 327
    Adrian K. Clear, Ross Shannon, Thomas Holland, Aaron Quigley,
    Simon Dobson, and Paddy Nixon

Methodologies for Continuous Cellular Tower Data Analysis 342
    Nathan Eagle, John A. Quinn, and Aaron Clauset

The Importance of Context

"It's Just Easier with the Phone" – A Diary Study of Internet Access
from Cell Phones 354
    Stina Nylander, Terés Lundquist, Andreas Brännström, and
    Bo Karlson

Does Context Matter ? - A Quantitative Evaluation in a Real World
Maintenance Scenario 372
    Kai Kunze, Florian Wagner, Ersun Kartal,
    Ernesto Morales Kluge, and Paul Lukowicz

On the Anonymity of Home/Work Location Pairs 390
    Philippe Golle and Kurt Partridge

Working Overtime: Patterns of Smartphone and PC Usage in the Day
of an Information Worker 398
    Amy K. Karlson, Brian R. Meyers, Andy Jacobs, Paul Johns, and
    Shaun K. Kane

Author Index 407

Display Blindness: The Effect of Expectations on Attention towards Digital Signage

Jörg Müller[1], Dennis Wilmsmann[1], Juliane Exeler[1], Markus Buzeck[1],
Albrecht Schmidt[2], Tim Jay[3], and Antonio Krüger[1]

[1] University of Münster
[2] University of Duisburg-Essen
[3] University of Bristol

Abstract. In this paper we show how audience expectations towards what is presented on public displays can correlate with their attention towards these displays. Similar to the effect of *Banner Blindness* on the Web, displays for which users expect uninteresting content (e.g. advertisements) are often ignored. We investigate this effect in two studies. In the first, interviews with 91 users at 11 different public displays revealed that for most public displays, the audience expects boring advertisements and so ignores the displays. This was exemplified by the inclusion of two of our own displays. One, the iDisplay, which showed information for students, was looked at more often than the other (MobiDiC) which showed coupons for shops. In a second study, we conducted repertory grid interviews with 17 users to identify the dimensions that users believe to influence whether they look at public displays. We propose possible solutions to overcome this "Display Blindness" and increase audience attention towards public displays.

1 Introduction

Due to continuously falling display prices, digital displays have started to be installed in many public spaces. Deployers of such technology often assume that public displays inherently attract attention and therefore that people will look at them. We had similar expectations when we deployed two different public display networks. We noticed that the iDisplays[8] network, installed at the university and showing information for students, received a fair amount of attention. By contrast, the MobiDiC[7] network installed in the city center and showing coupons for nearby shops, received much less attention. Huang et al. have shown that indeed most public displays are ignored by many users or receive only very few glances[4]. Despite the fact that this effect has been well established, explanations for this behaviour are still lacking.

In other areas, lack of attention for aspects of the environment has been explained by the fact that attention is highly selective. The world provides far too much information to be processed by an individual. This is especially true for urban environments, where Milgram[6] showed that many individuals experience *information overload*. Milgram identified six common reactions to information overload, among them the allocation of less time to each input and

H. Tokuda et al. (Eds.): Pervasive 2009, LNCS 5538, pp. 1–8, 2009.

disregard of low-priority inputs. In their survey on information overload, Eppler and Mengis[2] define the concept as follows: "Information overload describes the situation when too much information affects a person and the person is unable to recognize, understand or handle this amount of information." They conclude that when information supply exceeds information-processing capacity, a person has difficulties in identifying relevant information. He/she becomes highly selective and ignores large amounts of information, has difficulties in identifying relationships between details and overall perspective and needs more time to reach a decision.

One prominent example of such "disregard of low-priority inputs" has become known in the Web as "Banner Blindness". Burke et al.[1] have shown with eye-tracking experiments that people rarely look directly at banners and show low recall for banner content. They conclude that *"Participants in the present studies had an overriding incentive not to look at banners, and no amount of banner manipulation increased their pull. Longer exposure time, animation, and the presence of images did not make the task irrelevant ads more conspicuous. Connecting advertising to viewers goals may make ads more successful"*. In other words, people expected banners to lack task relevance and so ignored these areas.

In this paper, we pose the question of whether the effect of selective attention observed on the Web also applies to digital signage. In other words: What is the role of audience expectations regarding audience attention towards digital signage?

2   Study 1: A Comparison of Looking Behaviour at Public Displays

In order to investigate reasons why people look at public displays or not, we conducted interviews with people who passed by public displays. This provided us the ability to study both users and non-users of the technology in an ecologically valid setting.

2.1   Method

We created a corpus of all public displays we could find in the city of Münster, Germany. From these, we selected 11 different public display locations which we considered to provide a representative sample. These included three of our own iDisplays (two in front of lecture halls and one in the lobby of a university building) and one MobiDiC display installed in public telephones (see Figure 1). Others were located in shop windows (in a passageway, a mobile phone store and a credit store). Three displays showed television programs; one in a bank, one in a café and one in the waiting area of the citizen bureau. One was installed in a clothes store, showing fashion videos. Interviews were conducted on weekdays as well as weekends, between 9 am and 8 pm. Participants were selected on an opportunity basis, irrespective of gender, age (approx. 14-80 years) or other variables. In total 91 interviews were conducted. The procedure was as follows:

Table 1. Summary of interview results. It is shown how many participants stated to have seen the display, what they expected to be shown there and whether they consider the content they expect interesting.

Location	N	Looked	Expectations	Interesting
iDisplay Small Lecture Hall	15	15	Content known (15)	13
iDisplay Big Lecture Hall	15	13	Content known (13)	12
iDisplay Entrance	6	6	Content known (6)	5
Citizen Bureau	10	10	Content known (10)	8
MobiDiC	10	0	Advertising (5), Phone book (2), Telekom Information (2), City Information (1), City Map (1), Events (1), Internet (1), Manual (1), Emergency Numbers (1)	0
Shopping Mall	6	0	Ads for Fashion store nearby (5), Fashion (3), Videos (1)	0
Cafe	2	0	Soccer (1), Ads (1), News (1)	0
Credit Shop	11	0	Ads for credit shop (5), Credit conditions (4), Temperature (1), Time (1), Television (1), Ads for Cosmetics (1)	0
Bank	2	1	Ads (1), Special offers (1)	0
Phone Shop	9	2	Ads for Telephones (9)	0
Clothes Shop	5	0	Fashion (2), Music Videos (1), Ads (1)	0

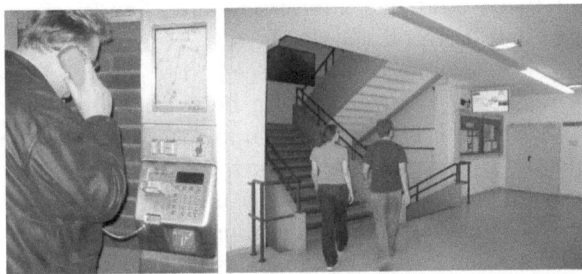

Fig. 1. The MobiDiC and iDisplays displays

First we let participants pass the displays and observed whether they looked at them. After they passed the display, we stopped them in a position where they could see the display but not the content shown. We showed them the display and asked whether they had seen it. Next, we asked what they expected the display showed right now. We then asked whether what they expected would be interesting to them. Finally, we asked what they would like the displays to show at that very moment.

2.2 Results

The results of the interviews (shown in Table 1) showed that, for the iDisplays and the citizen bureau, most participants said they had seen the displays, knew the content and considered the content to be interesting to them. For the other displays, relatively few participants said they had seen them; many expected advertisements to be shown and not one participant considered the expected content to be interesting for them. Some typical statements were: "You know, everything here is so full of advertisements, I don't look at these things anymore."(an older lady). A younger woman said "No, I'm not interested in technology. I don't look at displays." One younger man said: "I don't have time to look there. You got ads everywhere, I just ignore them." What kind of advertisements participants expected depended strongly on the context. Most participants tried to guess who the display owner was (the telecommunications company for the MobiDiC displays, and the shop for displays installed in shop windows) and expected that the displays would show advertisements for the display owner. No participant expected any of the displays to show general ads for different advertisers. Finally, we asked participants what they would like to see on the displays. For the citizen bureau and the iDisplays, 3 participants (7%) would like to see information on the building. At the iDisplays, 7 (19%) participants would like to see information about events at the university. For the other displays, 15 (45%) participants would like local information about the city, and 10 participants (30%) would like current (local) news. 5 (15%) participants would like to see sports news (e.g. football results) and 4 (12%) participants would like entertainment content and lifestyle news. Two (6%) would like to see the current temperature, and two would like advertisements. Two participants suggested that it would be best to turn off the displays.

From the interviews, two major factors seemed to affect whether participants looked at public displays. The comparison between the display in the citizen bureau and the café (which showed the same television channel) showed that displays where people wait and have nothing else to do received a lot more attention. A comparison between the iDisplays and the other displays showed that displays where participants expected something interesting (for them) received a lot of attention (in various locations). Displays where the participants expected nothing interesting on the other hand (like the MobiDiC displays and others) were largely ignored. Interestingly, the content that participants would have liked to have seen on public displays and what they expected seem to be diametrically opposed. While they would like to have seen local information and news, sports and lifestyle news as well as entertainment, they expected only advertisements to be shown. Another interesting observation was that while expectations were quite homogenous (ads), different people indicated they wanted to see very different content, implying it might make sense to personalize the content.

3   Study 2: Directions for Further Research on Top-Down and Bottom-Up Effects on Attention

After identifying the correlation of expectations with attention towards digital signage, we were interested in a broader view of potential factors that influence attention. We considered this would be useful both for placing the role of expectations in the context of other factors and for identifying promising areas for further research. To achieve this, we wanted to consider a corpus of very different signs, rather than investigate the impact of a few factors in detail. Therefore, we decided to study participants' responses to videos, showing a broad range of digital signs in the context of their surroundings. Using repertory grid interviews [5], it is possible to systematically elicit the dimensions that participants used to compare different digital signs. Using this method, each interview results in a number of dimensions used to compare different displays - these dimensions are then used by the participant to rate each display. In this case, Honey's content analysis[3] was used to analyse the grids, as this enables analysis of multiple grids (one from each participant) and comparisons of the importance of different dimensions by measuring the correlation of each dimension with the participants rating whether they would look at a display. The elicited dimensions are then categorized using affinity analysis[5] and for each category a mean correlation with the rating whether participants believe they would look at a display can be computed. This score can be used together with the number of dimensions in that category to estimate the relative importance of that category.

3.1   Method

We first collected videos (both our own and from YouTube) showing public displays including their surrounding context (e.g. people passing by, nearby buildings etc.), until no more videos dissimilar to those already in the corpus could be found. The resulting corpus contained 93 videos. A selection of these videos was then presented to a total of 17 participants on a computer screen. Seventeen participants were selected because the repertory grid technique dictates that one only adds participants until no new dimensions arise during the course of the interviews. The 17 people (7 female, 10 male, age 23-30 years) were selected among students of the institute and were not compensated for their participation.

In line with the repertory grid technique, each participant saw only a subset of the corpus. In this case, for each participant 10 videos were selected randomly and these were presented in random sets of three next to each other on a computer screen. For each group of three displays, participants were asked to state which two of them had something in common (emergent pole) that was different from the third (implicit pole). Exemplary poles were 'The display content is informative' versus 'The display content is not informative'. After providing a description, they were asked to rate each display from the 10 selected on a (5-point) Likert Scale on this new dimension. This process was repeated until no more dimensions arose. On average this occurred after 10 triples. Thus, each participant saw only 10 different videos but on average saw each video three times. At this point (when participants could not find any new dimensions), we asked them to rate each video on an additional, but key, dimension 'I would always look at this display in this situation' versus 'I would never look at this display in this situation' (overall dimension).

We computed the correlation of all elicited dimensions to the corresponding overall rating of whether they would expect to look at the display or not. Finally, we used the bootstrapping technique[5] to group the dimensions into categories. Dimensions were categorized by two independent raters. Each dimension was picked from the stack and compared by similarity to all the existing categories. It was decided whether the dimension would fit any of the existing categories, if existing categories needed to be split, or if a new category should be created. This process was repeated until all dimensions were categorized. The categorizations of the two raters were then compared and discussed. Then, the raters repeated the categorization process, until over 90% similarity was reached. For each category, the mean % similarity score (correlation to the overall likelihood that they would look) was then computed. Finally, the categories were sorted based on this value.

3.2   Results

The results of the repertory grid study are depicted in Table 2. In the first column, a descriptive name for the category, chosen by the raters, is given. In the second column, exemplary results are presented. The pole that correlated with the pole 'I would always look at the content' is printed in bold. Furthermore, the degree of correlation between this dimension and the overall dimension is given. In the third column, the number of dimensions in this category is given. In the fourth column, the mean similarity (correlation) of dimensions in this category to the overall dimension is presented. This mean similarity to the overall dimension

Table 2. Factors that correlate with whether participants believe they would look at public displays as resulting from the Repertory Grid interviews. A high similarity score means the rating for dimensions in this category strongly correlates with whether participants expect to look at a display, a low score means it correlates only weakly.

Category	Dimensions	n	mean % sim.
Conspicuity of display and content	does not attract attention ↔ **attracts attention** (90%sim.,H), **eye-catching** ↔ simple (65%sim.,H)	5	60
Colorful content	not so colorful ↔ **colorful** (75%sim.,H)	3	56.67
Interesting content	**interesting** ↔ boring (70%sim.,H), **informative** ↔ not informative (40%sim.,I)	6	54.17
Aesthetical content	not attractive ↔ **attractive** (65%sim.,H), not so beautiful ↔ **beautiful** (60%sim.,H), dull, sterile ↔ **aesthetic** (60%sim.,I)	9	42.78
Emotional content	not emotional appealing ↔ **emotional appealing** (50%sim.,H), emotional content ↔ **informative content** (40%sim.,I)	3	40
Long distance visibility	**visible on long distance** ↔ visible on short distance (25%sim.,H)	2	40
Content adapted to context	target audience undirected ↔ target audience directed (60%sim.,I), **differs from surrounding area** ↔ fits to surrounding area (45%sim.,I), isolated in surrounding area ↔ **integrated into surrounding area** (40%sim.,I)	7	39.29
Animated content	**animated** ↔ static (65%sim.,H), **fast content change** ↔ slow content change (55%sim.,H), **dynamic content** ↔ static content (55%sim.,I)	14	37.5
Waiting area	**time to watch the display** ↔ no time to watch the display (55%sim.,H), **waiting in front of the display** ↔ walking by the display (50%sim.,H)	4	36.25
Display size	**large** ↔ small (70%sim.,H)	10	35,5
Type of content	advertising ↔ **entertainment** (60%sim.,H), **news** ↔ video game (55%sim.,H), **news, commercials** ↔ planning, design (45%sim.,H), **advertising** ↔ information system (45%sim.,H)	17	32.94
Different content	varied content ↔ **identical varied** (55%sim.,H), varied commercials ↔ **single commercial** (35%sim.,I)	6	32.5

can be interpreted as an indication of how strongly this category correlates to whether people expect to look at displays.

The first category simply describes the likelihood of looking and as such is not a "factor". Thus "colourfulness" (second row) is the most important factor we found influencing whether people thought they would look at the displays. The expectation of participants as to whether the content shown on the displays would be interesting comes in third. If participants expect interesting or informative content on the displays, they say they are more likely to look. The category "Content adapted to context" is somewhat ambiguous. Participants said they would be more likely to look at public displays if they appealed to the general public rather than to specific groups and did not agree whether they would look more if the display contrasted with surrounding context or if it was integrated with it.

4   Discussion

A limitation of Study 2 is that it only asked what people *believed* they would do, instead of determining what they would actually do, so these findings demand further research. As a next step, these factors should be validated in experiments to explore whether people's behaviour really reflects their stated beliefs in this context. As can be seen from Table 2, there are a range of factors associated with peoples' belief that they would look at a public display. These include bottom-up factors like colourfulness or attractiveness of the display and, mentioned less often, the amount of time the display is potentially visible to a passer-by (mentioned as long distance visibility) and the size of the display. Other effects that have been mentioned in the literature, e.g. by Huang et al. [4], are the angle to walking direction, installation height, the distance of the display to a passer-by and level of distraction (e.g. by other displays). Notably, however, whether participants expect interesting content seems to be more important than other effects that could be naively assumed, like the display size (which scores relatively low in this list). This is very much in line with the results from Study 1, where audience expectations strongly correlated with audience attention. Thus, we propose that in addition to bottom-up effects, top-down effects - audience expectations - need to be considered. The combination of the two studies indicates that for public displays people may expect nothing interesting and so do not look at them. People's expectations appear to depend on the perceived context; in particular who they believe the display owner is. For certain owners (e.g. the university, the citizen bureau) people expect the content to be interesting, while for others (e.g. shops) they expect advertisements. People would like to see content interesting to them, like local city information, local news, sports and entertainment content. It is likely that none of these bottom-up and top-down factors alone determines how much attention a display receives, but rather it is more likely that it is combination of these factors at play depending on the particular context. The influence of expectations is indeed similar to the effect of Banner Blindness. An interesting departure from this analogy is that while for Banner Blindness, increased colorfulness and animation do not increase attention[1], for displays, such bottom-up factors do appear to correlate with whether participants believe they will look at displays.

5 Conclusion

We can conclude that indeed the process of selective attention that is known from the Web also applies to digital signage. Thus, similar to the effect of "Banner Blindness", there is an effect of "Display Blindness" meaning that expectations of uninteresting content leads to a tendency to ignore displays. The relatively short time for which public displays have existed seems to have been sufficient to build such negative expectations for many passers-by. In order to avoid an "arms race" for the audience's attention, display owners should investigate audience expectations for certain displays and take them seriously. Content should be designed to fit these expectations, and if recognizable displays provide interesting content, it may even be possible to influence audience expectations over time. It should be kept in mind however that expectations are not the only effect to influence attention, but are embedded in a complex interplay of bottom-up and top-down factors.

References

1. Burke, M., Hornof, A., Nilsen, E., Gorman, N.: High-cost banner blindness: Ads increase perceived workload, hinder visual search, and are forgotten. ACM Trans. Comput.-Hum. Interact. 12(4), 423–445 (2005)
2. Eppler, M.J., Mengis, J.: The concept of information overload: A review of literature from organization science, accounting, marketing, mis, and related disciplines. The Information Society 20, 325 (2004)
3. Honey, P.: The repertory grid in action. Industrial and Commercial Training 11, 452–459 (1979)
4. Huang, E., Koster, A., Borchers, J.: Overcoming assumptions and uncovering practices: When does the public really look at public displays? In: Indulska, J., Patterson, D.J., Rodden, T., Ott, M. (eds.) Pervasive 2008. LNCS, vol. 5013, pp. 228–243. Springer, Heidelberg (2008)
5. Jankowicz, D.: The Easy Guide to Repertory Grids, 1st edn. Wiley, Chichester (2003)
6. Milgram, S.: The experience of living in cities. Science, 1461–1468 (March 13, 1970)
7. Müller, J., Krüger, A.: How much to bid in digital signage advertising auctions? In: Adjunct proceedings of Pervasive 2007 (2007)
8. Müller, J., Paczkowski, O., Krüger, A.: Situated public news and reminder displays. In: Schiele, B., Dey, A.K., Gellersen, H., de Ruyter, B., Tscheligi, M., Wichert, R., Aarts, E., Buchmann, A. (eds.) AmI 2007. LNCS, vol. 4794, pp. 248–265. Springer, Heidelberg (2007)

Users' View on Context-Sensitive Car Advertisements

Florian Alt, Christoph Evers, and Albrecht Schmidt

Pervasive Computing Group
University of Duisburg-Essen
Schuetzenbahn 70, 45119 Essen, Germany
{florian.alt,albrecht.schmidt}@uni-due.de,
christoph.evers@stud.uni-due.de

Abstract. Cars are ubiquitous and offer large and often highly visible surfaces that can be used as advertising space. Until now, advertising in this domain has focused on commercial vehicles, and advertisements have been painted on and were therefore static, with the exception of car-mounted displays that offer dynamic content. With new display technologies, we expect static displays or uniformly-painted surfaces (e.g. onto car doors or the sides of vans and trucks) to be replaced with embedded dynamic displays. We also see an opportunity for advertisements to be placed on non-commercial cars: results of our online survey with 187 drivers show that more than half of them have an interest in displaying advertising on their cars under two conditions: (1) they will receive financial compensation, and (2) there will be a means for them to influence the type of advertisements shown. Based on these findings, as well as further interviews with car owners and a car fleet manager, we discuss the requirements for a context-aware advertising platform, including a context-advertising editor and contextual content distribution system. We describe an implementation of the system that includes components for car owners to describe their preferences and for advertisers to contextualize their ad content and distribution mechanism.

1 Introduction

Cars are ubiquitous in today's societies and, given their size and shape, offer potential space for placing highly-visible advertising. Vehicles are used and seen by people, and can therefore lead to many points of visual contacts between people and cars. For example, a pedestrian seeing cars driving by, a driver or passenger driving behind another car and looking at its rear, and a person walking by a row of parked cars at a parking lot and glancing at each of them. Until now, very few privately-owned cars have become advertising spaces and commercial vehicles have mostly static advertisements that promote the services and products of the company that bought them. Since the early 90s, vehicles used in public transport have been utilized as advertising space. One of the first examples is the Pepsi advertising campaign that started in 1993, where the Pepsi logo was painted on urban buses in the Seattle area. Since then, advertising on vehicles has remained mostly static, in the sense that the ads do not adapt to their immediate situation or context. The first approaches to offer dynamic advertising content involved mounting electronic displays on cars where the content

H. Tokuda et al. (Eds.): Pervasive 2009, LNCS 5538, pp. 9–16, 2009.

might be location dependent, e.g. taxis in the Boston area[1]. With advances in display technologies, we anticipate that there will be new ways for attaching or embedding electronic displays into car surfaces, such as car doors or the sides of vans and trucks. This will create new opportunities for advertising. We argue that context will play a major role for efficiency and acceptance of advertising displays.

We explore the potential of cars as dynamic and contextual advertising space in this paper. In section 2, we present the results of an online survey with 187 drivers summarizing their attitudes towards car-based advertising. Based on these findings, we describe in section 3 the requirements, concept and design space for a contextual advertising system for vehicles. In section 4, we describe an implementation of the central components and a platform of a dynamic advertising system prototype to show its feasibility. We then discuss related work in this domain and further opportunities and implications of our work in the conclusion.

The contribution of this paper is twofold:

- A detailed analysis of users expectations on context advertising on vehicles that reveals an interest in providing advertising space on privately-owned cars in return for an incentive (e.g. money) if there is an appropriate way for users to influence which advertisements are shown.
- A discussion of the design space and requirements for a platform that provides a means for creating and deploying contextual advertising to be shown on cars and a proof-of-concept implementation based on these requirements.

2   Car Owners' Expectations on Contextual Car Advertising

To understand users' expectations, motivations, and constraints for providing their car as a platform for advertising, we conducted an online survey and follow-up interviews with car owners and a car fleet manager. We hope that by looking broadly at potential target groups–from private car owners to large companies with company car fleets–, we can provide a more comprehensive picture of the requirements.

2.1   Privately-Owned Cars: Online Survey

In our survey, we were especially interested in 1) when users want to make their car available for advertising, 2) product categories users would be willing to show advertisements for, 3) the acceptance of different technological solutions, and 4) possible rewards for showing advertisements on private vehicles.

The survey was completed by 187 persons, 130 males and 57 females. The average age of the participants was 27.4 years (25.2 among males vs. 23.3 among females). The majority of participants were driving compact-sized vehicles (128) and medium-sized vehicles (41). Recruitment for the survey was done via email.

First, we asked for the preferred reward schema. We suggested different types of rewards, including monthly allowance, discount on car purchase, benefits from third parties (coupons for cinema/concerts, etc.), and coupons for free fuel. On a 5-point Likert scale (1=not interesting at all, 5=definitely interesting), 155 of 187 participants

---

[1] Taxis with roof mounted signs that change their content according to the GPS position, see http://www.clearchanneltaximedia.com/products/taxi-tops-digital-smart.asp for details.

were interested (gave a rating of 3 or higher) in a monthly allowance, 139 in a discount upon purchasing the car and 137 in coupons for free fuel. Only 50 subjects were interested in receiving coupons from third parties as a reward.

Then, we asked when users would like to provide their cars as advertising space. Users were asked to choose from several options: at any time, while driving, while I am not close to my car, and while I cannot see my car. We found that 105 subjects (56.1%) did not care when advertisements are displayed on their car, while 37 subjects (19.8%) preferred not to be close to their car while it was showing ads 24 subjects (12.8%) did not want to see their car while it was advertising.

Next, we evaluated the effect of the advertised products on the attitude of car owners towards advertisements. We provided users with a selection of different categories, including music, clothes, fast food, beverages, tobacco, firearms, politics, culture, eroticism, and sport events. We found that users preferred not to show advertisements on their cars for controversial categories. These included tobacco (124 subjects, 66.1%), eroticism (144 subjects, 77.0%), firearms (151 subjects, 80.8%), and politics (116 subjects, 62.3%). On the other hand, they strongly preferred showing advertisements on culture (173 subjects, 92.5%), music (167 subjects, 89.3%), sport events (161 subjects, 86.1%), and clothes (151 subjects, 80.7%).

Additionally, we looked into the acceptance of different technical solutions. We suggested four options to the users: painting, easy-to-detach foil, fixed labels, and roof-mounted equipment. Using a 5-point Likert scale (1=inacceptable, 5= acceptable), we found that the most popular solution among the participants was easy-to-detach foil (92.0% with a rating of 3 or better), followed by fixed labels (55.62%). Rather unpopular were painting (43.85%) and roof-mounted equipment (19.78%).

Finally, we wanted to know which parts of the car participants preferred to have the advertisements appear on. Participants used a 5-point Likert scale to evaluate the following locations: rear/trunk, rear windshield, rear side windows, sides/doors, roof, front lid. The result shows that most participants favor the sides/doors (83.9% with a rating of 3 or better), followed by the vehicle's rear/trunk (78.6%), roof (77.01%), rear side windows (67.4%), rear windshield (60.9%) and front lid (60.43%).

The survey revealed the following findings with regard to privately-owned cars:

(1) Incentives are essential for convincing users to provide their cars as advertising space. Preferred rewards for placing advertisements on cars are monthly allowance, discounts upon purchasing a car, and coupons for free fuel.

(2) More than half of the participants would not care if they were close to their car while ads were being shown. However, one third preferred not to have the ads shown when they were close by or could see the ads themselves.

(3) Users want to stay in control of the products advertised, especially those falling under controversial categories, such as eroticism, firearms, and tobacco.

(4) For technical solutions, easy-to-detach foil was rated by far the most preferred option while roof-mounted equipment was rather unpopular.

(5) The side doors, trunk and roof are among the favorite areas on the car for placing advertisements.

2.2 Privately-Owned Cars: Qualitative Interviews

In order to gain further understanding of the car owners' views on car advertising, we conducted 4 follow-up interviews with people who participated in the online survey. The subjects were two students (each 21 years old), a contractor (50), and a real estate administrator (45). First, we asked them about concrete incentives and values. All subjects would be interested in a monthly allowance (80€-120€ monthly), two of them would also like to receive coupons for free fuel (80€-100€ monthly). Next, we asked the subjects about how to select the products they would be willing to advertise on their cars. Two favored a category list, where they could select groups of products, and the other two favored positive lists for a more fine-grained selection. However, all of them wanted to be able to update the list at any time. Finally, we asked about privacy concerns regarding contextual and location information that might be transmitted about their car, e.g. GPS data. Three of the subjects were not concerned at all, adding that they were participating in payback programs anyway. However, all four subjects stated that their GPS data should not be provided to the advertisers.

2.3 Company-Owned Cars: Interview with a Car Fleet Manager

To complement the view of privately-owned cars with that of large companies that own cars for business use, we conducted a telephone interview with a fleet manager of a pharmaceutical company, who is responsible for a fleet of approximately 600 cars with a replacement rate of 200 cars/year. During the interview, we discovered issues and concerns regarding the hypothetical introduction of car advertisements in the fleet. One of the manager's primary concerns was the additional administrative workload. This additional workload would be partly due to the need to equip cars with the required technical gear and partly due to the necessity to define products that should not appear as ads on the fleet's vehicles. The manager also felt that a monthly allowance of about 10% of the monthly leasing rate of the cars would be an attractive incentive. A final and interesting finding was with regard to the acceptance among the employees using the cars. Since cars are a part of an employee's compensation, as well as a status symbol, the manager felt that it would be essential (especially among sales personnel) to allow the employees to have a strong influence on the decision of whether or not to allow advertising on their cars.

3 Contextualized Advertisements on Cars

In this section, we outline the design space for a system for context-aware car advertisements.

Means for expressing preferences. One central requirement is that car owners have the power and means to specify what advertisements are shown on their car. It is also essential that the system is easy to use. Potential solutions include giving the user a list of products and brands that she or he can accept or deny (not individually but as a whole). Asking the user to provide preferences for each product may lead to a lengthy and tedious process and asking for each product category may be too general to accurately specify their preferences. Therefore, we suggest creating a preferences editor based on a combination of product groups and brands. This approach scales well to large numbers of advertisements and appears to satisfy the users' concerns.

Simple technical solution. The technical solution for advertising must be very simple to apply and remove. A major change to the appearance of the car (e.g. display system on the roof) is also not welcomed. The required effort by the car owner for administrating, configuring, and maintaining such a display surface must be minimal. One potential technical solution is a self-contained display unit consisting of a magnetic foil (to make it easily attachable to the car door) with a bendable display (e.g. eInk display) on top and built-in communication and context-sensing. This unit would potentially be powered by energy harvesting.

Defining context. In the case of contextualized advertisements on vehicles, the advertising message would be broadcasted to all people who focus directly or indirectly on the car. Thus, the audience is not limited to a specific user but rather to a group defined by context. We identified the following crucial context information:

- **Spatial**: Location is an important factor when dealing with context-aware ads. Using knowledge of the geographic location of the display, ads can be targeted at the audience expected in that region. It may also be possible to estimate the potential view context (e.g. distance of the observer to the car).
- **Demographic**: From spatial information, demographical data can be derived, such as information on the people living, working, or staying in that area.
- **Temporal**: For the advertiser, knowing when ads are shown is important. The temporal context can either be date- and/or time-based or even further abstracted (e.g. during rush hour, lunch break, after work).
- **Weather**: For several product groups (e.g. cloth, drinks, food), the current weather conditions, temperatures and forecasts may be relevant to deciding which ads to display.
- **History**: Past information on displayed advertisements can facilitate the decision-making process when other context-information is unavailable.

Gathering context. There are two steps to gathering the context: (1) collecting objective data from the context using sensors embedded in the display, and (2) complementing the sensor data with external information. For the first step, location data via a GPS sensor is an obvious option but additional sensors that provide information about proximity (e.g. WLAN or Bluetooth scan) or weather conditions are also possible. Other types of context information (e.g. traffic, weather, demographics) may be gathered from external sources, e.g. of the Internet or from a database.

Matching context and content. One of the challenges of context-aware advertising systems is the matching of ad content to context (e.g. time, location, etc.). In a traditional advertising business, the marketing group decides when, where, and under what circumstances to display the content. With context-aware systems designed to make the decisions about content selection automatically, context information would have to be available in real time. Instead, we recommend giving the advertisers the freedom to specify the conditions in which an ad is shown. Then, the current context of a car and its display(s), the car owner's preferences, and the conditions specified by the advertiser could simply be matched to the context data.

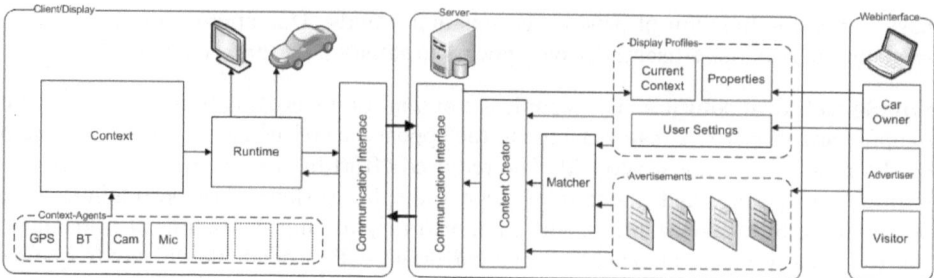

Fig. 1. Architecture of the advertising platform. (1) Client side: context agents gather context data (2) Communication Interface: data exchange between client and server side (3) Server side: front-end for advertiser and car owner.

4 A Platform for a Dynamic Contextual Advertising System

As a proof-of-concept, we implemented a prototypical platform that supports contextual advertising on dynamic displays. The system is set up based on a client-server architecture. In order not to rely upon the car's internal computational resources, we implemented a thin client, requiring as few computations and data storage as possible.

The two core functionalities of our car advertising system are 1) the matching of the current context in which the car is situated with the available advertisements, and 2) the smart and efficient exchange of information between multiple cars and the ad server. In the existing prototype, we follow an *exact matching approach*. Our implementation supports a push-pull communication model: cars send an update of their current context to the server periodically. At the same time, the server itself constantly calculates the best available ad campaign for every client's car. The best-fitting advertisement is then pushed to the client. This is technically realized by dynamically modifying the web page used by the client to display the content.

The front-end of our system consists of two web interfaces – one for the advertiser and one for the car owner. The advertiser's interface provides a way of specifying both the content of the advertising itself, as well as additional settings that define the context in which an ad has to be shown. The decision algorithm needs to know where, when, and under what circumstances to display the ads. The *web interfaces* were implemented using PHP and currently support the input of locations, date, and time, as well as weather and temperature conditions. The advertiser can select the desired area(s) on a map that is realized by the Google Maps API. The advertiser's interface also supports the upload of pre-designed content (e.g. images). In the car owner's interface, the people providing their cars as advertising space can specify their preferences and tell the system what ads they will and will not allow.

The client side includes three different context agents for gathering the context: (1) a GPS agent reporting the location of the client, (2) a time agent providing the time, and (3) a weather agent pulling current weather information and weather forecasts. The server keeps a record of which ads were shown where and on which car. This information can be used to make the car owners context-aware and give them incentives to change their parking behavior (e.g. parking in an highly frequented area will provide a higher revenue than parking in a private backyard), as well as to allow advertisers to assess how successful their campaigns are.

5 Related Work

In the following section, we provide an overview of recent work that relates to context-sensitive car advertisements. We focus on three main areas: (1) location-based mobile advertisements supporting the context-based selection of advertisements, (2) suitable technologies for distributing content among mobile devices, and (3) public displays as suitable media for targeting large audiences with advertisements towards.

Examples of mobile applications that support the selection of advertisements based on context are SMMART [7], Ad-me [4] and SmartRotuaari [8]. The first two approaches use Bluetooth, the third one uses WLAN in order to determine the location of a device. All systems support the concept of making decisions about the to-be-displayed content based on context information.

When it comes to distributing advertisements on mobile clients, two approaches prevail: push (sending information automatically to the client) and pull (sending the information only upon the client's request). Both approaches have been discussed by Ratismor [3] et al. and Varshney [10]. Another interesting approach has been taken by Castro [2]. Castro tries to migrate from a push to a pull model, where users can decide if they wanted to receive more ads after an initial notification.

Finally, research on context-aware advertisements for public displays is closely related to our work, since both types of advertising media try to attract large audiences with different personal backgrounds and interests. Work that focuses on maximizing the exposure of advertisements based on context has been presented by Karam et al. [5] and Kern et al [6]. Both approaches take the interests of the people in the vicinity as context in deciding which advertisements can be tailored to the target group.

6 Conclusion and Future Work

Our survey showed that cars have the potential to become an interesting, dynamic and context-aware advertising space in the future. However, it is essential that users stay in control of their vehicles, have an appropriate means for specifying their preferences and are provided with an agreed-upon compensation model. We recommend that location, demographics, time, weather, and history be considered as important contextual parameters. As a proof-of-concept, we implemented a client/server-based platform that provides a web interface for advertisers and car owners to enter content and preferences. On the client side, several context agents provide real-time data about the current context of the advertising medium. On the server side, a matcher and content creator assemble and prepare web pages for rendering.

Several challenges arise while trying to provide dynamic car advertisements. Although static deployments, such as labels or printed advertisements, are widely available, robust technological solutions for easily setting up context-sensitive advertisements on cars are still under development. National traffic regulations can also lead to additional challenges. For example, according to a telephone interview we conducted with a representative of the German *Technical Monitoring Association (TÜV)*, the use of light-emitting or light-reflecting materials is only allowed after passing a strict approval procedure, due to their high risk of distracting other drivers on the road.

As future work, we plan to extend the advertising platform and further develop the contextual displays to make them robust and easy to deploy. We also aim to conduct a larger-scale study to better understand the needs and preferences of advertisers and car owners and to create an appealing business model for both groups.

References

1. Aalto, L., Göthlin, N., Korhonen, J., Ojala, T.: Bluetooth and WAP push based location-aware mobile advertising system. In: Proceedings of the 2nd international Conference on Mobile Systems, Applications, and Services. MobiSys 2004, pp. 49–58. ACM, New York (2004)
2. Castro, J.E., Shimakawa, H.: Mobile Advertisement System Utilizing User's Contextual Information. In: Proceedings of the 7th international Conference on Mobile Data Management. IEEE Computer Society, Washington (2006)
3. Finin, T., Ratsimor, O., Joshi, A., Yesha, Y.: eNcentive: A Framework for Intelligent Marketing in Mobile Peer-To-Peer Environments. In: Proceedings of the. Fifth Int'l Conf. Electronic Commerce (ICEC) (2003)
4. Hristova, N., O'Hare, G.M.P.: Ad-me: Wireless Advertising Adapted to the User Location, Device and Emotions. In: Proceedings of the 37th Annual Hawaii International Conference on System Sciences (HICSS 2004) (2004)
5. Karam, M., Payne, T., David, E.: Evaluating BluScreen: Usability for Intelligent Pervasive Displays. In: Proceedings of ICPCA 2007 (2007)
6. Kern, D., Harding, M., Storz, O., Davis, N., Schmidt, A.: Shaping how Advertisers See Me: User Views on Implicit and Explicit Profile Capture. In: CHI 2008 Extended Abstracts, pp. 3363–3368 (2008)
7. Kurkovsky, S., Harihar, K.: Using Ubiquitous Computing in Interactive Mobile Marketing. Personal and Ubiquitous Computing 10(4), 227–240 (2006)
8. Ojala, T., Korhonen, J., Aittola, M., Ollila, M., Koivumäki, T., Tähtinen, J., Karjaluoto, H.: SmartRotuaari – Context-aware mobile multimedia services. In: MUM 2003: 2nd International Conference on Mobile and Ubiquitous Multimedia, pp. 9–18 (2003)
9. Tamminen, S., Oulasvirta, A., Toiskallio, K., Kankainen, A.: Understanding mobile contexts. Personal Ubiquitous Comput. 8(2), 135–143 (2004)
10. Varshney, U.: Location management for mobile commerce applications in wireless Internet environment. ACM Trans. Interet Technol. 3, 236–255 (2003)

ReflectiveSigns: Digital Signs That Adapt to Audience Attention

Jörg Müller, Juliane Exeler, Markus Buzeck, and Antonio Krüger

University of Münster, Münster, Germany

Abstract. This paper presents ReflectiveSigns, i.e. digital signage (public electronic displays) that automatically learns the audience preferences for certain content in different contexts and presents content accordingly. Initially, content (videos, images and news) are presented in a random manner. Using cameras installed on the signs, the system observes the audience and detects if someone is watching the content (via face detection). The anonymous view time duration is then stored in a central database, together with date, time and sign location. When scheduling content, the signs calculate the expected view time for each content type depending on sign location and time using a Naive Bayes classifier. Content is then selected randomly, with the probability for each content weighted by the expected view time. The system has been deployed for two months on four digital signs in a university setting using semi-realistic content & content types. We present a first evaluation of this approach that concentrates on major effects and results from interviews with 15 users.

1   Introduction

As display prices drop and cheaper display technologies are invented, digital signs are beginning to be installed everywhere in public spaces, gradually complementing and replacing paper signs. This leads to a radical change in the urban landscape, as can already be observed in places such as Times Square, New York or Shibuya Crossing, Tokyo. On the positive side, a new generation of information access is enabled, as digital signs have many properties and affordances that differ from that of their traditional paper counterparts (e.g. cheap dynamic updates, context adaptivity and interactivity). On the negative side, such signs may lead to visual clutter and information overload for audiences. In addition there are ecological costs by installation & maintenance, power use and recycling. Signage and its content is known to work differently in different contexts. As Mitchell states: "Literary theorists sometimes speak of text as if it was disembodied, but of course it isn't; it always shows up attached to particular physical objects, in particular spatial contexts, and those contexts—like the contexts of speech—furnish essential components of the meaning." [7], p.9. Traditional signs have been adapted to their context for a long time. However for contexts other than location, this has proven laborious (e.g. manually displaying an "Open" sign when a shop is open). When digital signage is equipped with

H. Tokuda et al. (Eds.): Pervasive 2009, LNCS 5538, pp. 17–24, 2009.

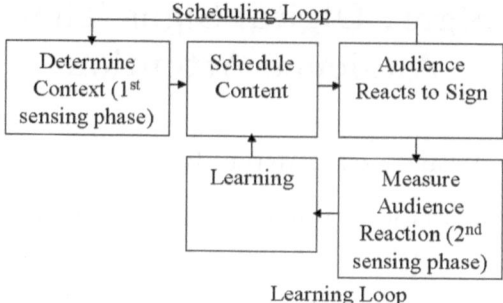

Fig. 1. Information flow in the proposed concept, consisting of the scheduling and learning loops

sensors, this process of adapting to context can be automated. However its been a difficult task for media planners to estimate how content works in different contexts and manually schedule content accordingly. We propose to automate this process by just using the audience as a laboratory. By simply presenting content to an audience, using appropriate sensors it can be observed how the audience reacts to the content shown in a particular context. Machine learning mechanisms (e.g. the Naive Bayes classifier), can then be employed to automatically learn scheduling strategies from these experiences. Thus, the proposed process consists of two feedback loops: A scheduling loop and a learning loop (see Figure 1).

2   Related Work

While bodies of research exist for public displays, ambient displays and context-aware systems, less study has been undertaken for context-aware public displays or learning public displays. GroupCast[5]is an example of public displays that identify the audience via a wireless badge and display content according to a pre-stored user profile. The Vision Kiosk[11] observes the audience with a camera and shows an animated face that looks in the audience's direction. The Interactive Public Ambient Display[12] observes the audience via a motion tracking system and adapts content to their distance and posture. BluScreen[10] is a system that identifies the present audience via Bluetooth-enabled mobile phones and employs auctions to select content the audience has not seen before. MobiDiC[8] is a system that distributes coupons to passers-by, measures advertising effectiveness with the coupons and optimizes content shown with auctions. Huang[3] presents an investigation of digital signage deployments *in the wild* and concludes that most digital signs receive relatively little attention. So, while some work has been done for showing different content in different situations, the task of automatically determining which content works best in various contexts has not been approached yet.

3   Adaptive Digital Signs through Sensing, Learning and Content Scheduling

The concept we propose for learning digital signage consists of two feedback loops (see Figure 1). First, the signs are context adaptive, i.e. they can automatically select content that fits the situation (the scheduling loop). Second, the signs are enabled to automatically learn how the audience reacts to different content in different situations (the learning loop). In the scheduling loop, the signs sense their context with any sensors that are available. Interesting context for selecting appropriate content could for example be the location, the time, weather, gender or age of the audience as well as audience profiles. Also, context that can be influenced by the audience, such as where they are looking, their distance from the sign or facial expression can also be used. From this information, the signs then decide which content to show in this situation. The audience hopefully reacts to it in some way (e.g. by watching, smiling, or interacting with it). Then, again, the context can be measured, and the loop begins anew. As long as users do not consciously understand this process, one would speak of incidental interaction[1], where the system reacts to the actions of the audience but no conscious interaction takes place. However, as soon as the audience understands this process, this loop potentially transforms to classical interaction. For example, the audience could notice that whenever they look sad, the signs would present some jokes. As soon as they start to look sad to make the sign present jokes, one would speak of classical interaction. One major difficulty in this scheduling loop is the creation of scheduling rules, e.g. the decision of which content to show in a certain context. The learning loop automates this process. After presenting particular content, the audience reaction to the content shown is measured with sensors. A learning mechanism can then be employed to learn which content provokes which audience reaction in a certain context.

In the presented prototype, the scheduling and learning mechanisms are designed to measure and maximize the time the audience looks at the signs. At the beginning of a content cycle, each sign determines autonomously which content to show. To determine context, instead of a set of active sensors, ReflectiveSigns currently only uses time and location. The sign uses the current time (in the categories night, morning, lunchtime, afternoon and evening) and its location, to retrieve the expected view time (an estimation of how long the audience is expected to look at the content) for each available content category. Then, a category is selected randomly, where the probability of each category to be selected is weighted by its expected view time relative to the other categories. Thus, content that attracts attention (in terms of time spent looking towards the signs) in a certain context is shown more often.

In the learning loop, currently the only sensor is the face detection that measures the audience's view time and then calculates the expected view time. Whenever some content is shown, the number of faces that are oriented towards the sign as well as the duration of time that these faces look at the sign are determined. The sum of these view times is then stored to a database. In order to be able to estimate the view time even with only few data, we used the

Naive Bayes approach to calculate the expected view time. The expected view time e is then calculated as $e = \sum_{i=1..\infty} p(v = i|l, t)i$, where v is the view time, l the location and t the current time. Under the assumption that location and time are conditionally independent, the Naive Bayes rule[6] is used to estimate $p(v = i|l, t) = \frac{p(l|v=i)p(t|v=i)}{p(l)p(t)}p(v = i)$. In practice, these parameters are simply estimated from historical data. As the system starts with no data, there is a problem of many probabilities being zero at the beginning. This problem is circumvented by applying the m-estimate[6] to individual probabilities. The effect of this is to give the system a set of values for a hot-start.

4   Implementation

The ReflectiveSigns prototype consists of four digital signs installed at a university department with approximately 60 employees. One sign is located at an entrance (see Figure 2), one in a sofa corner, one in a hallway and one in a coffee kitchen. Before being used for this project, the signs were used for the university information system iDisplays[9]. We measure the audience reaction to content shown via cameras installed on top of the signs. The system uses a face detection algorithm[4] that detects faces when they are oriented towards the signs. For the system, we aimed at providing very different kinds of content. Besides the iDisplays system, which has been designed in a user-centered design process [9], we collected videos as well as text and still images that would be eye-catching, interesting and appeal to different people. Such content was somewhat unusual for a research institute (although many comics can be found attached to walls, and employees have used the displays to show sports channels during olympics). This is reflected in the interviews. Content includes video categories such as animated movies, short films showing people who are cooking, football matches or funny animal videos. There are seven non-video categories including landscape photography, three comic strips, textual news, buzzwords and the iDisplays as a mixed information category. The videos are cut into pieces of 20 seconds, still images and text rest on the display for 20 seconds each. Graphics and photographs are scaled to full screen size. All contents are presented without audio. The scheduling algorithm does not decide on individual pieces of

Fig. 2. A user passing a ReflectiveSign, and exemplary content

content but only on categories. Every day there will be a new piece of content for each category. As a consequence, the same items will be displayed multiple times per day. The system consists of four components: face detection software, a MySQL database, a Java-based content scheduler and a Java-based content player. We use the real time face detector from Fraunhofer IIS[4] to analyze the video stream. This software is able to detect multiple faces within the camera image during runtime. The data that is collected by the face detection running on the different machines is stored in the database. The content scheduler decides on a new category to be played every 20 seconds applying the described scheduling mechanism. Based on the category determined by the content scheduler the content player displays one item from this category.

5 Noise

One of the most important problems for ReflectiveSigns is the amount of noise due to the face detection. In order to estimate the error rate, we collected 8 hours of video for two different display locations and hand annotated all view times. In total, 87 views towards the signs occured in this time. The face detection recognized 27 of these views, totaling a recognition rate of 32%. Looking more closely at the nature of the errors however reveals that the face detection only missed views with a duration of under 1 sec. All views with longer duration were correctly recognized. The face detection however also recognized 304 false positives, mostly faces recognized for a single frame in objects like the fire extinguisher. We implemented a filter for false positives by ignoring regions where many faces appeared at exactly the same position. Although methods for coping with people moving to fast or being present but not looking at the sign exist (e.g. high speed cameras and eye detection like Xuuk[1]), the further reduction of error rates is considered future work.

6 Data Collected

The system operated for two months 24 hours per day, seven days a week. The first month served to learn audience attention patterns, the second month to collect data. The data from the second month is analyzed. In total, 38612 views towards the signs were detected. There were obvious effects for different attention towards the signs depending on location, time and content shown. The display installed in the sofa corner received the most attention (mean $(\mu)= .323s$, standard deviation $(sigma) =1.383s$). All times are mean view times when content is shown for 20s. As often nobody is looking the mean values are quite small. However as so much data was collected, most differences are still significant. The sofa corner was followed by the coffee kitchen ($\mu = 0.312s, \sigma = 1.427s$), the hallway ($\mu = 0.229s, \sigma = 1.146$) and the entrance ($\mu = 0.146s, \sigma = 0.920s$). Not surprisingly, attention was highest during lunchtime ($\mu = 0.592s, \sigma = 3.876s$), followed

---

[1] www.xuuk.com

Table 1. ANOVA for location, hour, content, regarding view times. df are the degrees of freedom for that variable (e.g. 24 hours -1), Sum Sq is the summed square error for this variable, and mean sq is weighted by the degrees of freedom. These variables indicate how much variance in the view times can be explained by location, hour, and content, respectively. The last column shows that each of the variables has a significant influence on view times.

	df	Sum Sq	Mean Sq	F value	Pr($>F$)
Location	4	56996	14249	8577.836	$< 2.2e^{-16}$ $***$
Hour	23	3894	169	101.928	$< 2.2e^{-16}$ $***$
Content	18	1178	65	39.385	$< 2.2e^{-16}$ $***$
Residuals	291947	484967	2		

by the afternoon ($\mu = 0.523s, \sigma = 2.728s$), the morning ($\mu = 0.307s, \sigma = 1.424s$) and the evening ($\mu = 0.178s, \sigma = 0.894s$). More interestingly, different content received different degrees of attention. For example, whenever animal videos were shown, they were viewed for 0.287 seconds on average, whereas iDisplays were only viewed for 0.206 seconds. Resulting from this difference, animal videos were shown 28115 times in total, while iDisplays were only presented 21091 times. When we conducted interviews (Section 7), we asked interviewees to rate each content with grades on a scale from 1 (very good) to 6 (bad). Surprisingly, we found no strong correlation between the average viewtime for certain content and these grades (Pearson correlation=-.089, Significance .83). Apparently, user preferences do not significantly influence their attention to display content.

We were interested in how big the influence of the content on audience attention is compared to the influences of location and time. Therefore, we conducted a three-factor analysis of variance on the view times (see Table 1). The influence of all three factors, location, time and content, are all significant (which is no surprise given the large sample of 291,947 content slots of 20s each). More interesting is the relative ordering of the factors (see column for Mean Sq.). This indicates that in our data location has the biggest impact on view times, followed by time. The influence of content on view times is considerably smaller. This is not surprising given that viewing the signs is usually not planned. Nobody will pass a sign only because certain content is being shown. Instead, mere presence of people is obviously only influenced by time and location. Often, people who pass the signs will look at them (or not) regardless of content shown. Presenting the right content only has the opportunity to make users look longer.

7   Interviews

After running ReflectiveSigns for two months, we conducted semi-structured interviews with 15 employees and regular visitors of the institute (age 23-31, $\mu = 27$). The interviews were partially transcribed and evaluated using Grounded Theory[2]. The system was understood with mixed feelings. Five users perceived the system as one that would show random videos and comics. While four users liked this content, three experienced an information overload: "[there is] only trash, always changing videos, simply totally crazy, everything colorful and fast. It drives me crazy." (User 9). Three users critized the system as it was apparently not "useful" (as opposed to the iDisplays shown on the same signs before). Asked what they believed the system was for, three users experienced the display as agressively attracting attention: "The display cries: Hello, here I am!" (U 11). Still, five users liked the (static) comics shown ("It's like a noticeboard, its nice to look there and laugh a bit." (U 13)), the iDisplays, and the surfing videos ("There were surfing videos, sport videos. That was an eye-catcher!" (U 11)). Four users considered the content, especially the videos, annoying. One user stated that he considered videos without sound useless: "For most of the videos you need sound. Because there is no sound, it's not interesting. Videos would be better with sound, but—when you don't like to see it, the sound would be horribly annoying." (U 8). Regarding the observation through the cameras, there were mixed feelings. Four of 15 users were heavily annoyed by the cameras, mainly because they did not understand their functionality: "[the cameras] annoy me because I don't know what happens with the videos taken. I don't want others to know the ways I walk [...]" (U 3). Four said the cameras are OK because they know who put them up. Seven did not care at all about the cameras. There was an interesting effect where incidental interaction (i.e. looking at the sign) turned into conscious interaction. Two users said they tried not to look at the sign when they don't like the content: "I think the content is stupid but then I look there and you know that" (U 5). Asked, what other content they would find interesting, four users mentioned news (esp. regarding the university and the city) and three mentioned sports videos (if short and self-contained). Two users said that they prefer useful content to entertainment: "I consider the display to be more for information, less for entertainment." (U 12). The chosen content apparently annoyed some of the audience, and some were annoyed by the cameras. However, most of them found some of the content interesting.

8   Conclusion

In this paper we have presented ReflectiveSigns, a digital signage system that automatically learns the audience attention for certain content (depending on the context), and presents content accordingly. The system was deployed for two months and evaluated through analysis of the logging data and interviews with users. Somewhat to our surprise, the analysis of variance of the view times indicates that the influence of the chosen content categories on view time is relatively small. Apparently, the right choice of sign location bears a much greater potential than the right choice of content. This is an important finding for the use of public displays in Pervasive Computing scenarios. However, the audience was very homogenous for all locations. If signs attract very different audiences at different locations and times, the impact of the content may be much higher. It was also somewhat surprising to us that there seemed to be no strong correlation between view times and whether users liked the content. It may simply be the case, that users also look at content they don't like. Regarding the cameras, there are three kinds of users. Some disapproved of using cameras at all, some didn't care and for some it seemed OK as long as they trusted those who installed them. The approach presented opens many opportunities for future research. For example, it should be investigated whether a signage system that optimizes for audience attention indeed makes users look more and longer, and if so, how much. It should be further investigated how strongly attention towards different content in various contexts differs. Therefore, the noise in the system needs to be reduced, and more data collected. As such systems appear in urban spaces, visual spam and audience privacy are two major problems that need to be solved to not make them a harmful or annoying experience but beneficial for society.

References

1. Dix, A.: Beyond intention - pushing boundaries with incidental interaction. In: Building Bridges: Interdisciplinary Context-Sensitive Computing, pp. 1–6 (2002)
2. Glaser, B.G., Strauss, A.L.: The Discovery of Grounded Theory: Strategies for Qualitative Research. Aldine Pub. (1967) (2008 edition, 6)
3. Huang, E., Koster, A., Borchers, J.: Overcoming assumptions and uncovering practices: When does the public really look at public displays? In: Indulska, J., Patterson, D.J., Rodden, T., Ott, M. (eds.) Pervasive 2008. LNCS, vol. 5013, pp. 228–243. Springer, Heidelberg (2008)
4. Küblbeck, C., Ernst, A.: Face detection and tracking in video sequences using the modified census transformation. Image and Vision Computing 24 (6), 564–572 (2006)
5. Mccarthy, J.F., Costa, T.J., Liongosari, E.S.: Unicast, outcast & groupcast: Three steps toward ubiquitous, peripheral displays. In: Abowd, G.D., Brumitt, B., Shafer, S. (eds.) UbiComp 2001. LNCS, vol. 2201, pp. 332–345. Springer, Heidelberg (2001)
6. Mitchell, T.M.: Machine Learning. McGraw-Hill Science/Engineering/Math (1997)
7. Mitchell, W.J.: Placing Words. Symbols, Space, and the City. MIT Press, Cambridge (2005)
8. Müller, J., Krüger, A.: How much to bid in digital signage advertising auctions? In: Adjunct proceedings of Pervasive 2007 (2007)
9. Müller, J., Paczkowski, O., Krüger, A.: Situated public news and reminder displays. In: Proc. European Conference on Ambient Intelligence, pp. 248–265 (2007)
10. Payne, T., David, E., Jennings, N.R., Sharifi, M.: Auction mechanisms for efficient advertisement selection on public displays. In: Proceedings of European Conference on Artificial Intelligence, pp. 285–289 (2006)
11. Rehg, J., Loughlin, M., Waters, K.: Vision for a smart kiosk. In: Proc. IEEE Conf. on Computer Vision and Pattern Recognition, pp. 690–696 (1997)
12. Vogel, D., Balakrishnan, R.: Interactive public ambient displays: transitioning from implicit to explicit, public to personal, interaction with multiple users. In: UIST 2004: Proceedings of the 17th annual ACM symposium on User interface software and technology, pp. 137–146. ACM, New York (2004)

Realistic Driving Trips for Location Privacy

John Krumm

Microsoft Research
Microsoft Corporation
One Microsoft Way
Redmond, WA 98052 USA
jckrumm@microsoft.com

Abstract. Simulated, false location reports can be an effective way to confuse a privacy attacker. When a mobile user must transmit his or her location to a central server, these location reports can be accompanied by false reports that, ideally, cannot be distinguished from the true one. The realism of the false reports is important, because otherwise an attacker could filter out all but the real data. Using our database of GPS tracks from over 250 volunteer drivers, we developed probabilistic models of driving behavior and applied the models to create realistic driving trips. The simulations model realistic start and end points, slightly non-optimal routes, realistic driving speeds, and spatially varying GPS noise.

Keywords: location privacy, location-based services, false trips, GPS.

1 Trip Simulations For Privacy

Some location-based services require users to transmit location from their mobile device to a central server. These transmissions can be user-initiated and sporadic, such as a query to find nearby restaurants. Other location transmissions can be periodic and relatively frequent, like those querying for alerts about nearby friends, events, and advertising. These location transmissions and the responses from the server could be compromised by an attacker, resulting in a potentially sensitive privacy leak.

One approach to bolstering privacy is to anonymize the location transmissions by stripping away any identifying information. The server often still requires a pseudonym, however, in order to know how to respond and to whom. It has been shown in [6] that an attacker can find a person's home even with pseudonomized GPS tracks, and [10] shows how such an attack can go further and find the actual name of the victim based on publicly available street address listings. Even using completely anonymized tracks, with no pseudonym, [4] has shown how to find which location points belong together in the same track, effectively creating a pseudonym for each trip.

Another commonly proposed technique for improving location privacy is obfuscation. This approach degrades the transmitted location in some way that reduces the chance that an attacker can find the potential victim's true location. Obfuscation techniques include inaccuracy and imprecision, introduced for location privacy in [1]. Inaccuracy can be achieved by adding random noise to location measurements, and

H. Tokuda et al. (Eds.): Pervasive 2009, LNCS 5538, pp. 25–41, 2009.

imprecision can be achieved by snapping measurements to a grid. Unfortunately, [10] showed that the amount of obfuscation necessary to foil an attack can be very high, *e.g.* an identity attack still worked after adding noise with a 1-kilometer standard deviation. Gruteser and Grunwald [3] introduced *k*-anonymity for location privacy, in which point location reports are replaced by regions containing k-1 other people, another way of achieving imprecision. While obfuscation can be effective, it necessitates the degradation of the location data, which can be fatal for certain applications.

One little-explored but promising technique for location privacy is for the user to send several false location reports along with the real one. The server would respond to all the reports, and the user would ignore all but the response to their actual location. With enough false reports, the chances of an attacker picking the true one could be reduced to an acceptable level. This technique uses no obfuscation, meaning it would still work for location-based services that require accurate and precise point reports, such as alerts of nearby friends and location-based advertising. The only previous work exploring this idea appears to be that of Kido *et al.* [9] who explore an algorithm for reducing the inevitable increase in communication cost.

The effectiveness of false reports depends heavily on minimizing the ability of an attacker to determine which reports are false. Reporting completely random locations is risky, because they may fall at obviously unlikely locations like lakes, oceans, swamps, and rugged mountains. Furthermore, since locations from the true report will follow a plausible path, the false reports must also be plausible paths. Otherwise, the continuity of the true path would be easy to distinguish from the "twinkling" of the false reports.

The Kido paper concentrates on reducing communication costs, so its two proposed false path generation techniques are not emphasized. One of these techniques, "Moving in a Neighborhood", is essentially a random walk model, while the other, "Moving in a Limited Neighborhood", modifies the first to avoid clumping false reports near other users' true locations. However, Duckam *et al.* [6] point out sophisticated techniques that can be used to filter out false reports. For instance, they note that movement may be constrained to a graph, like a road network. Also, people normally move with a goal in mind. Thus, random walk models are likely to be easily identifiable by an attacker who could then strip away all but the true location report.

Related to our work is research on mobility patterns to model the use of wireless networks. For mobile networking, mobility simulations are important for wireless networking with both fixed base stations [11] and mobile peers [1, 3]. Because fixed base stations normally have a large range, the associated mobility models can work at the relatively coarse level of cells surrounding each base station, as in [18]. For mobile ad hoc networks (MANETS), however, finer grained simulations are necessary due to the short range of the participants' radios. Such models are used to help simulate a collection of wireless nodes, such as automobiles, forming a network with no central control. The Random Waypoint model [2] is one of the first simulations relevant to this situation. Here, a subject moves in a straight line toward a randomly chosen waypoint at a randomly chosen speed, then chooses another waypoint and speed, *etc.* Other such mathematical models have been developed since, all aimed at increasing realism. For the case where the mobile nodes are vehicles [8], as in this paper, one of the more sophisticated models constrains the vehicles to a road network, either random or from a real map [20]. These mathematical models fall short of reality, however, because they lack the degrees of freedom to faithfully simulate real drivers. Maximum realism comes from trace-based models that use actual path traces played back from real subjects. These are limited, however, because measuring traces is relatively expensive, especially for high volumes of traffic in cities.

We also note that simulated trips for privacy *vs.* wireless networking have different goals, and therefore different criteria. For instance, mobility simulations for wireless networking often try to account for group behavior and interactions among mobile nodes, because this can affect loads on base stations and present opportunities for messages to hop between peers. For privacy, however, our goal is to fool an attacker, which means we can give many isolated, false trips that do not need to show any regard for each other.

This paper presents simulated traces based on an actual road network. Our method approaches the realism of actual traces by using probabilistic models of driving behavior abstracted from real traces. Our simulated driving trips exhibit these realistic characteristics, all derived from a statistical analysis of actual driving traces:

- Realistic starting and ending points
- Goal-directed routes with randomness
- Random driving speeds
- Spatially varying GPS noise

We can generate an arbitrary number of these traces, all of which adhere to the statistical behaviors we see for actual drivers.

The following sections describe how we model each of these characteristics, preceded by a description of our measured driving data.

2 Multiperson Location Survey

Our statistical behavior models are based on observations of where drivers drive measured from GPS receivers. We have been gathering GPS data from volunteer

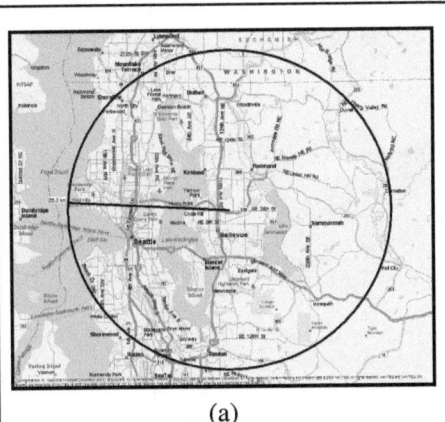
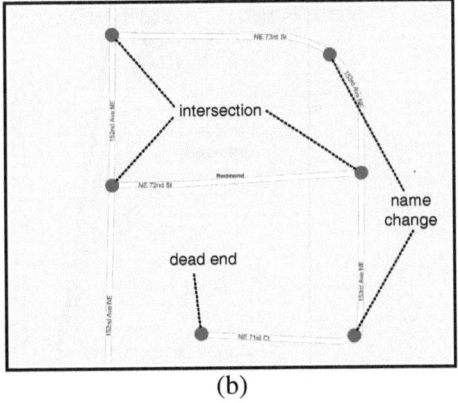

(a)   (b)

Fig. 1. (a) We analyzed and generated trips inside the 20 kilometer radius circle covering the area around Seattle, Washington, USA. (b) Our road network is anchored by nodes that occur at intersections, dead ends, and road name changes.

drivers in our Microsoft Multiperson Location Survey (MSMLS) starting in March of 2004. Volunteer drivers are loaned one of our 55 Garmin Geko 201 GPS receivers, capable of recording 10,000 time-stamped latitude/longitude measurements. The GPS receivers are set to an adaptive recording mode that records more points when the vehicle is moving and accelerating. The median interval between recorded points is 6 seconds and 62 meters.

For this study, we used data from 253 subjects. From these subjects, we have approximately 2.3 million time-stamped latitude/longitude points comprising about 16,000 separate trips. We split the sequence of points into individual trips at gaps of more than five minutes and at apparent speeds of more than 100 miles per hour. We also eliminate trips with fewer than 10 measured points. High apparent speeds and unusually short trips often come from random, noise-induced measurements while a vehicle is parked.

Approximately 80% of our GPS data is contained in a 20 kilometer radius circle centered in the Seattle, Washington, USA region, so we limited our analysis to this area, shown in Figure 1(a).

3 Simulating Trip Endpoints

The first step in our simulation is choosing start and end points of a trip. Vehicle trips normally start and end near a road, and some parts of a geographic region are more popular than others. We attempt to model this behavior, first, by constraining starting

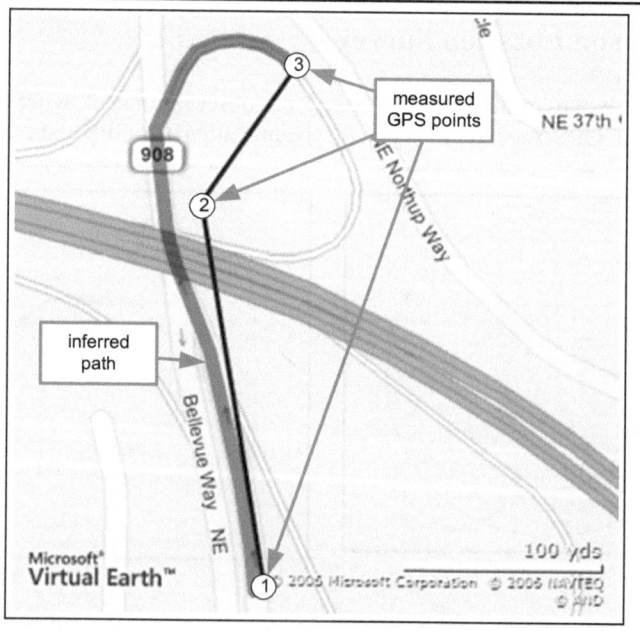

Fig. 2. We used a map-matching algorithm to determine which roads correspond to noisy GPS points. From [11]

and ending points to nodes in a road network. The road network is a graph, in a mathematical sense, where roads are edges and nodes occur at intersections, dead ends, and changes in the road name, as shown in Figure 1(b). Our analysis region (Figure 1(a)) contains 51,637 nodes and 65,549 edges with an average length of 131 meters. While actual trips could start or end almost anywhere, our nodes give a convenient spatial sampling of the geographic space. An attacker may notice that the false trips start only on nodes, but this is mitigated somewhat by the random GPS noise we add, described in Section 7.

Our goal is to compute a probability for each node governing the chances that a trip will start or end there. Toward this end, we first examine our GPS data to find the node nearest to the start and end of each actual trip. In subsequent sections, we need to know the *entire* sequence of nodes for each trip, which we compute with a probabilistic map-matching technique [11], illustrated in Figure 2. This algorithm takes as input a sequence of time-stamped latitude/longitude points and produces a sequence of nodes that best represents the trip. The map-matching algorithm uses a hidden Markov model to produce a route that simultaneously minimizes the GPS error and accounts for the GPS time stamps in light of the road network's connectivity and speed limits. After processing each GPS trip, we have a time-stamped sequence of nodes and edges for each one, including the start and end nodes.

We examined a variety of features of the nodes to compute the probability $P(n_i)$ that a node n_i will be a start or end point of a trip. The features are shown in Table 1. All except the "USGS" (United States Geological Survey) and "Roads attached" features are actually features of road edges, not nodes. To compute the corresponding node feature, we let the attached edges vote for the feature value and take the plurality. For instance, one of the features is called "Autos allowed". This will be true if most of the node's connected roads allow cars to drive on them. The meaning of the binary features is obvious from their names in Table 1. For these features, Table 1 also gives the fraction of the endpoint nodes whose corresponding feature value was "true". For instance, of all the endpoints extracted from the GPS data, a fraction of

Table 1. These are the features that determine the probability of a node being chosen as an endpoint of a trip

Feature	Values	"true" probability
Autos allowed	true/false	1.000
Ferry route	true/false	0.000
Paved road	true/false	0.997
Private road	true/false	0.050
Roundabout	true/false	0.001
Through traffic	true/false	0.965
Toll	true/false	0.000
USGS	21 ground types	--
Roads attached	1,2,3,4,5,6	--
Number of lanes	1,2,4	--
Road type	7 road types	--

0.997 of them were on nodes whose plurality of attached edges was paved. Similarly, no routes started or ended on nodes whose plurality of attached edges were toll roads or ferry routes, which makes intuitive sense.

The "USGS" feature pertains to the ground cover at the node, *e.g.* urban, grasslands, *etc.* The USGS makes available free, digital maps of the U.S. giving a ground cover type for each 30m x 30m square of ground [7]. The 21 ground cover types and the associated probability of an endpoint node landing on them are shown in Figure 3.

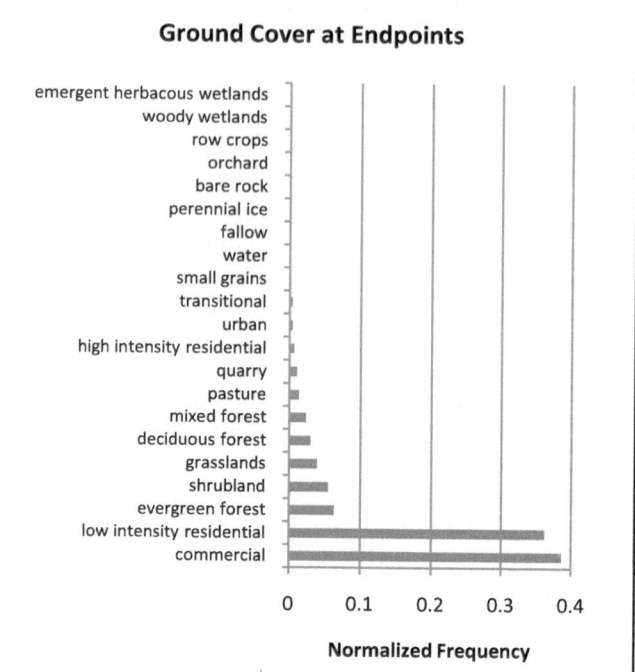

Fig. 3. The relative popularity of trip endpoints varies depending on the ground cover

The "Roads attached" feature counts the number of roads attached to the node. The number of roads attached and associated probabilities of endpoint nodes occurring there are 1 (0.010), 2 (0.152), 3 (0.436), 4 (0.295), 5 (0.014), 6 (0.001).

"Number of lanes" is the plurality of the number of road lanes on the node's connected edges. The number of lanes and probabilities are 1 (0.751), 2 (0.241), 4 (0.009). End points most often occur on single- and double-lane roads.

"Road type" gives the plurality vote of the type of road connected to the node. The probabilities, shown in Figure 4, indicate that highways, ferries, and ramps are unpopular places to start or end a trip.

To compute the probability of a given node being an endpoint, we use a naïve Bayes formulation for the 11 features f_j from Table 1 that says

$$P_{endpoint}(n_i|f_1, f_2, \ldots, f_{11}) = \prod_{j=1}^{11} P_{endpoint}(n_i|f_j) \qquad (1)$$

We take the $P_{endpoint}(n_i|f_j)$ values from the feature probabilities described above. Using naïve Bayes carries a risk of overweighting some features that have correlations with each other, but it has been shown to work well in practice [13]. With this technique, we find that the least popular endpoints are grouped along highways and also appear at the ends of unpaved or private roads.

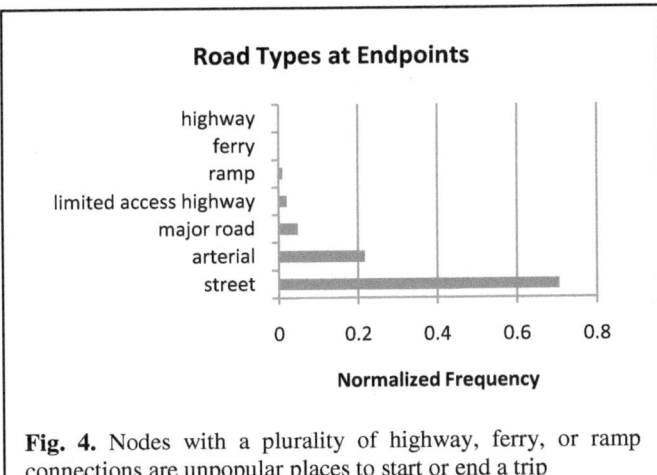

Fig. 4. Nodes with a plurality of highway, ferry, or ramp connections are unpopular places to start or end a trip

The preceding ysis does not guish between the start and end point of a trip, based on the observation that each point normally serves both roles for a typical driver. However, having chosen a random *starting* point based on the probabilities in Equation (1), the *ending* point should not be chosen at an arbitrary distance away. Intuitively, we know that most car trips are measured in minutes, not hours, which limits the range of likely destinations. To quantify this intuition, we used data from the U.S. 2001 National Household Transportation Survey (NHTS) [8]. The NHTS collected data on daily and longer-distance travel from approximately 66,000 U.S. households based on travel diaries kept by participants. A histogram of trip times from this study is shown in Figure 5.

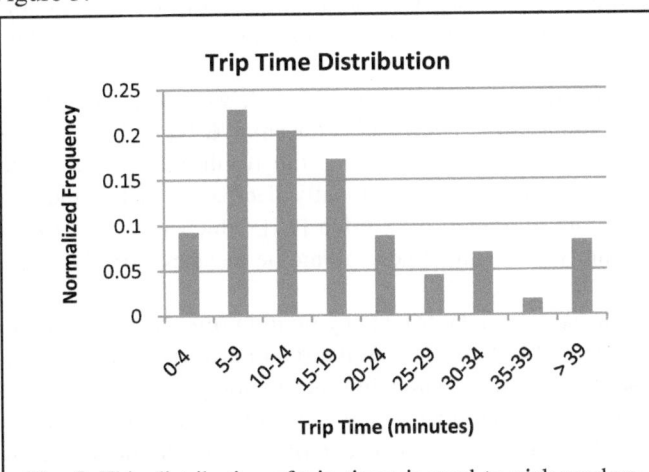

Fig. 5. This distribution of trip times is used to pick random trip destinations

We designate this distribution as $P_{trip\ time}(t)$, where t is the trip time. Having chosen a random starting point n_{start} from Equation (1), we compute the driving times to all the other nodes, designated as $t(n_{start}, n_i)$., using a conventional path planner. The probability of picking a destination node n_i is then

$$
\begin{aligned}
&P_{destination}(n_i | f_1, f_2, \dots, f_{11}, n_{start}) \\
&= P_{endpoint}(n_i | f_1, f_2, \dots, f_{11}) P(n_i | n_{start}) \\
&= P_{endpoint}(n_i | f_1, f_2, \dots, f_{11}) P_{trip\ time}(t(n_{start}, n_i))
\end{aligned}
\tag{2}
$$

This is simply the endpoint probability of the candidate destination node multiplied by the distribution governing trip times, evaluated at the time it would take to drive to the candidate destination. This gives the false trips the same distribution of trip times as the NHTS study suggests.

We use Equations (1) and (2) to randomly chose a start and end point of the trip, respectively. We note that the driving times used for computing $t(n_{start}, n_i)$ are based on the speed limits in the road network database, which may or may not be realistic driving speeds. For choosing a route and ultimately making a time-stamped, simulated trip, we need probability distributions governing the speeds that drivers actually drive. This is the topic of the next section.

4 Speeds at Nodes

For simulating routes and eventually time-stamped location traces, we compute probability distributions of actual driving speeds at every node from our measured GPS data. For each trip from our GPS loggers, the map-matching algorithm generates a sequence of time-stamped locations along the road network. From this, we compute a sequence of time-stamped distances along the trip, (t_i, x_i), $i = 1 \ldots N$. Here there are N points on the trip measured at times t_i. The variable x_i represents the accumulated distance along the trip, with $x_1 = 0$, x_N as the total length of the trip, and x_i monotonically non-decreasing with i. We note that the (t_i, x_i) representation is different from the more obvious, and ultimately less convenient, choice of representing our recorded trips as time-stamped latitude/longitude pairs.

Since we need to sample locations at an arbitrary interval, we interpolate (t_i, x_i) with a one-dimensional cubic spline, which gives $x(t)$ for any $t \in [t_1, t_N]$. A conventional cubic spline is not necessarily monotonic, thus the resulting wiggles in the spline could have the accumulated distance occasionally decreasing with time. We chose the monotonic cubic spline presented by Steffen [15], which is simple to implement and ensures monotonicity with time. Speed along the measured trip is simply $\dot{x}(t)$.

While $\dot{x}(t)$ approximates the speed on the trip at any point in time t, we still do not know which values of t correspond to the nodes in the road network along the driver's route. We need speed samples at these points in order to compute speed distributions at all the nodes. We solve this by computing the accumulated distance along the trip to each node encountered. From this, we can compute which particular spline section pertains to that part of the trip and then find t at that point by solving a cubic equation. Thus, each measured trip gives a sample of the drivers' speeds at each node along the way.

Using a time and distance representation (i.e. (t_i, x_i) as above) proved to be a good alternative over using time-stamp coordinates like $(t_i, \mathrm{lat}_i, \mathrm{long}_i)$. The time and distance representation made it relatively easy to interpolate points along the route without the worry of the interpolant wiggling off the road. It also made it easy to compute speeds with a simple derivative and, in Section 6, to solve a differential equation for filling in simulated locations between nodes.

With sampled speeds at each node, we can compute a histogram of speeds for each node that was encountered in actual driving by our GPS subjects. However, we want speed distributions for all the nodes in our region of study, not just the ones we measured. Toward this end, we abstract away the particular node, replace it with node features, and compute a speed histogram as a function of the feature values. The features we choose for each node are the seven possible road classifications (listed in Figure 4) and the seven possible speed limits of the approach and departure edges. With these features, we can abstract speed distributions from particular, measured nodes into all the nodes in our region of study. These features are intentionally sensitive to the characteristics of the edges used to approach and depart from the node, because we expect speeds to be sensitive to the context surrounding the node. Therefore, the same node could have multiple speed distributions depending on the roads connected to it. An example speed distribution is shown in Figure 6, which shows the speed distribution on a node that connects a limited access highway to an off ramp.

With a four-dimensional feature vector (approach road classification, approach speed limit, departure road classification, departure speed limit), and seven possible values for each dimension, there are $7^4=2401$ possible features vectors. We observed only 434 (18%) in our GPS

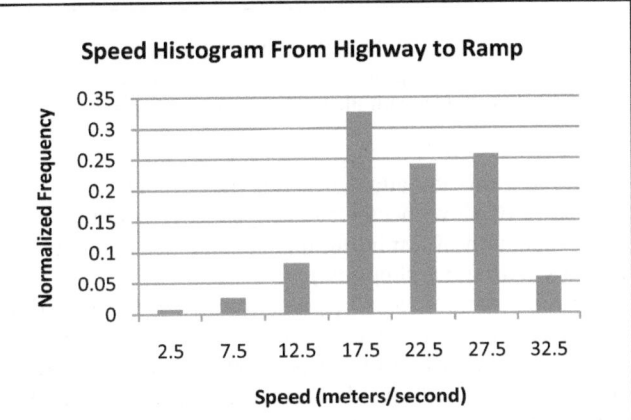

Fig. 6. Speed distribution observed going from a limited access highway (speed limit approximately 26 meters/second) to a ramp (speed limit approximately 11 meters/second). The average value is 21.4 meters/second, showing that drivers are generally slowing down from the highway's maximum speed limit. This is based on 377 observations at intersections of this type.

data. We explain in the subsequent sections how we actually generated random speeds for a node depending on the intended purpose. For each feature vector, we had an average of 2257 observations from our GPS data.

5   Random Routes

Given random start and end points from Section 3, we could use a conventional path planner to find a reasonable, simulated route between them. However, we know from previous research that drivers do not always take the optimal route from the route planner's point of view [12]. Drivers may be unaware of the "optimal" route, they may know a better route, or they may have preferences that go beyond the route-planner's idea of optimal, *i.e.* minimum time. We want our simulated trips to appear

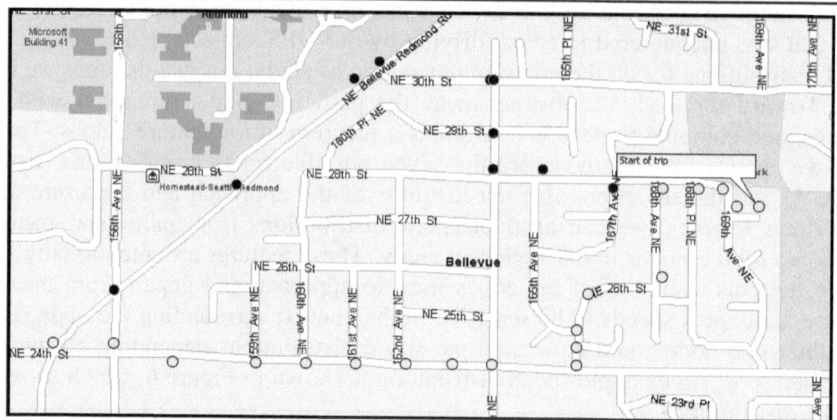

Fig. 7. A trip starts at the right of the figure and moves toward the lower left. The black dots show nodes along the standard path planned using the roads' speed limits. The yellow (lighter) dots show a path planned with random speed limits.

realistic to a privacy attacker. Thus they cannot always be optimal from a route planner's point of view, because that optimality could be easily detected, even for partial trips. (A section of a minimum cost route is still the minimum cost route between the section's start and end points.)

We inject randomness into our routes by injecting randomness into the cost of all the edges. We do this by computing random speeds for the road edges from the speed distributions described in the previous section. Specifically, for each node, we draw randomly generated speeds from the speed distributions using all the possible approach/depart pairs for roads connected to the node. If we have not observed a particular approach/depart pair, we skip it. The speed assigned to the road between a pair of nodes is the average of the random speeds drawn for each of the two nodes. We generate new, random speeds before planning each route, which helps to differentiate different trips between the same start and end points. With these random speeds, we apply a standard A* search algorithm to find the minimum time route.

Figure 7 shows parts of two routes, one generated with the road network's built in speed limits and the other with random speeds as described above. Both appear reasonable.

6 Points along Route

The routes from the previous section demonstrate start points, end points, and routes that are reasonable but random. The next element is the time stamps and locations of points along the route. We want to simulate a GPS taking measurements at any frequency along the route. One simple alternative would be to take speed limits from the original road network representation and apply them to get distance along the route as a function of time. However, drivers do not drive at constant speeds along edges, they do not undergo step changes in speeds at changes in speed limits, and their behavior varies over time.

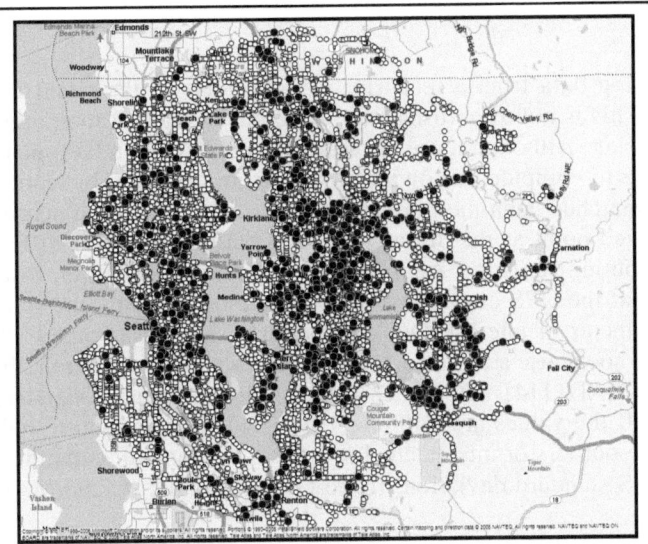

Fig. 8. GPS noise varies with location. The white dots show all the points where we estimated the standard deviation of GPS noise. The black dots show the 5% of points with the largest standard deviation.

We use our random speed distributions again to generate random speeds at each node encountered on the random route. For each node, we know the characteristics of the approach and departure edges, so we use the applicable speed distribution if we have it. If not, the computed speed for the node is the average of the nominal speeds limits on the approach and departure edges. This gives a speed at each node, and we do a linear interpolation of speed between nodes, resulting in a specification giving speed as a function of distance along the route. For example, at nodes i and $i + 1$, the (distance, speed) pairs along the route are (x_i, s_i) and (x_{i+1}, s_{i+1}). We linearly interpolate on distance to get the speeds between the two nodes.

With speed as a function of distance, we have to solve a differential equation to get distance as a function of time, which is what we need to generate points along the route. For example, with linear interpolation along an edge, we have this relationship between speed dx/dt and distance x:

$$\frac{dx}{dt} = mx + b \tag{3}$$

With the initial condition that $x = 0$ when $t = 0$, the solution in terms of t is

$$x = \frac{1}{m}\left(e^{mt + \ln b} - b\right) \tag{4}$$

We move along the route in x as we increment t with whatever Δt we choose. For a computed x along the route, we convert to latitude/longitude using our knowledge of the lengths and coordinates of the route's constituent edges. The result of this step is a sequence of time-stamped latitude/longitude pairs along the route, sampled at whatever frequency we chose. Figure 9 shows the result of this step, where points have been filled in at one per second according to randomly chosen speeds. These points represent the locations where the simulated driver made GPS measurements.

7 GPS Noise

As a final step in simulating data from a real trip, we add noise to the simulated latitude/longitude points. This is not to obfuscate the data, but to make it look more realistic to a potential attacker. Although there are statistics published on GPS inaccuracy, *e.g.* [16], we chose to compute our own statistics from our data. In section 3, we explained how we matched each measured GPS point to a point on a nearby road. We regard the matched point as the driver's actual location, giving us differences in distance for computing statistics. Adopting the Gaussian assumption from [16], we further assume that the GPS errors have zero mean, leaving only the standard deviation as the parameter of interest. Our observations show that some GPS measurements are outliers, so we use a robust estimate of the standard deviation, the median absolute distance (MAD) [14]. The MAD gives a valid estimate of standard deviation even if up to half the values are outliers. This is why, even if up to half our GPS measurements are outliers or mismatched to a road, we can still compute a reasonable estimate of GPS standard deviation. If the GPS errors are d_i, the MAD formula is

$$\sigma = 1.4826 \cdot \text{median}|d_i - \text{median}(d_i)| \tag{5}$$

Here, since we assume that GPS error has zero mean, we replace $\text{median}(d_i)$ with zero. The factor of 1.4826 makes the estimate consistent for Gaussian distributions. We computed $\sigma = 7.65$ meters using data from all our subjects.

Our observations also show that GPS error varies with location, with higher errors perhaps coming in areas with more obstacles to prevent a clear view of the GPS satellites. With this in mind, we compute a separate GPS error standard deviation for each node we observed in our GPS data. Specifically, for each GPS point matched to a road, we associate that error to the nearest road node and compute each node's standard deviation from its associated errors.

Figure 8 shows in black the 5% of nodes with the highest GPS error. Although there is no obvious pattern, there are several clear clusters of points, indicating areas of extended disruption, caused possibly by trees or buildings.

To add realistic GPS noise to our traces, for each point, we first generate a random direction with a uniform distribution, $\theta \sim U(0, 2\pi)$. We then find the σ associated with the nearest node and generate a random magnitude $d \sim N(0, \sigma)$. The point is then moved by $(\Delta x, \Delta y) = (d \cos \theta, d \sin \theta)$. Figure 9 shows a section of one of our traces, with and without added noise.

Adding noise is the last step of our process. We note that this is the only step that does not abstract away the specific training region. Our simulated start and end points, routes, and speeds are based on generic features that could be extracted from any city without taking GPS data there (*i.e.* the road network and USGS ground cover data). A simple alternative to site-specific training for GPS noise would be to use the same value of σ everywhere. A more interesting alternative would be to learn a model that infers σ as a function of relevant features, perhaps USGS ground cover and the density of nearby buildings.

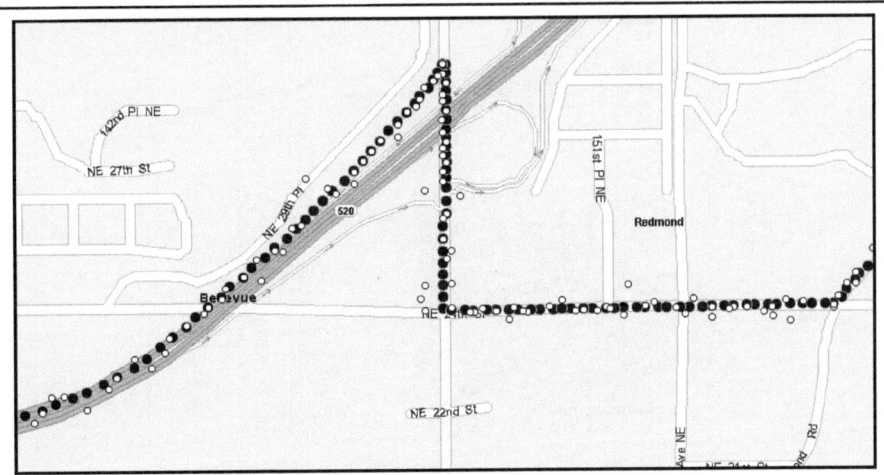

Fig. 9. The larger black dots show points sampled at a rate of one per second, filled in along edges according to randomly generate speeds. The smaller, white dots have had spatially varying, GPS noise added to them.

8 Summary

To summarize, this is the list of steps used to generate a false trip:

1. Trip endpoints – Use features from Table1 to compute the probability of each node serving as a trip endpoint. The start of the trip is chosen according to these probabilities. The end of the trip is chosen according to the same probabilities, augmented with the probability distribution of trip times given in Figure 5. This gives realistic starting and ending points and realistic trip times.

2. Trip speeds – Based on simple learning from GPS traces, compute probabilistic speed distributions for each node as a function of the posted speed limits and types of road approaching and departing each node. For example, Figure 6 gives a speed distribution for going from a limited access highway with a certain speed limit to a ramp with another speed limit.

3. Random routes – Given a random start and end of a trip, generate a route. Instead of using posted speed limits to compute the minimum time route, we use speeds randomly drawn from the speed distributions in the previous step. This makes the routes somewhat random and unpredictable, but still reasonable.

4. Points along route – Draw another set of random speeds at nodes along the computed route. Linear interpolation gives the speed at any point along the route, and solving a simple differential equation gives distance along the route as a function of time.

5. GPS noise – Add spatially varying GPS noise to the previously computed points on the route. The spatial variation was computed based on our sampled GPS data.

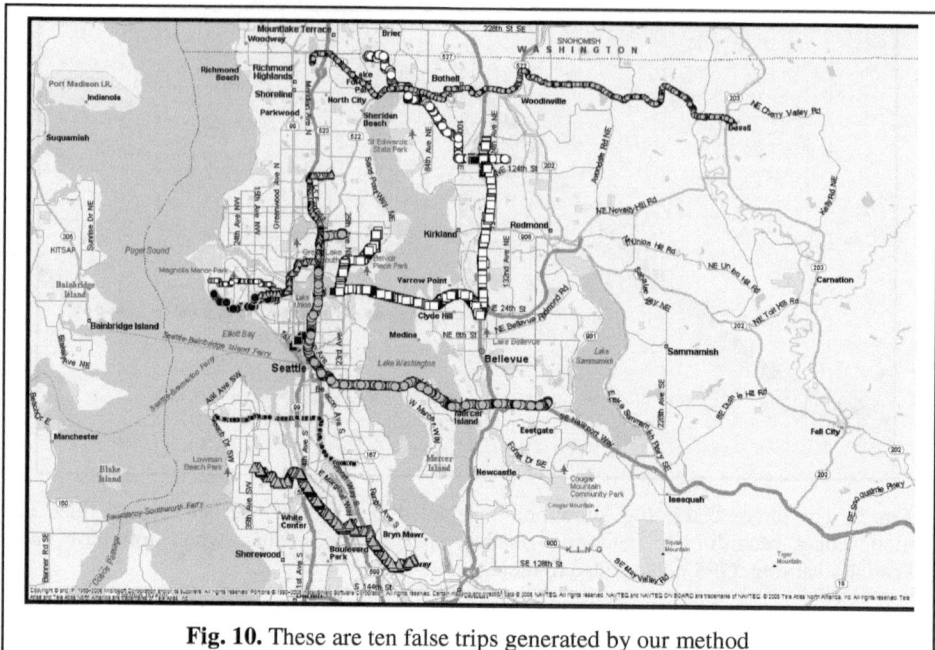

Fig. 10. These are ten false trips generated by our method

9 Discussion

The steps outlined in the preceding sections constitute a method for generating realistic, false trips for location privacy. Some false trips generated from the method are shown in Figure 10. As a way to enhance privacy, the technique's ultimate utility comes in whether or not an attacker could distinguish the false trips from real ones. The likely attack method would be to find some characteristic of real trips and test to see which trips pass the test. The current method incorporates the major characteristics of everyday trips.

Techniques like this should be subjected to scrutiny from unbiased researchers posing as attackers. If they find an unmodeled characteristic that distinguishes false trips from true trips, that characteristic should be incorporated into the simulation. Toward this end, we have made available 1000 simulated trips and 10 real trips from our test area available on a public Web site[1]. The simulated trips come from the technique described in this paper. This site also contains a movie showing the progress of the 1000 false trips on a map. The movie shows that most trips start and end in more urban areas, with fewer in less populated regions.

While ours is one of the first efforts to produce realistic trips for location privacy, there are published criteria for trip simulation. One list of criteria comes from a survey of vehicular simulation techniques for mobile ad hoc networks [5]. Their five "macro-mobility" criteria apply to our technique:

---

[1] http://research.microsoft.com/en-us/um/people/jckrumm/RealisticDrivingTrips/data.htm

- **Graph** – Vehicular models that move on a map-derived graph, like ours, are considered more realistic.
- **Initial Destination and Position** – Our endpoints are not random. They are restricted to the graph and represent characteristics of the endpoints we observe in data.
- **Trip Generation** – Endpoints can be generated based on likely activities of drivers (*e.g.* shopping, entertainment). Our models do not account for this.
- **Path Computation** – Our computed routes are based on random, but plausible, road speeds and thus demonstrate variability similar to actual drivers.
- **Velocity** – We take driving speeds from probability distributions based on our GPS data.

Another list of criteria, for a related purpose, comes from Duckham *et al.*'s [2] speculation on how a privacy attacker might attempt to refine obfuscated location data. The same refinement techniques could be applied to filter out false reports:

- **Maximum/minimum/constant Speed** – Road speeds that deviate significantly from normal are suspicious. Our trips use speeds derived from observations.
- **Connected Refinement** – An attacker would check that a sequence of location reports adheres to a connected graph of locations. Our false trips are consistent with the road network.
- **Goal-directed Refinement** – A trip that wanders aimlessly is unlikely. Our trips move toward a goal, but they do not always follow the optimal path according to published speed limits, thereby enhancing realism.

The benchmark for privacy-related, false trips is the random walk methods in Kido *et al.* [9], which is the only previous attempt we know of. Our trips are sensitive to the road network, the locations where drivers start trips, their destinations, the randomness of their routes, and the speeds they drive. While this is a significant improvement over previous work, there are more trip features to consider:

- **Time Sensitivity** – All our models disregard the time of day, day of the week, *etc.* It is likely that trip characteristics vary with time. For instance, commuters normally leave residential areas in the morning to drive to commercial areas. However, our goal is to simulate plausible trips, not aggregate traffic flows, so time sensitivity is not critically important. It would be easy to retrain our driver behavior models with different time slices.
- **Stops** – Without knowledge of the locations of stop signs, stop lights, and traffic slowdowns, we could not adequately model stops during a trip. While our speed distributions do admit very slow speeds, we do not explicitly model stops nor their durations.
- **GPS Outliers** – We know that GPS receivers occasionally produce outliers, sometimes repeatedly whenever they return to a certain place. We do not attempt to model this.

Increasing realism is not the only way to improve the effectiveness of false reports. It is also worth considering making the true report look more like a false one in order to confuse an attacker. For instance, if the false reports lack fidelity on a micro scale (*e.g.* lane selection before a turn, brief stops), it may be easier to simply add more noise to the false *and* true reports to cover minor infidelities. Decreasing precision and accuracy of location reports is an acknowledged method for protecting privacy [1], and it can make it more difficult for an attacker to distinguish real trips from false ones. Likewise, instead of adding outliers to the false reports, it may be easier to filter outliers from the real reports.

Still unresolved is *when* a privacy-minded client would report false trips – continuously, only while the client is actually moving, random times? It would be possible to build a higher level process that invokes our realistic trips at realistic times of the day to simulate movement and stop patterns over extended periods of time.

10 Conclusion

Generating false trips is one way to enhance location privacy. We generate false trips by abstracting probabilistic models from real trips and using these probabilities to generate random start and end points, random routes, random speeds, and random GPS noise.

References

1. Duckham, M., Kulik, L.: A Formal Model of Obfuscation and Negotiation for Location Privacy. In: Gellersen, H.-W., Want, R., Schmidt, A. (eds.) Pervasive 2005. LNCS, vol. 3468, pp. 152–170. Springer, Heidelberg (2005)
2. Duckham, M., Kulik, L., Birtley, A.: A Spatiotemporal Model of Strategies and Counter Strategies for Location Privacy Protection. In: Raubal, M., Miller, H.J., Frank, A.U., Goodchild, M.F. (eds.) GIScience 2006. LNCS, vol. 4197, pp. 47–64. Springer, Heidelberg (2006)
3. Gruteser, M., Grunwald, D.: Anonymous Usage of Location-Based Services Through Spatial and Temporal Cloaking. In: First ACM/USENIX International Conference on Mobile Systems, Applications, and Services (MobiSys 2003), pp. 31–42. ACM Press, San Francisco (2003)
4. Gruteser, M., Hoh, B.: On the Anonymity of Periodic Location Samples. In: Hutter, D., Ullmann, M. (eds.) SPC 2005. LNCS, vol. 3450, pp. 179–192. Springer, Heidelberg (2005)
5. Harri, J., Filali, F., Bonnet, C.: Mobility Models for Vehicular Ad Hoc Networks: A Survey and Taxonomy, Institut Eurecom, Department of Mobile Communications: Sophia-Antipolis, FRANCE (2007)
6. Hoh, B., et al.: Enhancing Security and Privacy in Traffic-Monitoring Systems. In: IEEE Pervasive Computing Magazine, pp. 38–46. IEEE, Los Alamitos (2006)
7. http://landcover.usgs.gov/ftpdownload.asp
8. Hu, P.S., Reuscher, T.R.: Summary of Travel Trends, National Household Travel Survey, U. S. Department of Transportation, U.S. Federal Highway Administration. p. 135 (2001)
9. Kido, H., Yanagisawa, Y., Satoh, T.: An Anonymous Communication Technique Using Dummies For Location-based Services. In: IEEE International Conference on Pervasive Services 2005 (ICPS 2005), Santorini, Greece, pp. 88–97 (2005)
10. Krumm, J.: Inference Attacks on Location Tracks. In: LaMarca, A., Langheinrich, M., Truong, K.N. (eds.) Pervasive 2007. LNCS, vol. 4480, pp. 127–143. Springer, Heidelberg (2007)

11. Krumm, J., Letchner, J., Horvitz, E.: Map Matching with Travel Time Constraints. In: Society of Automotive Engineers (SAE) 2007 World Congress, Detroit, MI USA (2007)
12. Letchner, J., Krumm, J., Horvitz, E.: Trip Router with Individualized Preferences (TRIP): Incorporating Personalization into Route Planning. In: Eighteenth Conference on Innovative Applications of Artificial Intelligence (IAAI 2006), Boston, Massachusetts USA (2006)
13. Rish, I.: An Empirical Study of the Naive Bayes Classifier. In: IJCAI 2001 Workshop on Empirical Methods in AI (2001)
14. Rousseeuw, P.J., Croux, C.: Alternatives to the Median Absolute Deviation. Journal of the America1 Statistical Association 88(424), 1273–1283 (1993)
15. Steffen, M.: A Simple Method for Monotonic Interpolation in One Dimension. Astronomy and Astrophysics 239(II), 443–450 (1990)
16. van Diggelen, F.: GNSS Accuracy: Lies, Damn Lies, and Statistics. In: GPS World (2007)

Enhancing Navigation Information with Tactile Output Embedded into the Steering Wheel

Dagmar Kern[1], Paul Marshall[2], Eva Hornecker[2,3], Yvonne Rogers[2], and Albrecht Schmidt[1]

[1] Pervasive Computing and User Interface Engineering Group,
University of Duisburg-Essen, Germany
dagmar.kern@uni-due.de, albrecht.schmidt@uni-due.de
[2] Pervasive Interaction Lab, Open University, Milton Keynes, UK
p.marshall@open.ac.uk, y.rogers@open.ac.uk
[3] Dept. of CIS, University of Strathclyde, Glasgow, UK
eva@ehornecker.de

Abstract. Navigation systems are in common use by drivers and typically present information using either audio or visual representations. However, there are many pressures on the driver's cognitive systems in a car and navigational systems can add to this complexity. In this paper, we present two studies which investigated how vibro-tactile representations of navigational information, might be presented to the driver via the steering wheel to ameliorate this problem. Our results show that adding tactile information to existing audio, or particularly visual representations, can improve both driving performance and experience.

1 Introduction

Humans are limited in what they can simultaneously perceive. This is particularly noticeable when driving a car and trying to do something else at the same time, such as talking and changing the radio channel. In-car navigation systems are now making new demands on a driver's attention [3]. Extensive research has been carried out to investigate how this cognitive demand might be reduced through the provision of various kinds of collision detection systems and on how best to warn the driver of possible collision using different modalities and also representing information in multiple modalities (see [13]). Our research focuses on a less safety-critical aspect of driving, although one that is still affected by the multiple stresses on a driver's attention: navigating when using an in-car navigation system.

In-car navigation systems are common and many drivers use them regularly. Typically, three types of systems are in use: (1) built-in systems offered by the manufacturer (2) specific navigation add-on devices offered by third party companies, and (3) navigation applications on mobile phones which include GPS. Sales trends show that these devices are increasingly being used and that it will be the norm to use a navigation system within the next few years[1]. In our work we are investigating how vibro-tactile

---

[1] http://www.telematicsresearch.com/PDFs/TRG_Press_Jan_08.pdf

H. Tokuda et al. (Eds.): Pervasive 2009, LNCS 5538, pp. 42–58, 2009.

output, as an additional channel, can help to provide navigation information without interfering with the overall user experience and without distracting the driver.

From a technical perspective, the devices of built-in systems are more tightly integrated with the car's sensors, displays, and speaker system. The main navigation screen in these systems is often of high fidelity and commonly only shared with other information and entertainment systems in the car. Hence, when using one of the other functions (e.g. browsing the music collection or looking up weather or news) this screen is not used for navigation. In many built-in designs there is an additional display of smaller size (e.g. in the dashboard or the head-up-display) that shows only the next action for the driver. The audio output of built-in systems is linked to other audio sources in the car and, hence, it can be prioritized over entertainment content. However, if the user listens to the radio, to music or information, the interruption is disruptive and interferes with the user experience.

In contrast, add-on devices typically provide an additional single screen that can be exclusively used for the navigation task. The audio output is provided by additional speakers, but which compete with the in-car audio system for the user's attention. Some of these devices can be linked to the in-car audio system via Bluetooth for a tighter level of integration. Navigational applications on mobile phones are similar to add-on devices with regard to their output capabilities, with the exception that the output channels may be shared with other applications on the phone (e.g. SMS, music player, calling) and hence the output channel is not exclusive to the navigation application.

All of these systems provide visual and audio output to convey information about the recommended driving direction to the user. The complexity of the information presented varies from simple directional indicators (e.g. an arrow that indicates the driver should turn right or left at the next crossing) to complex 3D scenes (e.g. a first person view of the geographical surrounding with an added arrow indicating driving directions) and map views. The additional audio information can also vary in complexity, ranging from simple commands (e.g. "turn right") to longer explanations (e.g. "take the next exit and continue towards highway 7").

If visual and audio output are present and the user concentrates on the driving task then current systems work very well. However, this optimal scenario often fails to occur in real driving scenarios as drivers engage in many tasks while driving, ranging from social conversation with passengers, talking on the phone or consuming entertainment such as music or audio books. These additional tasks are important to the driver and contribute significantly to the user experience. For example, engaging in a conversation or listening to an audio book can keep the driver alert and may make a trip seem shorter. The audio output of current navigation systems fails to integrate well with these practices and hence can negatively affect the user experience.

Answers given by participants in our user studies indicated that audio output is problematic for many users of these navigation systems. They deal with this issue in different ways. A common approach is to mute the navigation system while in conversation or listening to the radio or music, and to rely exclusively on visual information. However, people reported that this can lead to missing turns as the audio doesn't prompt them to look at the display. In this situation, the driver either has to focus on the navigation system or risk missing important information.

These considerations, and previous work on tactile driver warning systems, e.g. [6] motivated us to look at different modalities for presenting navigation information to the driver. Our hypothesis was that vibro-tactile signals might be less intrusive than audio signals and interfere less with other activities. Our study therefore explores the design space of different modalities for presenting information to the driver. We created a prototype to explore the utility of vibro-tactile feedback in the steering wheel both for transmission of simple information and as an additional modality that supplements the conventional channels.

2 Prototype and Design Space

To build our prototype navigational system, we first assessed potential locations in which to present vibro-tactile output in terms of feasibility and user experience. To make vibro-tactile output useful as an additional modality a central requirement is that the actuators are in constant contact with the user. This leaves three potential options for integration: steering wheel, pedals and floor, and the driver seat.

We decided to explore the design space for the steering wheel. Some car manufactures have recently added vibration output to their steering wheels for warning signals e.g. Audi[2]. The whole steering wheel vibrates to provide binary information. There has also been initial work on providing tactile information in the steering wheel to communicate more specific information that inspired our prototype [4]. The seat has been used to provide coarse tactile information, e.g., for warnings[3] or other information [10, 14].

The steering wheel is used with hands and fingers, which are very sensitive to tactile information. Additionally, in contrast to the body (driver seat) or feet (pedals), fingers are usually bare, making it easier to provide rich tactile information. To explore the design space we created a prototype steering wheel with integrated tactile actuators. An advantage of integrating the signal into the steering wheel is that the signal itself might intuitively prompt the driver to turn the wheel using a direct physical mapping [8], nudging and tugging the driver in the correct direction. This approach has been successfully employed, for example with a shoulder-tapping system for visually impaired people [11] which was preferred over and engendered better performance than audio feedback. According to research on stimulus-response compatibility (see [9]) spatially corresponding mappings yield better performance than non-corresponding mappings, and matching modes of stimuli and response (e.g. manual responses to visuo-spatial stimuli). This further motivates investigation of vibro-tactile cues in the steering wheel.

The system consisted of a microcontroller (PIC 18F252), 6 power drivers, 6 vibration motors, and a Bluetooth communication module (Linkmatik). The microcontroller ran a small application that received commands from the serial line (via Bluetooth) and controlled the vibration motors using a pulse-width-modulation via power drivers. Via the Bluetooth module, the prototype can be connected to a test application or the

---

[2] http://www.audiworld.com/news/05/naias/aaqc/content5.shtml
[3] http://www.citroen.com.hk/tech/sec_04.htm

navigation system. Each vibration actuator could be controlled individually with regard to intensity and duration of tactile output. The minimal perceptible duration for the on-time of the motor is about 300ms and about 5 levels of intensity could be discriminated. Any software that can send command strings over the Bluetooth serial link could generate the control commands. In our experimental setup we used Flash and Java on a PC to control the hardware.

The physical design was a steering wheel the same size as that found in cars. The vibration motors (6 x 3.5 cm) were integrated on the outer rim of the wheel under a layer of rubber (see fig 1). It was attached on top of a gaming steering wheel used to control car racing games (logitec). This acted as controller for our simulated driving task.

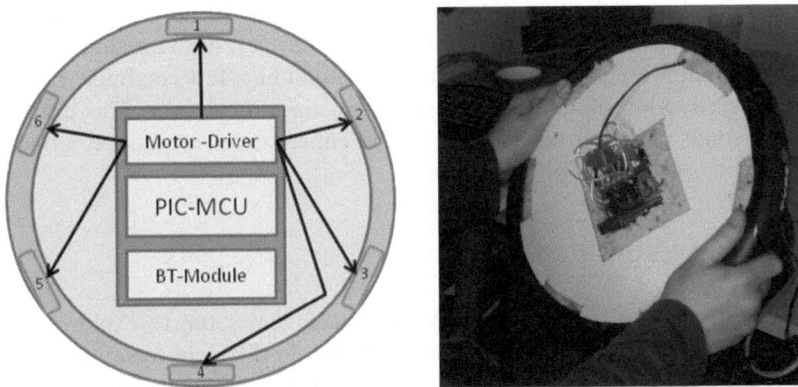

Fig. 1. The steering wheel: concept and internal data flow and photo of the prototype used in the study with the elements exposed

In the design of the tactile output we were able to use the following dimensions:

1) number of actuators: each of the six actuators could be used independently; 2) intensity: the intensity of each actuator could be controlled independently from an off-state up to level 5; and 3) timing of the signal: the actuators could receive signals at any time. This enabled us to create static output (e.g. switching on the left side of the steering wheel with a medium intensity for 2 seconds) as well as dynamic patterns (e.g. activating vibration in a circular pattern moving clockwise, with 1 actuator always on and a brief overlap during transitions).

For our comparative studies, we mainly focused on static patterns because our current setup with only six distinct locations (actuators) for the signal limited the fidelity of dynamic patterns and the speed of the traveling signal. Our static pattern consisted of two different vibration signals: 1) vibration on the right side (actuators 2 and 3 turned on) indicating that the driver should turn to the right; and 2) vibration on the left side (actuator 5 and 6 turned on) indicating a left turn.

However, we also used the study as an opportunity to probe the general feasibility of dynamic patterns. We introduced a dynamic circular pattern, where the vibration signal moves along the wheel (i.e. a vibration signal starts at actuator 1 with full intensity, then after 300ms the vibration stops and starts immediately at actuator 2 for the same time with the same intensity and so on). The idea is to lead the driver to turn the wheel in the correct direction by following the moving signal, i.e. when it moves from left to right the driver should turn to the right and vice versa. Dynamic patterns are also an interesting alternative, since they are not affected by extreme turns of the steering wheel and could transmit more complex information. Integrating many small actuators into the wheel would allow the signal to quickly move between adjacent actuators, enabling the user to, for example, feel the vibration move along the fingers of one hand.

In the studies described below we concentrate on simple static vibration signals. This was feasible because our test situation required no extreme turns. Thus, there was no risk of the wheel being turned around to a degree where a vibration on the left side of the wheel might be felt at the driver's right hand. Participants were instructed to keep both hands on the wheel. To ensure that they felt the vibration regardless of where their hands were located (the next motor might be a few centimeters away from the hand) the vibration signal had to be put on maximum intensity. This unfortunately resulted in some vibration transmitting to the entire wheel, negatively affecting the ease of distinguishing left/right vibration.

3   Setup and Experiments

We ran two studies using an almost identical technical setup to explore the design space. Variations were due to the studies being run in different locations and lessons learned from the first study. Both studies utilized the steering wheel prototype and vibration signal (see fig 2).

The first study compared three conditions: a spatially localized audio beep (provided via headphones), a tactile-only condition, and an audio+tactile condition. The second study investigated spoken audio instructions, visual instructions (arrows), and multimodal instructions (visual+audio, audio+tactile, visual+tactile). While the first study aimed at a comparison of signals of similar length and informational content, the second study was designed to closer emulate current navigation systems which employ spoken instructions.

For the simulated driving task we chose a deliberately simple road layout, inspired by the Lane Change Task layout [7]. Our road consisted of three straight lanes. The participants had to drive on the middle lane of the road and to change to the left or right lane whenever they received a corresponding instruction and then return to the middle lane again. They also had to keep to the speed limit indicated by the road signs they were passing. Order and timing of direction instructions were randomized.

The chosen road layout offered the opportunity to easily measure direction recognition and driving performance without the risk that the drivers might turn the steering wheel to an angle where the actuators were not at the left or the right side. Recommended speed limits alternated between 30 and 50 km/h at varying distances. Participants also had to carry out a distractor task. The setup is depicted in fig. 2.

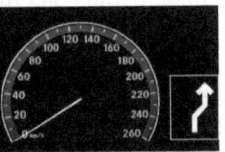

Fig. 2. Setup in the first study with the control panel on a laptop (left), setup in the second study with control panel on a 8" display (middle) and a close-up of the control panel with an direction arrow used in the second study (right)

3.1 Software and Equipment

Participants were seated on a chair in front of our prototype steering wheel. The logitec driving game pedals were located on the floor, taped to the ground, and augmented to provide some resistance to being pressed.

The physical setup can be seen in figure 2. A 42" display behind the steering wheel emulated the view through the front window, showing the road ahead. As a driving simulator we employed CARS[4] , run on a PC. The CARS software was adapted to send messages to the vibration actuators using UDP over a Bluetooth connection. In the first study we utilized a laptop located towards the side of the driver behind the steering wheel to show the speedometer on a control panel, (see fig 2 right). Due to the design of our wheel prototype (with electronics filling the inside of the wheel) the control panel could not be placed directly behind the wheel. In the second study we used an 8" display to show the control panel, this time including navigation instructions for the visual information conditions (see fig 2, middle and right).

The drivers were equipped with a headset that delivered audio information, distracter information and tasks (background music emulating a radio show in the first study and spoken questions in the second study) and additionally shielded off audible noise from the vibration actuators. In the first study a Sennheiser HD 280 pro 64 Ω was used, and in the second study a Philips headset.

3.2 Study 1: Driving with Audio, Tactile or Combined Directional Information

In the first study we utilized spatially localized audio (a beep) as the most direct equivalent to a vibration signal for the audio condition. The audio signal was given by a 140 ms beep following guideline 7 from Green [5] about the duration of signal bursts. In the vibration condition two actuators were activated for 300 ms on the left or right side of the wheel (much shorter signals are not noticeable). The third

---

[4] https://www.pcuie.uni-due.de/projectwiki/index.php/CARS

condition combined audio and vibration. 16 participants took part in this study, with the order of conditions counterbalanced.

As a distractor task participants heard music through their headphones made to resemble a radio station playing standard easy-listening pop music, and were instructed to tell the experimenter when they hear a specific jingle. All music tracks were about a minute long, and the jingle lasted three seconds.

To investigate the general viability of a dynamic vibration pattern for conveying directional information, we presented the participants with a final task after they had completed the three conditions. The actuators were turned on one after another to create a signal moving along the steering wheel either clockwise or anticlockwise. Holding the wheel without any driving task, participants had to recognize and tell the experimenter either verbally or using gestures in which direction the signal was moving. We researched two different conditions: in the first one the signal made one circle of the steering wheel, meaning that each actuator was turned on only once; in the second condition, the signal made two circles of the steering wheel. In each condition they were presented with 16 instances, half running clockwise and the other half anticlockwise in random order.

Design

A within-subjects design was employed, with each subject performing the task in all conditions (in counterbalanced order). Participants were first introduced to the simulator and to the task. The three modalities of directional information were demonstrated: audio, tactile and combined audio+tactile). They were then given six minutes to drive the simulator in order to get used to it with signs on the road giving left-right instructions.

Each condition then lasted six minutes, during which subjects received 18 instructions (nine left and nine right) in random order. The time between instructions was randomly between 15 and 24 seconds. Subjects were instructed to drive in the middle lane and to switch to the left or right lane according to the signal and to come back to the middle lane immediately after having reached the respective lane. At the end, participants were given a questionnaire and asked to rate the conditions according to their preferences (e.g. being annoying or pleasant). Further open-text explanations (e.g. why it was annoying) for their statements were collected, as well as demographic data.

As dependent variables we assessed driving performance, measured in terms of lane keeping (mean deviation from the race line) and compliance to the suggested speed and correctness of lane-shifts in both studies.

As a measure of lane keeping we examined the position of the car on the street in comparison with an ideal race line that we assume participants should drive along (cf. [7]). Every 20 millisecond the standard deviation of the mean distance of the car from the ideal race line was calculated up to this point. To make the calculation of the curves to the left and right lane easier we approximate them also with straight lines, see figure 3.

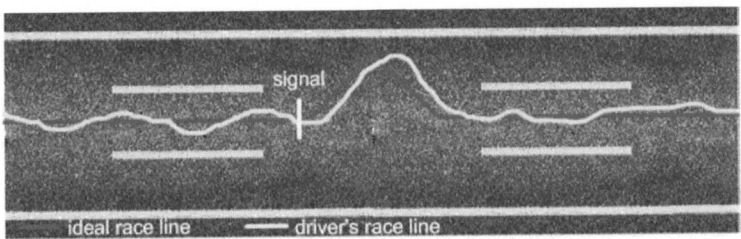

Fig. 3. graphical representation of calculating the driving performance by measuring the standard deviation of the mean distance to the ideal race lane

Participants

16 participants took part in the study: 9 female and 7 male, aged 25 to 52 (mean of 36). All were administrative or research staff from the Open University. Driving experience varied from having held a driving license from 1 year up to 36 years (mean of 15.3 years) . Only 2 people had less than 6 years driving experience. The majority (nine people) drove more than five times per week and only five drove less than once a week. Only one used a navigation system, but reported that they frequently turned off the audio when listening to radio or talking with other passengers.

3.3 Results of User Study 1

Analysis of driving performance data

The effects of representing directional information in different modalities (audio, tactile or audio+tactile) were compared for three measures of driving performance using repeated-measures ANOVAs: likelihood of moving in the correct direction, average speed and mean standard deviation from the race line.

There was an effect of interface condition on participants' accuracy in choosing whether to steer left or right, $F(2, 28) = 14.25$, $p < .001$. Planned comparisons showed that participants were correct less often in the vibration condition ($M = 16.4$) than in either the audio ($M = 17.9$), $p < .01$, or combined condition ($M = 17.9$), $p < .005$. There was no significant difference in accuracy between the audio and combined conditions, $p > .05$. There was no significant effect of the modality of directional information on the average driving speed, $F(2, 30) = 2.42$, $p > .05$. There was also no effect of the modality of directional information on the standard deviation from the race line, $F(2, 30) = 1.04$, $p > .05$.

Therefore, we can conclude that the tactile information led to decreased driving performance compared to the audio and there was no improvement in providing both together. There were however interesting qualitative responses to the different modalities from participants' responses to the questionnaire. These are outlined in the next section.

All participants were able to distinguish the direction of the dynamic vibration signal in the follow-up experiment. The variation of having the signal run twice around the wheel was preferred, as this enabled a confirmation of the initial judgement after the first round. The fidelity of the signal (due to our setup with only six actuators) was not high enough to be easily detected.

Analysis of questionnaire results

The questionnaire asked participants to rate the output modality variation they preferred and to what extent they found each pleasant, annoying or distracting. There was no significant effect of interface condition on the rating of pleasantness, annoyance or distraction. However, 11 participants were found to prefer the audio-only interface and 5 preferred the combined interface. No participant preferred the vibration interface. A Friedman's ANOVA found differences in participants' preferences for different navigational systems to be statistically significant $\chi^2 = 11.37$, p < .005. Post hoc Wilcoxon tests indicated that the audio condition was preferred to the combined condition (p<.05), which was in turn preferred to the tactile condition (p<.05).

All of the participants in the first study gave extensive answers to the free text questions in the questionnaire. These asked them to explain what they liked and disliked, and what was annoying, pleasant or distracting in the three conditions as well as what they thought were advantages or disadvantages. These open text answers as well as remarks by participants during the study indicate that the preference for the audio condition was mostly due to difficulties in distinguishing the direction of the signal and the limitations of our prototype: for example, ensuring that vibration could be felt regardless of how or where participants held the wheel required a maximum intensity signal, which resulted in vibration transmitting across the entire wheel).

Almost two-thirds of the participants mentioned difficulties in distinguishing direction and location of the tactile vibration signal, possibly due to the insufficiencies of our current hardware implementation. Vibration on its own was considered to be less clear or comprehensible than the audio signal. Several participants thought there was a risk of confusing the signal with road vibrations or it being masked. As a practical issue, several participants mentioned that it might be hard to notice if only one hand is on the wheel (although this issue might be alleviated by a more sophisticated setup with many actuators). One possibility is that integrating more actuators into the steering wheel, increasing signal fidelity and reducing its intensity for a more localized signal, would provide a remedy to most of these issues except for the potential interference of road vibration.

Many general problems were listed for the audio-only condition, confirming our hypothesis that alternative modalities would be useful. Half of the subjects mentioned that background noise, conversation and radio could interfere, mask the signal or distract the driver. As a practical issue, hearing impairments were mentioned. The utility of spatially localized sound instead of verbal instructions was questioned, e.g. the audio signal could be masked by other sounds (although it should be noted that the beep signal was used more for reasons of experimental parity than practical utility). Participants furthermore wondered whether it would be feasible without headphones, as they would not want to wear headphones while driving, and were concerned that turning one's head around could lead to a mismapping of directions. Several participants commented on the audio being annoying. Thus, it seems that verbal instructions are superior to more abstract sound, even if they might feel tedious to listen to.

Overall, problems in one condition mirrored advantages of the other: Several people who mentioned background noise/radio as a problem for audio signals listed as an advantage of vibration that it would not be masked by surrounding noise, while the audio signal was listed by the majority as being "easier to notice" and to "distinguish direction".

Participants almost unanimously liked the multimodality of the audio+tactile condition. Its main advantage was seen in providing confirmation and reinforcement of the signal perceived in the other modality, and a backup in case one signal was missed, for example: "alerting more than one sense not to miss it"; "the sound reinforced the vibration"; and "the sound will confirm the vibration if the driver was not sure". A few people were concerned that an inconsistency in the combined signal would be highly confusing and that the combination of two modalities might become overwhelming or distracting when experienced over an extended time.

The questionnaire results led us to continue to explore the design space and to focus on the utility of vibration as auxiliary information. Results and user feedback indicated that this might be a likely avenue for finding benefits. That performance measures for speed and race-line for the vibration-only signal were comparable to the other conditions despite of the limitations of our prototype was encouraging. User feedback confirmed our hypothesis that audio information on its own is felt to be problematic in driving practice due to interference with the radio and passenger chat). Vibration-only might be useful, but needs much better prototypes (better resolution of signal) to be evaluated fairly. Further research in this direction will need to keep in mind users' concerns about one-handed driving and the possibility of road vibrations masking the signal.

3.4 Study 2: A Comparison of Different Forms of Multimodal Directional Information

The questionnaires in the first study revealed a range of concerns regarding spatialized audio (use of headphones while driving, danger of confusing directions when turning head during the audio signal). Furthermore a spatially localized beep sound is too restricted in terms of the information it can convey to be useful for complex driving instructions. The second study therefore investigated a more realistic scenario emulating existing navigation systems. This study investigated whether multimodal information improves performance and whether an auxiliary vibrotactile signal would outperform the existing combination of audio and visual information.

Design

A within-subjects design was again employed: participants took part in 5 conditions in counterbalanced order: audio information alone, visual alone, audio+visual, visual+tactile and audio+tactile.

Information was presented via spoken audio instructions ("please change to the left/right lane") by a female computer voice, and in the visual conditions through an arrow next to the speedometer indicating the direction. The vibration signal, again, was given for 300 ms by two actuators on the left or right side of the wheel.

An audio distractor task was designed to emulate distractions from passenger conversations that interfere with audio navigation information. It consisted of mathematical questions, asking participants to calculate (e.g. "Peter and Paul are 16 together, Paul is nine, how old is Peter?"), with a ten second interval between questions. The volume of questions was lower then the audio instructions. Participants also had to pay attention to visual information by looking out for signs indicating the speed limit and making sure they did not go to fast or slow. All other aspects of the design were identical to the first study.

Participants

17 master's students from the University of Duisburg-Essen participated in the second study: 2 female and 15 male, aged between 23 and 35 (mean 26). Driving experience varied from having held a license for between one and 12 years (mean of 7.8) years. 6 typically drove less than once a week, another 6 between one and four times a week and 5 five to seven times. Half (9 people) used a navigation system. Six reported that they found voice output inappropriate or disturbing when talking with passengers or listening to the radio. One reported turning it off while talking to people and another when listening to the radio. Three never turned it off. Those participants who used a navigation system were asked to specify on a scale from 0 (very often) to 5 (never) how often they miss turns while the voice output is turned off: the mean was 2.96 (standard deviation 0.94).

3.5   Results from User Study 2

Analysis of driving performance data

Participants' driving performance with each of the five representations of directional information (audio, visual, visual+audio, audio+tactile, visual+tactile) were compared using repeated-measures ANOVAs. Modality of the information had no effect on the number of correct lane changes $F(1.9, 30.3) = 2.45$, $p > .05$. There was also no effect of the modality on the average speed, $F(1, 16) = 1.21$, $p > .05$. However, there was a significant effect of information modality on the standard deviation from the race line, $F(4, 64) = 3.40$, $p < .05$. Mean standard deviations from the race line are shown for each condition in fig. 4.

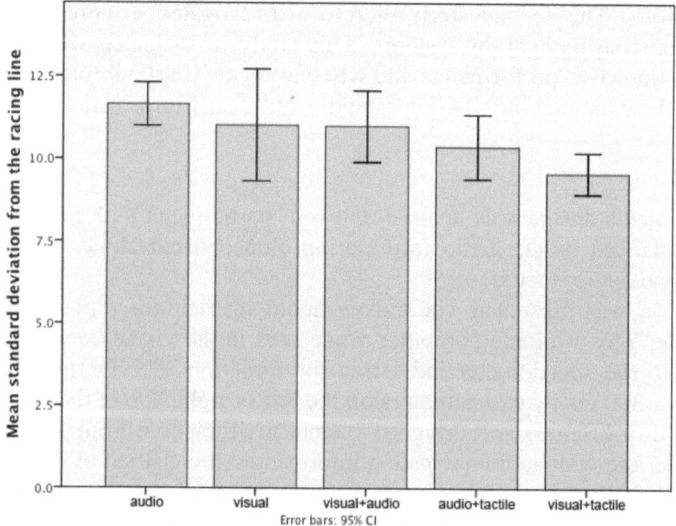

Fig. 4. Mean standard deviation from the race line by condition. The combined tactile and visual condition has the lowest mean standard deviation.

Pairwise comparisons revealed that there was a significant improvement in performance when coupling audio with tactile information compared to audio alone (p < .05); however, there was no improvement when coupling audio with visual information compared to audio alone (p > .05). There was also an improvement in coupling visual and tactile information over visual information alone (p < .05), but no improvement over visual alone when coupled with audio (p > .05). There was no significant difference in performance between the audio+tactile and visual+tactile conditions (p > .05).

Questionnaire Data: Preference ratings

Participants were asked to rate each of the five navigational system configurations in terms of preference from 1 (most preferred) to 5 (least preferred). Preference scores were compared using Friedman's ANOVA. A significant effect of the type of navigational system was found on participants' preferences ($\chi^2(4) = 43.77$, p < .001). Wilcoxon tests were carried out to follow up on this finding. A Bonferroni correction was applied, so all effects are reported at a p<.007 level of significance. Both the visual+tactile (Mdn = 1, T = 3.71, p=.001) and visual+audio (Mdn = 3, T=2.81, p=.005) configurations were preferred to the visual alone (Mdn = 5). Similarly both the audio+tactile (Mdn = 3, T =2.76, p=.006) and visual+audio (T=3.10, p=.002) configurations were preferred to the audio alone. The visual+tactile configuration was also preferred to the other two multi-modal configurations: visual+audio (T= 3.70, p=.001) and audio+tactile (T=3.25, p=.001). There was no significant difference in preference for the audio+tactile and visual+audio configurations. Therefore to summarize, multi-modal are preferred to single modal navigational system and the most preferred multi-modal configuration uses visual and tactile representations.

Questionnaire Data: Ratings of Pleasantness and Annoyance

Participants were asked to score how pleasant and annoying each of the navigation systems were to use, indicating their preference by crossing a line. The distance along the line was then measured and translated into a scale ranging from 0 (not at all) to 5 (very). Mean ratings are shown in figure 5 for both pleasantness and annoyance.

Mauchly's test indicated that the assumption of sphericity had been violated for the pleasantness scores ($\chi^2(9) = 27.6$, p<.05), therefore degrees of freedom were corrected using Greenhouse-Geisser estimates of sphericity (ε=.49). A significant effect of navigational system was found on pleasantness ratings, $F(2.0, 31.4) = 12.3$, p<.001. Planned contrasts revealed that visual+tactile was found to be more pleasant than visual alone (p<.001), visual+audio (p<.001) and audio+tactile (p<.005). No significant differences were found between the audio and visual+audio (p>.05) or audio+tactile (p>.05).

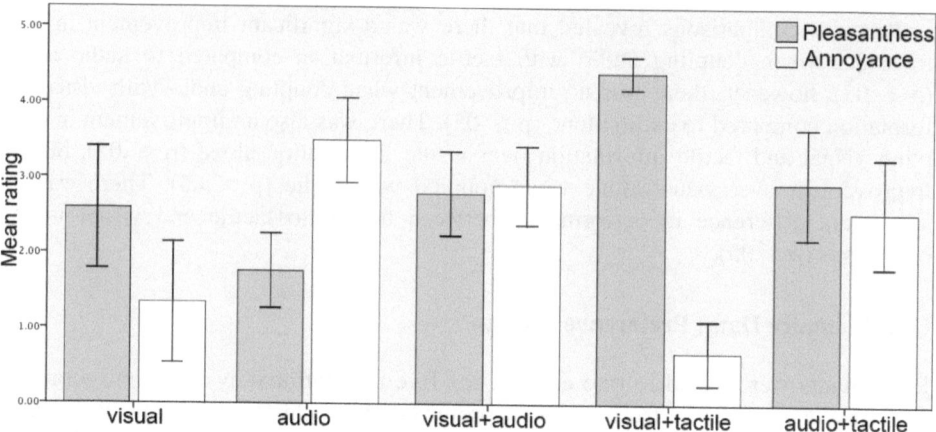

Fig. 5. Mean rating of how pleasant and annoying the conditions were perceived to be (0 = very unpleasant, 5 = very pleasant)

Mauchly's test also indicated that sphericity had been violated for the annoyance ratings ($\chi^2(9) = 31.7$, p<.05)), therefore degrees of freedom were corrected using Greenhouse-Geisser estimates of sphericity (ε=.54). A significant effect of navigational system was again found, $F(2.2, 34.6) = 16.7$, p<.001). Planned contrasts revealed that participants found no difference between visual+tactile and visual alone in terms of how annoying they were (p>.05), but found the visual+audio to be significantly more annoying than either visual-alone (p<.005) or visual+tactile (p<.001). Adding vibration (p<.01) or visual representations (p<.05) to audio were found to make it significantly less annoying. Audio+tactile was found to be significantly more annoying than visual+tactile (p<.001).

In summary, participants tended to find the visual+tactile representations both most pleasant and least annoying. The audio navigational system was found to be particularly annoying and unpleasant. This effect was ameliorated somewhat by combining it with another representation: either tactile or visual.

Questionnaire Data: Ratings of distraction

Participants were also asked to rate how distracting they found each of the navigational systems, again by crossing a line between the extremes of 'very' and 'not at all'. Mean ratings of distraction are represented in figure 6.

Mauchly's test indicated that the assumption of sphericity had been violated for the distraction ratings ($\chi^2(9) = 19.3$, p<.05), therefore degrees of freedom were corrected using Greenhouse-Geisser estimates of sphericity (ε=.58). A significant effect of navigational system was uncovered, $F(2.3, 37.0) = 4.8$, p<.05. Planned contrasts revealed that participants perceived the visual alone system to be more distracting than the visual+tactile (p<.001), but no more distracting than the visual+audio system (p>.05). The audio system was perceived to be neither more nor less distracting than the audio+tactile system (p>.05), or the visual+audio system (p>.05) The visual+tactile system was perceived to be less distracting (p<.05) than the visual+audio system.

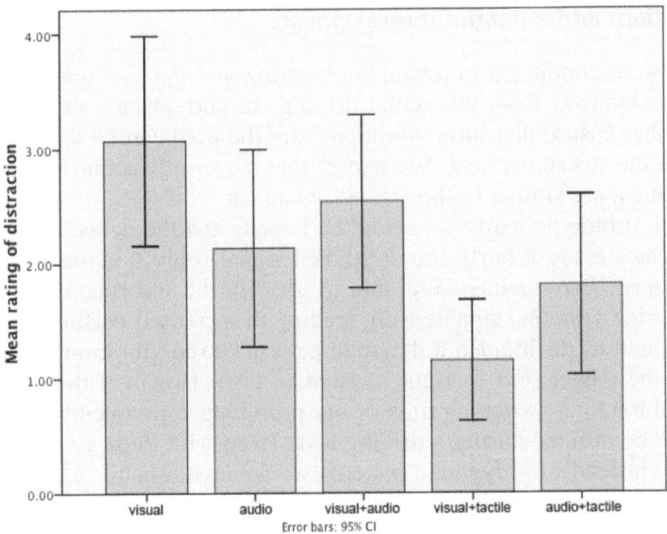

Fig. 6. Mean rating of perceived distraction (5=most, 0=least distraction). The visual-only and the visual+audio condition are considered most distracting.

Questionnaire Data: Summary

The navigational system that combined visual and tactile information came out as a clear winner in participants' questionnaire responses. In the preference ratings, it was preferred to all other modality variations. Multi-modal systems were also preferred generally to the single modality systems. The visual+tactile system was found to be the most pleasant system to use, the least annoying and the least distracting.

The most frequently listed advantage for the audio condition was that audio information allows the driver to keep their eyes on the road (7 times) and that it is very salient (4 times). As a disadvantage, interference with conversation was listed, and that it can quickly become annoying. Participants seemed to perceive as an advantage of the visual information display that it does not distract from driving or listening to passengers. Four people mentioned that its biggest advantage is that one can look a second time and therefore do not need to remember the information. Visual information was considered useful as a back-up and confirmation for another signal that has not been well understood or clearly perceived, in particular since it does not disappear and can be looked up again. The back-up/confirmation function was listed frequently for all of the multimodal conditions.

The biggest disadvantages of visual information, listed most often, are that it requires the driver to look away from the road (listed 10 times) and can be missed as it does not attract attention unless glancing at the display. An auxiliary channel, either audio or vibration, was felt to provide a remedy to both disadvantages. Few people listed any disadvantages for the visual+tactile condition, while visual+audio was listed by a some people as having the 'disadvantages of both'. Vibration was valued as more ambient and less distracting by a few people and also listed as being fast and providing the least distraction from traffic or conversation.

3.6 Limitations and Potential Improvements

The studies were conducted in a simulator setting and not in a car, hence there were no vibrations induced from the actual driving. In current cars there are suspension mechanism that ensure that little vibration from the road can be felt in the car and in particular on the steering wheel. We expect that the results acquired with the simulation environment are similar to those in an actual car.

Due to our prototype hardware setup we have tested the general viability of using vibration signals using a fairly rough-grained signal (only 6 actuators and switching times of 300ms). Participants were able to identify the information from static (left side or right side vibrates) signals well, leading to increased performance. They were furthermore able to distinguish a dynamic pattern of the vibration moving directionally around the wheel (left to right or right to left). However the small number of actuators and the long switching time of our prototype consequently made the pattern too 'slow' to be utilized during a driving task. Even with these limitations of using a static signal (instead of a dynamic pattern) we achieve a better user experience. We expect that with more actuators distributed throughout the steering wheel and a faster-moving signal, the experience could be further improved with vibration being felt to move between the fingers of one hand on the wheel, supporting one-handed driving.

4 Discussion and Conclusions

Presenting information to users during a driving task is challenging. The central goal is to communicate useful information in a timely fashion without creating distraction and without increasing the cognitive load. Navigation devices provide just-in-time information for drivers on upcoming decisions, such as turning at the next corner or changing lanes. Providing this information in small pieces at the time the driver needs it to decide where to go eases the navigation task and hence reduces cognitive load and distraction. However, how this information is provided remains crucial as it is typically presented to the driver in situations where the primary task requires additional caution (e.g. taking a turn or driving off a motorway). The modality in which this information is represented can be critical, especially given the limitations to what the human cognitive system is able to simultaneously perceive. There are multiple potential demands on a driver's attention: talking with passengers, telephone conversations, looking out for potential dangers and in the car's mirrors, to name just a few.

In the research described in this paper we investigated the effects of presenting vibro-tactile information to the driver [6, 14]. In particular, we looked at the effect of presenting navigational cues with vibration output embedded into the steering wheel. Our hypothesis was that as most driver distractions are either visual or auditory, by presenting tactile information, we might minimize the cognitive load associated with navigation. The result of the first study indicated that vibro-tactile information display may not be as beneficial as more conventional auditory display of information in a distracting environment. This was because participants found it more difficult to perceive the direction represented by the tactile information and thus made more directional mistakes. Largely because of this, the participants preferred an auditory interface. We predict that tactile output in our prototype could be improved upon to increase the perceptibility of information (e.g. by using tactons [1]). However, based upon our user feedback, we chose to pursue the different approach of investigating whether representing redundant information in the tactile modality might be beneficial and favoured over single modality setups. In the second study we investigated whether multimodal representation of directional information would be associated with improved driving performance compared to single modality visual and audio representations. We also compared users' qualitative impressions of the different systems using questionnaires.

As predicted, we found the best driving performance in the conditions where there was redundant multi-modal representation of information. However, this performance improvement was only found in the two conditions where audio and visual representations were coupled with vibro-tactile representation and not where visual and audio representations were combined. As the task carried out by the participants was highly demanding of visual and auditory attention, one plausible explanation for this finding is that the participants were able to use the tactile information as a pointer to tell them when to attend to the other forms of information being presented, thus enabling them to offload the cognitive work associated with monitoring for navigational information in the auditory or visual modalities and allowing them to concentrate on the driving and auditory distracter tasks (cf. [12]). Some participants indicated in the questionnaire that they relied primarily on the tactile representation for navigational information, but were able to use visual or auditory information as a backup where they were unsure which direction had been indicated.

This finding is supported and augmented by the questionnaire findings: participants showed a strong preference for the multimodal navigational interfaces, and in particular visual information coupled with tactile information. Participants reported finding audio information on it's own distracting when they were trying to concentrate on speech. This led to an unpleasant experience and annoyance, which was somewhat ameliorated through the simultaneous provision of tactile information.

Our research suggests that the current design of in-car navigational systems, where both visual and audio output are combined, is acceptable for users, but inferior to the combination of visual output and embedded vibration suggested in our work. Our observations suggest that users rely on the vibro-tactile output as a trigger and use the visual display for confirmation and to gain additional information. The main advantage over audio as second modality is that vibration is unobtrusive, does not hinder ongoing conversation, and does not interfere with music or media consumption.

Overall the design recommendation drawn from the results are to present navigational information multimodality combining visual and tactile output. Our results, found that despite using a quite crude form of tactile interface, such a design improves the driving experience and might make it saver.

In further work we will investigate further how vibro-tactile presentation influences driving performance and overall user experience. In particular we are interested what effects spatial distribution, fidelity of tactile output, and timing of the actuators have. A potential way of increasing the fidelity of tactile information might be to use tactile icons or 'tactons' [1,2], where directional information might be associated with a particular tactile pattern.

We also plan to use more sophisticated measures to quantify changes in visual attention when tactile feedback is introduced, using an eyetracker. Here we expect that the driver will look significantly more at the road. Our current hardware includes an acceleration sensor that provides information about the steering angle; in a car similar information could be obtained from the can-bus. Making use of the measured angle of the wheel we plan to compare the effect of output that is relative to the wheel or relative to the car. This is important when the information is presented while the wheel is turned far out of its normal position, e.g. while turning. In the first case output on the left side would always be on the same (originally left) part of the wheel (which may then be on top if turning right) and in the second case output on the left will be always on the left side of the car. From a technical and systems perspective we are currently improving the output actuators (allowing faster switching and greater spatial resolution) and looking at options how to integrate this in an actual car – as built-in solution as well as an add-on device.

References

1. Brewster, S.A., Brown, L.M.: Non-Visual Information Display Using Tactons. In: Extended Abstracts of ACM CHI 2004, pp. 787–788. ACM Press, Vienna (2004)
2. Brown, L.: Tactons: Structured Vibrotactile Messages for Non-Visual Information Display. PhD Thesis, Department of Computing Science, University of Glasgow, Glasgow (2007)
3. Burnett, G.E.: On-the-Move and in Your Car: An Overview of HCI Issues for In-Car Computing. Int. Journal of Mobile Human Computer Interaction 1(1) (2009)
4. Enriquez, M., Afonin, O., Yager, B., Maclean, K.: A pneumatic tactile alerting system for the driving environment. In: Proceedings of the 2001 workshop on perceptive user interfaces, pp. 1–7 (2001)
5. Green, P., Levison, W., Paelke, G., Serafin, C.: Suggested Human Factors Design Guidelines for Driver Information Systems. University of Michigan Transportation Research Institute Technical Report UMTRI-93-21 (1993)
6. Ho, C., Spence, C., Tan, H.Z.: Warning signals go multisensory. In: Proceedings of HCI International 2005, 9, Paper No. 2284, pp. 1–10 (2005a)
7. Mattes, S.: The Lane Change Task as a Tool for driver Distraction Evaluation. In: IHRA-ITS Workshop on Driving Simulator Scenarios (2003)
8. Norman, D.: The Psychology of Everyday Things. Basic Books, New York (1988)
9. Proctor, R.W., Tan, H.Z., Vu, K.-P.L., Gray, R., Spence, C.: Implications of compatibility and cuing effects for multimodal interfaces. In: Proceedings of. International Conference on Human-Computer Interaction, vol. 11. Lawrence Erlbaum Associates, Mahwah (2005)
10. Riener, A., Ferscha, A.: Raising awareness about space via vibro-tactile notifications. In: Proceedings of the 3rd European IEEE Conference on Smart Sensing and Context (EuroSSC 2008) (2008)
11. Ross, D.A., Blasch, B.B.: Wearable Interfaces for Orientation and Wayfinding. In: Proceedings of ASSETS 2000, pp. 193–200. ACM Press, New York (2000)
12. Scaife, M., Rogers, Y.: External cognition: how do graphical representations work? International Journal of Human-Computer Studies 45, 185–213 (1996)
13. Spence, C., Ho, C.: Multisensory warning signals for event perception and safe driving. Theoretical Issues in Ergonomics Science 9(6), 523–554 (2008)
14. Van Erp, J.B.F., van Veen, H.: Vibro-Tactile Information Presentation in Automobiles. In: Proceedings of EuroHaptics 2001, pp. 99–104 (2001)

Landmark-Based Pedestrian Navigation with Enhanced Spatial Reasoning

Harlan Hile[1,2], Radek Grzeszczuk[2], Alan Liu[1,2],
Ramakrishna Vedantham[2], Jana Košecka[2,3], and Gaetano Borriello[1]

[1] University of Washington
[2] Nokia Research Center
[3] George Mason University

Abstract. Computer vision techniques can enhance landmark-based navigation by better utilizing online photo collections. We use spatial reasoning to compute camera poses, which are then registered to the world using GPS information extracted from the image tags. Computed camera pose is used to augment the images with navigational arrows that fit the environment. We develop a system to use high-level reasoning to influence the selection of landmarks along a navigation path, and lower-level reasoning to select appropriate images of those landmarks. We also utilize an image matching pipeline based on robust local descriptors to give users of the system the ability to capture an image and receive navigational instructions overlaid on their current context. These enhancements to our previous navigation system produce a more natural navigation plan and more understandable images in a fully automatic way.

1   Introduction

Mobile phones provide users with highly portable and connected computing devices. Moreover, the trend towards increased performance and inclusion of new sensors such as GPS and cameras in mobile phones make them a compelling platform for location-based services. In particular, navigation is emerging as a critical application for the mobile phone industry. We extend our previous work [1] on automatically generating landmark-based pedestrian navigation instructions with improvements on multiple fronts. In addition to improving landmark selection to provide more natural directions, we utilize computer vision techniques to improve both image selection and the quality of arrows augmenting the image. This extension also allows us to support the live annotation of images as the user follows a path.

Consider the situation of a visitor attending a talk on a university campus. A user can use their GPS enabled mobile device to navigate the campus by entering their desired destination. In the simplest case, this navigation aide may just be calculating a path and displaying it on a map along with the current GPS location. Matching the physical environment to the map may still be challenging, even if there are landmarks labeled on the map, as in Figure 1A. It is also

H. Tokuda et al. (Eds.): Pervasive 2009, LNCS 5538, pp. 59–76, 2009.

Fig. 1. Examples of landmark-based instructions. A (left): A map client with landmarks labeled. B (middle): Using text and canonical images C (right): Our system, utilizing reconstructed camera pose to accurately augment images.

possible to generate text-based instructions referencing landmarks, and provide accompanying images (Figure 1B). This makes it easier for a user to match to the physical environment, but without camera pose information for the images, an image may be chosen that is a significantly different perspective than the user sees. To lower the cognitive load further, we utilize the reconstructed camera pose to choose an image that is similar to the expected view of the user, and to automatically draw accurate arrows on the image, as in Figure 1C and Figure 2. This makes it easier for the user to orient themselves with respect to the images and the path. As a user walks along the path, the GPS location can be used to automatically show the next direction.

Many studies have shown that landmark-based navigation instructions provide significant benefits over map or distance-and-turn based directions [2,3,4]. Landmark-based navigation instructions are easier to follow, shorten the navigation time, and reduce confusion by providing visual feedback on the correctness of a navigation decision. Our previous work addressed the challenge of automatically creating landmark-based navigation instructions by leveraging an existing collection of geotagged images [1]. This work demonstrated the possibility to produce a set of navigation instructions utilizing these images. It also showed that users are able to follow these instructions and preferred them over other types of directions. The user studies from this system guided us toward the multiple improvements presented here.

Our previous system made decisions about which landmarks and images to use on a local basis. Here we improve on this by including higher-level reasoning to choose landmarks across larger regions of the path. This ties into user comments about the prior system indicating that text directions to accompany the images were important. While generating text corresponding to a single image is relatively straightforward, we aim to produce an entire set of directions that fits naturally with the way people navigate. For this reason, we have developed a set of heuristics to guide landmark choice. We also optimize landmark choice over larger sections of the path to provide a smooth flow. In addition, we provide support to fallback on map-based directions when appropriate landmarks or images are not available.

The previous system rendered arrows onto the chosen images using rough estimates for viewing direction and camera tilt. While this worked well in many

Fig. 2. A sample client view of a generated instruction. The current direction is displayed prominently, but a preview of the next step is shown in the top right corner.

cases, it would occasionally produce confusing augmentations due to high GPS error or landmarks and camera poses that violated the standard assumptions. For example, see the uncorrected case in Figure 5. To address these problems we run an automated reconstruction algorithm to solve for full 3D camera poses of the images in the database, where a camera pose is described as a 3D location and a 3D orientation. This corrects for GPS error and provides accurate camera pose information, allowing us to improve low-level spatial reasoning. This information improves both image selection and augmentation, leading to more realistic arrows without any manual labeling. It also allows us to solve for camera pose of a new image, giving us the ability to augment a live image a user has just taken. Live augmentation is done by matching the image provided by the user with the images in our database and computing its pose from the poses of the matching images [5]. Matching is done using a mobile phone implementation of an image matching pipeline based on robust local descriptors [6]. This enhancement opens the possibility of rendering the image as part of a larger context. In summary, our proposed framework provides more compelling landmark-based navigation instructions in a fully automated way.

1.1   Prior Work

Requirement studies and user surveys have shown that landmarks are by far the most predominant navigation cues and should be used as the primary means of providing directions for pedestrians [7]. Goodman *et al.* [8] showed that landmarks are an effective navigation aid for mobile devices—they shorten the navigation time, reduce the risk of getting lost and help older people lower the mental and physical demand required to navigate. See our previous publications for more background on the advantages of using landmarks in navigation on mobile devices [1].

Recently mobile phones are becoming a popular platform for Augmented Reality type applications. Kähäri and Murphy [9] demonstrate such an application running on the newly released Nokia 6210 Navigator mobile phone. It uses an embedded 3D compass, 3D accelerometer, and assisted GPS unit for sensor-based pose estimation. Viewfinder images are augmented with information about the landmark at which the user is pointing the camera. However, sensors are often not accurate enough to allow for precise augmentation. A closed loop approach based on robust image matching supported by our system alleviates this problem. Recently, Google has announced Street View for mobile [10]. The software downloads a panoramic view of the current location and supports walking directions. Unfortunately, it is being reported that data transfer speeds are limiting the usefulness of that system.

Efforts to build a guidance system based on online collections of geotagged photos include the work by Beeharee and Steed [2]. They built a prototype database of geotagged photos by manually entering the location and the viewing direction of each photo. The study found that with landmarks the users finished the route significantly faster than without them, and that users found the landmarks helpful and informative. Since the authors did not use any augmentation of images, the users found some of photographs that were not taken exactly along the navigation path confusing.

We previously extended their work by making landmark images of primary importance in generating directions and automatically augmenting images with navigation directions, which greatly increases the confidence of the users and their understanding of the navigation instructions [1]. The technique of leveraging collections of photographs is a good approach for mobile devices since it is lightweight and does not require special hardware, which are drawbacks of some other systems [11,4]. While this previous work was designed to choose landmarks with important features (good *advance visibility* and *saliency* [12]), it would occasionally result in confusing images due to a lack of quality orientation information. This was partially alleviated by manually labeling a subset of images with orientation information, but it unfortunately limited the number of images available for annotation. The previous approach also reduced landmark choice to a local decision, resulting in a set of instructions that was not very natural for the user.

1.2   Contributions

This work extends pedestrian landmark-based navigation into a number of new areas:

- We propose a set of heuristics to optimize landmark choice over a larger area in order to produce a more comprehensible set of instructions.
- We present an approach to automatically solve for camera orientation and correct poor GPS readings by leveraging computer vision techniques. Our system automatically reconstructs camera pose by using the Photo Tourism system [13]. In addition to providing orientation information for more pictures than the previous approach of manually labeling a subset of images, it also provides complete camera pose information with refined location. The

reconstruction step also serves as a filter for quality images, removing images with poor exposure, excessive clutter, or mislabeling. We extend Photo Tourism's computation by automatically aligning the reconstructed geometry to the world by using (possibly noisy) GPS data associated with the images.

- We support automatic live augmentation of images taken by the user, which was deemed important in the user feedback on our previous system. Instead of using a system based on matching to a 3-D reconstruction as is done in Photo Tourism and other previous work [13,14], we propose a new method based on image-to-image matching that can work with a system designed to run in real-time on a camera phone [6].

- Additionally, since our technique automatically aligns the 3D reconstruction of a landmark with the world data, the system can be further extended to support other features that users asked for in our initial study: features such as zooming out to give users additional context information (for example street information), warping images so that they appear as if they were taken from the current location, highlighting a portion of an image that contains the landmark, etc.

2   System Overview

This section describes the details of our landmark-based pedestrian navigation system with enhanced spatial reasoning. Figure 3 shows the block diagram of the system. Our system supports two types of user interaction: following a set of cached navigational instructions and augmenting a live image taken while the user is en route. Our navigational instructions are built using images from a database of geocoded and labeled images, as described in Section 2.1. In order to produce a set of navigational instructions from an input path, we first apply high-level spatial reasoning to optimize landmark selection over longer sections

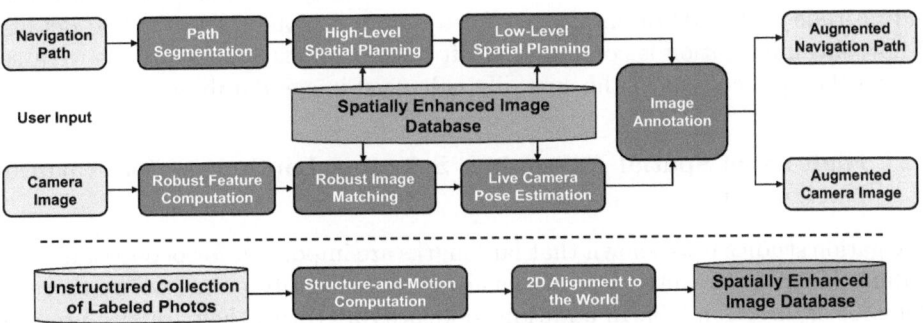

Fig. 3. Block diagram for landmark-based navigation supporting two modes: one that produces a set of augmented images for a user-supplied navigation path and one that augments a user-supplied image. Below the dotted line we show construction of the spatially enhanced image database from an unstructured collection of labeled photos.

of the path (Section 2.2). Next, we apply lower level spatial reasoning to optimize image selection locally and augment the images, as described in Section 2.3. In the case of the live image, we use a robust image matching pipeline adapted to mobile phones to find matching images in our database. We then use the poses of those images to compute the pose of the live image and augment it directly on the device. This is discussed in Section 2.4. Finally, in Sections 2.5 and 2.6 we describe how the unstructured database is processed to produce the spatially enhanced image database that includes 3D reconstruction of the landmark shapes and 3D camera poses for the images aligned to the real world. In each section we also present the results of each step.

2.1   Image Database Organization

We are currently using a database of landmarks and images from an existing outdoor augmented reality project [6][1]. It was populated by many users taking pictures with GPS-enabled camera phones over a period of time. Each image was tagged during capture with the names selected from a list of nearby landmarks. Additionally, most (but not all) of the images were tagged with GPS location. The phones used in collecting the data did not have a built-in compass, hence no orientation information was recorded. In addition, the GPS accuracy is also limited, ranging between 10-100m. We choose to use this data instead of data from Flickr or other photo sharing services because the image tagging found on those sites is generally of poor or inconsistent quality.

The landmarks stored in the database also have an associated GPS location— a single point placed somewhere within the geometric extent of the landmark. This data may come from a mapping service or may be manually entered. As was shown in the past, this crude approximation to landmark location is often not sufficient for estimating accurate camera orientation, which is critical for a correct image augmentation [1]. To deal with this problem, we utilize additional information: a 3D camera pose calculated using computer vision algorithms. The camera pose is defined as a 3D location described in terms of longitude, latitude and altitude and a 3D orientation defined as a unit sphere vector. The details of how this information is computed is left until Sections 2.5 and 2.6, as we first discuss the applications of the spatially enhanced image database.

2.2   High-Level Spatial Reasoning: Selecting Landmarks for Natural Navigation

Navigation studies have shown that landmarks are important for pedestrian wayfinding. Most importantly, landmarks are used to identify points where there is a change of direction [15]. In addition to identifying turns, they are also used to confirm travel in the correct direction [7]. Our system is given a path consisting of a series of GPS coordinates as input and outputs a complete set of navigation

---

[1]   The readers can access our databases through a web-based interface available from this URI: http://mar1.tnt.nokiaip.net/marwebapi/apiindex.

A. "Walk past Physics and Astrophysics and make a right turn"

B. "Make a right turn before reaching Moore Materials Research"

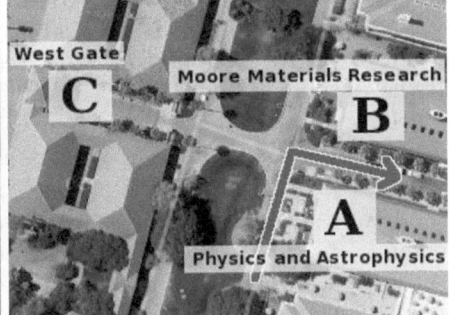

C. "Make a right turn, West Gate will be on your left."

D. Map view of the same instruction, showing landmarks

Fig. 4. Example showing degrading choices of landmark at a corner. *A* uses a landmark inside the corner, producing a natural description. *B* uses a landmark on the inside half. *C* uses a landmark on the opposite side of the turn and is less optimal. *D* shows an option to show a map of the turn, with landmarks used in *A,B* and *C* labeled.

instructions. The previous system made local decisions about landmark choices based on visibility and saliency. We aim to incorporate the ideas of how people naturally navigate by choosing landmarks based on larger regions of the path. We have developed a set of heuristics that we believe will support this.

Our heuristics focus first on turns, since these are the important decision points along a path. Figure 4 shows the landmarks and images chosen for a sample corner using various options. When navigating a turn, it is most natural to reference a landmark on the inside corner of that turn, for example, "Walk past *landmark* and turn" (Fig. 4A). To achieve this, we look for landmarks in the inside quadrant of the turn that have images taken along the path approaching the turn (to measure visibility). If a landmark is not available in this region, we search the inside half of the turn ("Turn before *landmark*," Fig. 4B), and lastly fall back to landmarks on the outside of the turn (Fig. 4C). If there are still no appropriate landmarks, we can produce a map representation of the turn (Fig. 4D). If a landmark exists but has no appropriate images, it can still be referenced in the text directions and labeled on the map.

We also note that a turn consists of two parts: the path before the turn and the path after the turn. The view before the turn gives an indication of where to turn, and the view after the turn serves to confirm that the correct turn was made. For this reason, the navigation client shows the current step and the next step, so the user knows what to expect next, as seen in Figure 2.

When choosing landmarks for straight segments, it is important to choose landmarks with good visibility. However, choosing the landmark with the best visibility at each point along a route can result in a user seeing several different landmarks along a straight segment, even if a single landmark might be visible throughout the entire length. It is preferable to minimize this switching between landmarks in order to provide a more coherent navigation experience. To accomplish this, we define a cost function over the set of landmarks used at each point in a straight segment. We assign a small penalty for using less visible landmarks (in proportion to its visibility rank) and a large penalty for switching landmarks (adjustable to the desired level of landmark stability). We then find the optimal (least costly) set of landmarks using dynamic programming. We believe this produces more natural directions that allow users to navigate using landmarks as waypoints, rather than present a navigation experience that consists of precisely following a series of "micro-steps." Reducing the number of landmarks associated with a route also increases each landmark's significance and can promote learning the path. Although we believe these heuristics to align with desired properties in navigational instructions, we plan to carry out user studies to evaluate their effectiveness.

2.3   Low-Level Spatial Reasoning: Using Reconstruction for Image Selection and Augmentation

The previous step in planning the navigation instructions only selects which landmarks to use at different portions of the path. The next step is to select an appropriate image of that landmark at each location. This is accomplished by using the reconstructed 3D camera poses stored in the database. The computer vision reconstruction also serves as a filter for quality images: images with poor exposure, excessive clutter, or mislabeling are not likely to be reconstructed. This inherent filtering allows us to simply pick a reconstructed image that is close to the path and well aligned with the path. This is an improvement over our previous approach which only selected from a small set of images with high computed saliency, some of which were manually tagged with camera direction information. Having a larger set of images to choose from increases the likelihood of finding a good match to the current path.

Once an image is chosen, the reconstructed camera information can be used to augment the image with navigational instructions. Figure 5 shows an example of how this information is used to improve the quality of image augmentation. Without reconstruction, the camera orientation is estimated by simply using the direction from the GPS location of the camera to the GPS location of the landmark. This will produce poor results when the landmarks are large, when an image capture location is very close to a landmark, or when there is high GPS

Fig. 5. An example of correcting image augmentation using reconstruction data. The goal is to specify a path which passes alongside the building. Compare the uncorrected case (left), which relies on GPS location of the camera and the landmark, to the case which uses the automatically computed camera pose (right). The center map shows the difference in camera angle between the two techniques.

error. This case can be seen in the left image of Figure 5, with the estimated camera direction shown in the center map. In addition to providing a more accurate orientation and corrected GPS coordinates, the camera pose also includes an estimate of tilt. This allows the arrow to be drawn in the correct orientation and also the correct perspective, making it seem better integrated with the image. The resulting image is shown in the right image of Figure 5. Another benefit of having full camera pose information is it opens the possibility of rendering the image in a new view with additional context, as discussed in future work.

2.4   Pose Estimation and Annotation of Live Images

We use the image matching pipeline for mobile phones developed by Takacs *et al.* [6] to support annotation of live images. First we compute the camera pose of the user-provided image from the poses of the matched images stored in our spatially enhanced image database. Next we use the computed camera pose to augment the live image using the same methodology as was described in the previous section for annotation of images from the database. An alternative approach is to try to register the user supplied image with the reconstructed 3D geometry using the structure-and-motion computation discussed in the next section. However, the approach we have chosen allows us to compute the camera pose and annotate the live image directly on the handset, thus reducing latency, bandwidth and computation. The whole process of finding matching images, computing the camera pose and augmenting the live image takes less than 3 seconds on a typical smart phone available today.

Figure 6 shows an example of a live image (shown in the middle) annotated using the camera pose computed from the poses of the two best matching images found in the database (shown on the sides). Given a new query view, we match it to the database using the existing system described by Takacs *et al.* [6]. From the returned images we select the top k images with the largest amount of geometrically consistent matches. Out of these we select the top two views (denoted (R_1, t_1) and (R_2, t_2)) which have been successfully registered in the stage described in Section 2.5 and with locations and orientations in the global world

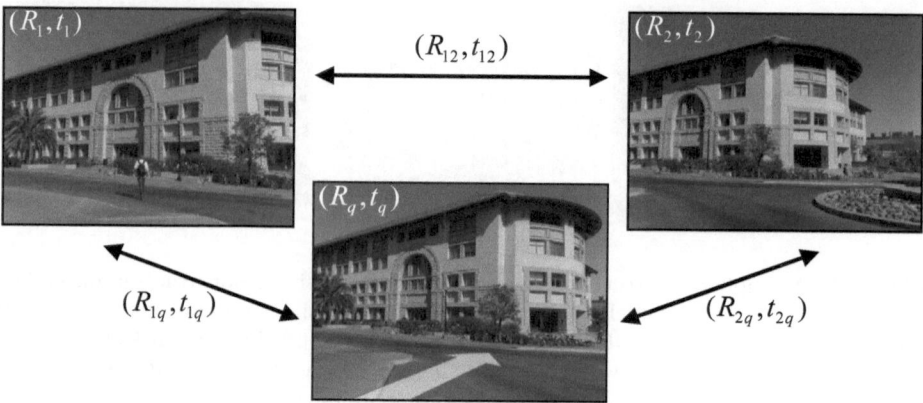

Fig. 6. An example of a live image (shown in the middle) annotated using the camera pose computed from the poses of the two best matching images found in the database (shown on the sides). The camera pose computed using this lightweight technique closely matches the result of the more complex structure-from-motion computation.

coordinate frame derived using the global pose alignment process described in Section 2.6. We denote the relative displacement between these two reference views to be (R_{12}, t_{12}). The position and the orientation of the query view can then be determined by triangulation using the following algorithm. Using the information about the focal length of the query view from the Exif (Exchangable image file format) tag stored in each image, we can compute the essential matrix E and consequently the motion between the query view and the two reference views. Let (R_{1q}, t_{1q}) be the motion between the first reference view and the query view and (R_{2q}, t_{2q}) be the motion between the second reference view and the query view, with both translations computed only up to scale. Note t_{12} and t_{1q} are with respect to the coordinate system of the first view, while t_{2q} is not. For the triangulation to proceed, we need t_{2q} in the coordinate frame of the first reference view $\tilde{t}_{2q} = R_{12}^T t_{2q}$. All three translation vectors are then projected to the ground plane. The three translation vectors in the coordinate system of the first reference view form a triangle. Knowing the absolute scale of t_{12} and the two angles between t_{12} and t_{1q} and t_{2q} we can use simple trigonometry to compute the correct location of the query view. The remaining orientation of the query view in the world coordinate frame is then $R_q = R_1 R_{1q}$.

We tested our pose estimation algorithm by comparing it to the camera pose computed using the structure-and-motion algorithm described below. For each query image we tested, the camera pose computed using our algorithm was nearly identical to the camera pose obtained using the more complex computation.

2.5   Structure-and-Motion Reconstruction

The navigation system enhancements detailed above require a way of automatically computing spatial image details from an unstructured set of images. We

Fig. 7. Structure-and-motion reconstruction results for one of the landmarks in the database. 3D structure shown as a gray point cloud in the back and the reconstructed camera poses shown in red in the front.

propose an algorithm for computing the camera orientations and propagating other meta-data, such as GPS location, to the images in the database that do not contain this information. Additionally, we show that when meta-data such as GPS information or compass orientation exists, we can correct for the error in sensors. This can be accomplished by means of full 3D registration of available views using visual information only, thanks to robust large scale wide base-line matching using scale invariant image features and a structure-and-motion estimation algorithm. At this step we use an open source package developed by Snavely *et al.* [13] which facilitates fully automatic matching and full 3D registration of overlapping views. The core of the incremental and final pose registration algorithm is done by a modified version of the sparse bundle adjustment package of Lourakis and Argyros [16][2].

The structure-and-motion pipeline is used to improve the quality of the meta-data associated with the images in a collection of user contributed photos. Processing is done independently for each landmark. First, we extract from the database all images labeled with the given landmark. The camera registration pipeline detects SIFT features [17] in all the images. Features are matched between images and the resulting matches are pruned by enforcing geometric consistency.

Geometrically consistent views are incrementally registered together using the bundle adjustment algorithm after selecting an initial starting image pair. For more information on this algorithm, refer to the Photo Tourism paper [13]. This algorithm requires significant computation and can take hours to run. However, this is an offline process and the results are easily cached for later use. The

---

[2] The software is available from this URI:
http://phototour.cs.washington.edu/bundler

resulting reconstructed 3D points $\mathbf{X}_j$ and the registered camera poses (R_i, t_i) are given in the reference frame of the initial camera pair.

Figure 7 shows a result of structure-and-motion computation for one of the landmarks in the database. This reconstruction is one of the most detailed, since it was computed from 150 images. Typically our reconstructions are done using between 5 and 15 images, resulting in much sparser 3D structure. However, since we are mainly interested in the camera poses, which are always reconstructed very accurately, this is not an issue. Figure 8 shows results of aligning the reconstructed camera poses to the world.

Often the images representing a landmark will form disconnected clusters (images in separate clusters will not have any features in common). This is particularly common when we are dealing with large landmarks visible from dissimilar viewpoints. Due to this disconnection, it is often not possible to register all images of a landmark in a single stage. Instead of dealing explicitly with clustering of views prior to reconstruction, we deal with this issue by running the bundler registration incrementally. In the first stage we register as many images as possible and reconstruct the first cluster. We then rerun the reconstruction pipeline on the images which were not successfully registered in the previous stage and repeat these steps until no more poses can be reconstructed successfully. Although computationally not optimal, this naturally enables us to keep initializing the registration process with new views, which may have no overlap with the starting image pair. This process typically converges after at most three iterations.

2.6   Aligning 3D Reconstructions to the World

In order to utilize the computed camera pose information, the reconstruction must be aligned to the real world. The structure-and-motion algorithm result is expressed in the reference frame of the first selected image pair, and it is ambiguous up to the similarity transformation comprised of rotation, translation and scale. In order to align these results to the world, we use available GPS information extracted from the image tags. We use the grid-based UTM (Universal Transverse Mercator) coordinate system, since it makes alignment to the metric 3D reconstructions simpler than the latitude/longitude coordinates.

First we compute the gravity to ensure that the y-axis of all camera coordinate frames is perpendicular to the ground plane of the world. Using the formulation described by Szeliski [18], we estimate a global rotation of the entire reconstruction, which minimizes the deviation of the perpendicularity for all camera coordinate frames. All the camera poses can then be projected to the ground plane, where the 2D similarity transformation is then estimated. Since some of the GPS coordinates of the reconstructed images have large errors, we proceed to estimate the 2D similarity transformation $T_s = (R_s, t_s, s)$ in a robust way similar in spirit to the RANSAC algorithm.

The minimal number of poses and corresponding GPS locations needed to estimate the 2D similarity transformation is two. Given two reconstructed camera locations Cp_i and Cp_j and two corresponding GPS sensor readings L_i and L_j,

Fig. 8. Two examples of a pose alignment using our algorithm. High GPS sensor error is recorded in many cases indicated by long red lines connecting green (GPS reading) and red circles (reconstructed position). Yellow lines indicate reconstructed camera direction. Dark blue triangles are synthesized GPS location for images that did not have a GPS reading. The right image shows an alignment computed from the 3D reconstruction shown in Fig. 7.

the scale s can be estimated as the ratio of the two distances $s = \frac{\|L_i - L_j\|}{\|Cp_i - Cp_j\|}$. The translation t_s is then taken as the displacement vector between two mean locations of the chosen image pair $t_s = \bar{C}_p - \bar{L}$, and the rotation angle α_s is the angle between two lines connecting the reference GPS locations and the reconstructed image locations after the translation alignment t_s.

With a few GPS locations that are well distributed (not all close to each other), we can find a good alignment. From the set of n GPS locations we pick randomly 2 GPS locations that are relatively far away from each other and use them to compute a translation, rotation and scale hypothesis. Given the obtained hypothesis, we transform all reconstructed camera poses to obtain their GPS positions. We then evaluate the hypothesis by computing the total alignment error, which is the sum of distances between the original GPS location and the reconstructed GPS location for each camera pose for which GPS information was available. We repeat this hypothesis selection process k times and select the hypothesis which generated the smallest total alignment error. This optimization allows us to correct for errors in GPS positioning.

Figure 8 shows the results of this algorithm for two examples. The green dots indicate the original GPS locations and the red dots indicate the reconstructed camera poses. The original and the reconstructed GPS locations are connected by red line and the the yellow lines show the reconstructed camera orientations. The two cyan pins indicate the two GPS locations that were used to compute the similarity transformation.

It is worth noting that many original GPS locations have a high error (green dots showing on top of a building, or in the middle of a street). Relatively high error is characteristic of the GPS sensors in today's mobile devices. Our algorithm is able to identify this error and correct for it. The reconstructed poses tend to be very accurate, to the point that they can be used as the reference locations. The reconstructed orientations are also computed appropriately, since they are all pointing in the direction of the building facade.

It is difficult to determine the accuracy of this algorithm because ground truth information is not easily obtainable. In order to evaluate the reconstruction and alignment, we performed the following experiment. For a number of reconstructions, we confirmed by visual inspection that the reconstructed camera poses project to the locations where the images were taken from. We then computed the mean/max GPS reprojection error. The average correction across our examples is approximately 6 meters, with a maximum of 47 meters. For a number of images for which we had correct GPS locations, we manually labeled the approximate orientation. We then compared that orientation with the orientation we got from the reconstruction. The typical error was small, on the order of 10 degrees, and well within the error of our manual labeling. We also calculated how much the reconstruction estimate of angle changed over the previous estimate that relied on GPS location of the camera and landmark. For our examples, the average angle correction was 38 degrees, and the maximum observed was 170. This indicates the reconstruction has significant impact on the generated instructions.

An added benefit of computing the structure-and-motion reconstruction is that after the alignment, we can synthesize the GPS locations and the world camera orientations for the images that did not have this information originally. The reconstructed GPS poses for the images with no GPS sensor reading are shown as dark blue triangles in the right image of Figure 8. We can see by visual inspection that these positions are also reconstructed with good accuracy.

Alternative Alignment Method. There are situations where the above described algorithm fails; for example, when all images for a single landmark are

Fig. 9. Two examples of pose alignment using our alternative method for small landmarks. Some GPS locations have very large errors–on the order of a hundred meters. Meanwhile, the reconstructed poses correctly identify that all images were taken from roughly the same location. The orientation and scale are also correctly estimated.

taken from roughly the same location. This may happen when we are dealing with small landmarks, or landmarks with a single interesting feature. In this case, the GPS error starts to dominate the distances between the different GPS samples, which prevents us from having a robust rotation alignment. In those situations, we use a different technique for aligning the poses.

We first compute the mean GPS location $\bar{L}$ and the mean reconstructed camera location $\bar{C}_p$. We then search for those GPS locations that are far away from $\bar{L}$ (the distance is bigger than the mean distance) and consider those locations to be the outliers. We re-estimate $\bar{L}$ and $\bar{C}_p$ using only the inliers. As before, the translation t_s is taken as the displacement vector between the two mean locations: $t_s = \bar{C}_p - \bar{L}$.

We compute the mean orientation by averaging the direction computed from the GPS locations of the inliers to the landmark location (since this algorithm is used only for small landmarks, a single point approximation of the landmark location works quite well). We also compute a mean orientation of the camera viewing direction obtained from the reconstruction. We use the difference in those two directions to determine the alignment rotation R_s. The scale s is determined by computing the ratio of the mean distance to $\bar{L}$ and the mean distance to $\bar{C}_p$. Once we have the proper alignment, we can propagate the correct pose reconstruction to those cameras that were considered to be outliers (or those that do not have the GPS information available).

Figure 9 shows two examples. The green dots indicate the original GPS locations and the red dots represent the reconstructed GPS locations. The yellow lines indicate the reconstructed viewing directions. The two cyan pins connected by a line correspond to the landmark location and the mean GPS location $\bar{L}$. The direction of the line connecting them was used to estimate the rotation angle of the alignment. As we can see, some of the GPS locations have very large errors—on the order of a hundred meters. Meanwhile, the reconstructed poses correctly identify that all images were taken from roughly the same location.

Fig. 10. Left image shows what happens if we apply the original algorithm to the landmark in the left image of Fig. 9. Large GPS error leads to bad pose estimation. Right image shows what happens if we do not perform outlier detection. We get proper orientation but scale is not estimated correctly. This demonstrates the need for our alternative alignment method.

If we were to apply the original alignment algorithm to the landmark shown in the left image of Figure 9, the GPS locations with a large error would be used to estimate the pose orientation (since they are far away from each other), and that would result in an incorrect orientation estimate. This is shown in the left image of Figure 10. The image on the right shows results for the same landmark if we do not eliminate outlier GPS locations. This leads to correct pose estimation, but incorrect scale estimation. This indicates the need for both corrections used in our alternative alignment method.

3   Future Work

Although we have made significant improvements over our previous system, we still have several areas of future work to explore.

Currently we have only used the detailed camera pose information for better augmentation of the images already present in the database or the live images taken by the user while navigating. However, this information could be further leveraged to produce zoomed-out and/or warped views showing more context and better viewing angle. Instead of rendering a directional arrow in the camera space of the image (which often do not contain a view of the desired path), we can render a view of the image using a virtual camera at the path location. This should provide more understandable views and can also include additional context. This approach is still lightweight enough for mobile devices since it uses few images, in contrast to full panoramas or complete 3D models. It also preserves the ability to highlight important landmarks along a route by not showing extraneous information. A conceptual view of this is shown in Figure 11.

Furthermore, the ability to register landmark geometry with the world could be used to increase visual fidelity of the navigation system. For example, the reconstructed geometry could be used to compute a landmark shape proxy. The database images of the landmark could then be projected on the shape proxy, resulting in a simple, compact and photo-realistic representation of the object akin to a surface light field [19]. This full 3D representation of the world would allow us to move away from giving navigation instructions as sequences of static

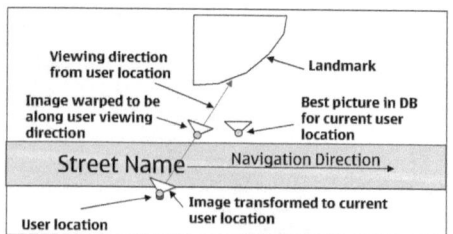

Fig. 11. This shows the concept of leveraging detailed camera pose information to produce zoomed-out view showing more context. The diagram on the left shows the camera location for the available image with respect to the desired camera position for the user. The right shows how this information is put together with additional context to create a representation for the new camera position.

images taken from fixed viewpoints to a continuous unconstrained 3D animation of the navigation route with total freedom in camera path selection.

We also currently take a complete navigation path as input, leaving the path planning to another system. We would like to include path planning in our system to produce a route between two points taking landmark information into account. Instead of simply generating the shortest route, we could generate the easiest to follow route or the most interesting route. Because individuals may have different preferences, we are also working on an adaptive framework to create a personalized user model for selecting better routes that contain more appropriate landmarks [20]. Finally, we plan to conduct user studies to evaluate the enhancements introduced in this work and those planned for the future in terms of qualitative improvements in user experience and clarity of navigation.

4 Conclusion

We have presented a pedestrian landmark-based navigation system with enhanced spatial reasoning. This work extends prior results in the area with new techniques to compute better choices of landmarks and more realistic augmentations of images, resulting in more natural navigation instructions. We enhance the system with a live-matching mode that augments the images of landmarks taken by the user while navigating.

Underlying these improvements are key computer vision technologies. The structure-and-motion reconstruction pipeline for computing the 3D landmark geometry and camera poses is necessary for better spatial reasoning and improved realism of image annotation. The robust image matching pipeline for mobile devices allows for a quick and reliable pose estimation of live images directly on the mobile device. A combination of these technologies leads to significant improvements in the quality of image selection, realism of their augmentation, and novel user-directed functionality. This work also shows a clear path for further improvements involving landmark-based navigation systems, some of which we discussed in the future work section.

References

1. Hile, H., Vedantham, R., Liu, A., Gelfand, N., Cuellar, G., Grzeszczuk, R., Borriello, G.: Landmark-Based Pedestrian Navigation from Collections of Geotagged Photos. In: Proceedings of ACM International Conference on Mobile and Ubiquitous Multimedia (MUM 2008). ACM Press, New York (2008)
2. Beeharee, A.K., Steed, A.: A Natural Wayfinding Exploiting Photos in Pedestrian Navigation Systems. In: MobileHCI 2006: Proc. of the 8th Conf. on Human-Computer Interaction with Mobile Devices and Services, pp. 81–88. ACM Press, New York (2006)
3. Millonig, A., Schechtner, K.: Developing Landmark-Based Pedestrian-Navigation Systems. IEEE Transactions on Intelligent Transportation Systems 8, 43–49 (2007)
4. Kolbe, T.H.: Augmented Videos and Panoramas for Pedestrian Navigation. In: Gartner, G. (ed.) Proceedings of the 2nd Symposium on Location Based Services and TeleCartography (2004)

5. Zhang, W., Košecka, J.: Image Based Localization in Urban Environments. In: International Symposium on 3D Data Processing, Visualization and Transmission, 3DPVT 2006, North Carolina, Chapel Hill (2006)
6. Takacs, G., Chandrasekhar, V., Gelfand, N., Xiong, Y., Chen, W.C., Bismpigiannis, T., Grzeszczuk, R., Pulli, K., Girod, B.: Outdoors Augmented Reality on Mobile Phone using Loxel-Based Visual Feature Organization. In: ACM International Conference on Multimedia Information Retrieval (MIR 2008) (2008)
7. May, A.J., Ross, T., Bayer, S.H., Tarkiainen, M.J.: Pedestrian Navigation Aids: Information Requirements and Design Implications. Personal Ubiquitous Computing 7, 331–338 (2003)
8. Goodman, J., Gray, P., Khammampad, K.: Using Landmarks to Support Older People in Navigation. In: Brewster, S., Dunlop, M.D. (eds.) Mobile HCI 2004. LNCS, vol. 3160, pp. 38–48. Springer, Heidelberg (2004)
9. Kähäri, M., Murphy, D.: Sensor-fusion Based Augmented Reality with off the Shelf Mobile Phone. In: ISMAR 2008 Demo (2008), http://ismar08.org
10. Shankland, S.: Google Street View Goes Mobile. CNET Download (2008), http://www.download.com
11. Reitmayr, G., Schmalstieg, D.: Scalable Techniques for Collaborative Outdoor Augmented Reality. In: ISMAR 2004: Proceedings of the Third IEEE and ACM International Symposium on Mixed and Augmented Reality (ISMAR 2004) (2004)
12. Winter, S.: Route Adaptive Selection of Salient Features. In: Kuhn, W., Worboys, M.F., Timpf, S. (eds.) COSIT 2003. LNCS, vol. 2825, pp. 349–361. Springer, Heidelberg (2003)
13. Snavely, N., Seitz, S.M., Szeliski, R.: Photo Tourism: Exploring Photo Collections in 3D. In: SIGGRAPH Conference Proceedings, pp. 835–846. ACM Press, New York (2006)
14. Hile, H., Borriello, G.: Information Overlay for Camera Phones in Indoor Environments. In: Hightower, J., Schiele, B., Strang, T. (eds.) LoCA 2007. LNCS, vol. 4718, pp. 68–84. Springer, Heidelberg (2007)
15. Look, G., Kottahachchi, B., Laddaga, R., Shrobe, H.: A Location Representation for Generating Descriptive Walking Directions. In: IUI 2005: Proc of the 10th International Conference on Intelligent User Interfaces, pp. 122–129. ACM Press, New York (2005)
16. Lourakis, M., Argyros, A.: The design and implementation of a generic sparse bundle adjustment software package based on the levenberg-marquardt algorithm. Technical Report 340, Institute of Computer Science - FORTH, Heraklion, Crete, Greece (2004), http://www.ics.forth.gr/~lourakis/sba
17. Lowe, D.: Distinctive Image Features from Scale-Invariant Keypoints. International Journal of Computer Vision 60, 91–110 (2004)
18. Szeliski, R.: Image Alignment and Stitching: A Tutorial 1. Technical Report MSR-TR-2004-92 (2005)
19. Chen, W.C., Bouguet, J.Y., Chu, M.H., Grzeszczuk, R.: Light Field Mapping: Efficient Representation and Hardware Rendering of Surface Light Fields. In: SIGGRAPH 2002: Proceedings of the 29th Annual Conference on Computer Graphics and Interactive Techniques, pp. 447–456. ACM, New York (2002)
20. Liu, A., Hile, H., Borriello, G., Kautz, H., Brown, P., Harniss, M., Johnson, K.: Informing the design of an automated wayfinding system for individuals with cognitive impairments. In: Third International Conference on Pervasive Computing Technologies for Healthcare, PervasiveHealth 2009 (to appear, 2009)

The Acceptance of Domestic Ambient Intelligence Appliances by Prospective Users

Somaya Ben Allouch, Jan A.G.M. van Dijk, and Oscar Peters

University of Twente, Drienerlolaan 5,
7522 NB Enschede, The Netherlands
{s.benallouch,j.a.g.m.vandijk,o.peters}@utwente.nl

Abstract. Ambient intelligence (AmI) is a growing interdisciplinary area where the focus is shifted towards users instead of merely emphasizing the technological opportunities of AmI. Different methods are employed to understand the adoption of AmI appliances by users. However, these are often small-scale methods that are focused on specific subgroups. Large scale quantitative studies to understand the adoption of AmI appliances are scarce. In this study, a questionnaire was designed to examine how the Dutch people (n = 1221) perceive AmI appliances for domestic settings. Findings show that intention to adopt AmI appliances was low and that respondents had a negative to neutral attitude towards AmI appliances. On the basis of structural equation analysis, results suggest that adoption of AmI appliances could be explained by outcome expectancies of AmI appliances. The potential implications of the findings are discussed.

Keywords: ambient intelligence, pervasive technologies, technology acceptance.

1 Introduction

Enhanced computing power and convergence of technologies make it possible for embedded systems and appliances in the environment to adapt to and anticipate users' needs [30]. The integration of these systems and appliances in everyday lives of people is a particular vision of the future called Ambient Intelligence. Ubiquitous computing, pervasive computing and calm computing are synonyms of AmI, which refer to visions of people surrounded with embedded computing which is mostly invisible to the user [46, 6, 17]. Different names emphasize different aspects of this vision, but, Abowd and Sternbenz [1] have noted, they all have one thing in common, namely the desire to create a more symbiotic relationship between humans and their environment. Therefore, in the rest of this paper we will use the term Ambient Intelligence (AmI).

In the ubiquitous computing field, various methods such as ethnographic studies [29], scenarios and risk assessment [25], historical analysis [47] and interviews combined with diary studies [19] are used to get a better understanding of ubicomp and its possible consequences in different domains. However, large-scale quantitative studies of the acceptance of AmI appliances are scarce. This study aims to, first, increase the diversity of formative evaluation results for AmI appliances by using a large-scale,

H. Tokuda et al. (Eds.): Pervasive 2009, LNCS 5538, pp. 77–94, 2009.

survey-based, quantitative study and, second, to empirically investigate the perception of a large group of prospective users toward AmI appliances. Here we present results from a large scale quantitative study exploring people's attitudes and intentions towards AmI appliances for domestic settings. Furthermore, we attempt to predict which variables influence the future adoption of AmI appliances. We conclude the paper reflecting on the anticipated adoption and use of domestic AmI appliances.

2 Related Work

Researchers from different backgrounds try to gain understanding of AmI in various settings to inform future design and to evaluate what kind of implications AmI can have for prospective users. In this section, we discuss the diversity of methods in relevant studies that used some form of formative evaluation of AmI appliances.

Health care is seen as a potential area where AmI could provide many benefits for both the patient and the care giver and where different research methods are used to explore this area. Interviews are widely applied in this area, for example to address the physical and cognitive needs of elderly to support their daily activities [36], to investigate the needs of technologies for elderly [7] and combined with a two-week phone diary study to explore the needs and implications of eldercare technologies [14].

Studies have also been undertaken which focus not only on health care for elderly but on the general needs and expectations of people regarding AmI. Venkatesh *et al.* [44] used photographs and illustrations of smart homes and appliances during interviews to gain insight into the attitude and potential interest of American household members towards the home of the future. To explore the requirements that people have for domestic AmI technologies workshops were used in a European study. Pictures of emerging technologies were shown to residents of five homes to trigger future scenarios [4]. Interviews combined with dairies were used for 47 people from different European countries (Norway, Finland, Hungary, and UK) to gain insight into user receptions of AmI [19]. Here, the respondents were mainly recruited through the researchers' social networks and therefore white collar workers were over-represented. Garfield [23] investigated the acceptance of the pc tablet as an example of ubiquitous computing in an organizational context. She conducted a longitudinal, qualitative study based on interviews with participants who voluntarily used the tablet for a three-month period. As mentioned earlier, survey-based studies are scarce in the ubicomp field. Only recently a survey-based study has taken place in Germany to measure the experiences of people with ubicomp technologies, specifically focusing on privacy issues [42].

This brief overview of studies shows that different methods such as small-scale questionnaires, focus groups, interviews, diary studies, and cultural probes studies are the most frequently used methods to elicit responses from users regarding AmI technologies. However, these are often small-scale methods that focus on specific subgroups. In many other fields, from the pharmaceutical industry to technology product development, large-scale survey methods are commonly used [13]. As AmI technologies will ultimately be woven into society and into the everyday lives of many people [18], large-scale quantitative studies can be a valuable addition to current methods to provide an overall picture and understanding of the anticipated adoption of AmI technologies by a large, diverse group of people. Therefore, in this research a large-scale survey was adopted to examine the anticipated adoption of AmI technologies.

3 Research Questions

Precursors of AmI appliances are entering the public domain and research activities worldwide have been employed to realize AmI. However, not enough knowledge is available about people's perceptions of domestic AmI appliances to understand and inform the future development of AmI. Therefore, the following research questions are addressed:

RQ1: How are the benefits and disadvantages of domestic ambient intelligent appliances perceived by potential users?

RQ2: What are the attitudes and intentions of potential users regarding ambient intelligent appliances?

Another aim of this study is to explore the variables which could explain and predict the anticipated adoption of ambient intelligent appliances. Therefore, a third research question is proposed:

RQ3: Which variables explain and predict the attitudes and intentions for adopting ambient intelligent appliances in domestic settings, and what are their relationships?

Existing user acceptance theories and models of technology such as the technology acceptance model (TAM) [15, 16] or the unified model of acceptance and use of technology (UTAUT) [45] could offer insight into the adoption process of ambient intelligent appliances. Only, they are usually applied to technologies which are fully developed and already in use. Furthermore, in TAM and UTAUT, performance expectancy and effort expectancy play an important role as predictors of technology acceptance intentions. However, because these predictors are very specifically operationalized at a level of detail that is not possible for technologies that do not exist yet, these predictors are only meaningful when people have at least some experience with the technology to be able to reflect on its performance. This is not yet the case with AmI appliances; they are not widespread and used by people. Therefore, using more general statements in the form of outcome expectancies [34] that people could have towards AmI technologies was more meaningful in this case. Furthermore, applying these models to a technology which is in its development phase means that only the anticipation of adoption and use can be investigated. For this purpose a new model has to be constructed. From the existing user acceptance theories and models of technology a number of relevant constructs are selected to investigate the anticipated adoption of domestic AmI appliances by prospective users. These constructs and their hypothesized relations form the basis of a conceptual model which will be used to explore the anticipated adoption of AmI appliances. This model will be tested in the user survey. In the next section, the conceptual model will be discussed in more detail.

4 Adoption of Ambient Intelligent Appliances: A Conceptual Model

Several factors influence the adoption of new technologies. Previous research on user acceptance of technologies has shown that factors such as social influence [41], performance expectancy, effort expectancy [45], attitudes, behavioral intentions [2], and outcome expectancies [34] play an important role in the adoption process of new technologies. Fishbein and Ajzen's theory of reasoned action (TRA) postulates that behavioral intentions are the most immediate determinant of behavior. Thus, we expect a strong correlation between people's intentions and their actual behavior. Therefore, we hypothesized that the anticipated intention to adopt AmI appliances will also strongly correlate with people's actual behavior to adopt these technologies and, therefore, intention to adopt AmI appliances will be included in the conceptual model.

We hypothesized that the specific characteristics of AmI will also play an important role in the adoption process. The specific characteristics of AmI such as its unobtrusiveness, invisibility, adaptability and pro-active anticipation of user behavior, are supposed to bring ease and convenience to everyday domestic life [40]. However, next to these potential positive benefits negative outcomes are also related to AmI, such as loss of privacy, loss of control, less reliability, and a low social acceptance of these technologies [32, 33, 37, 40, 8]. Loss of privacy and loss of control are often mentioned as potential negative outcomes of AmI in daily life or, in other words, as the "dark side" of AmI [43]. If users also have these concerns, this will probably have a negative effect on the adoption process of AmI appliances. McCullough [35] argues that we should pay considerable attention to privacy aspects in the development process of AmI. The loss of privacy and control are included in the conceptual model as perceived disadvantages because they can be seen as important potential barriers to the widespread adoption of AmI appliances. The potential positive benefits of AmI such as convenience, easiness, and personalization will be included in the model as perceived advantages of AmI appliances. In this study, we focused on the advantages and disadvantages to the ones currently dominating the literature, though we recognize that there are more and other benefits and disadvantages related to AmI, for example, having too much information, providing false information and using energy.

In addition to the perceived benefits and perceived disadvantages of AmI appliances, we hypothesized that attitude towards AmI appliances will also play an important role in the adoption process. Attitude towards a behavior is defined as "the degree to which performance of the behavior is positively or negatively valued" [21]. It refers to the desirability of the behavior, which is considered to be a function of the sum of the perceived values of the expected consequences of the behavior. We hypothesized that perceived benefits and perceived disadvantages of AmI appliances can influence people's attitude and, therefore, in the conceptual model perceived benefits and disadvantages will strongly correlate with people's attitude towards AmI appliances. Furthermore, we hypothesized that attitude strongly correlates with outcome expectancies because outcome expectancies are more specifically presented to future users (specified in specific items such as "I expect this technology to make everyday life easier") than the more general attitude concept (specified in general items such as "I think that using ambient intelligent appliances is good vs. bad"). It is also hypothesized that the more specific outcome expectancies will have a direct influence on users' intentions to adopt AmI appliances. Thus, we hypothesized that attitude to-wards AmI appliances will influence the outcome expectancies of people and these expectancies will probably have a direct effect on intentions to adopt AmI appliances.

The variables of the conceptual model are not independent of each other. We hypothesized that if people have a negative attitude towards AmI appliances, they will probably perceive fewer benefits of them, and vice versa. Therefore, a reciprocal relationship is expected between the perceived benefits and the perceived disadvantages concerning attitudes towards AmI appliances. Figure 1 shows the proposed conceptual model, including its proposed relationships among predictive variables.

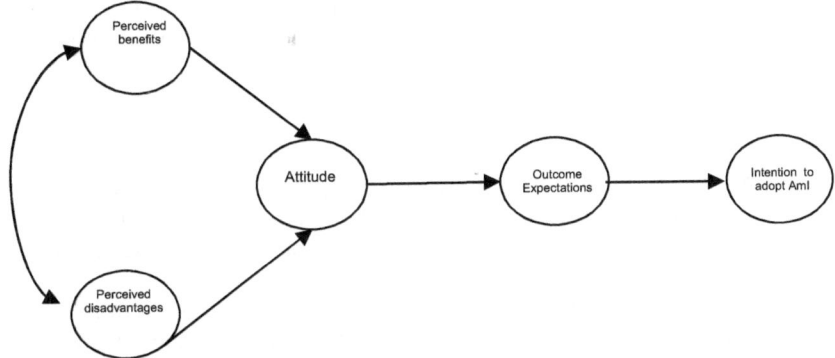

Fig. 1. Proposed path model

5   Method

5.1   Sample and Procedure

Members of a national panel ($N = 1539$) which is supposed to represent the Dutch population and is administrated by a research and consultancy company were invited via email to voluntary participate in the online survey. The survey was pretested by 25 people with ages ranging from 18 to 63 years on vocabulary, understanding of sentences, irregularities and length of time. Adjustments to the survey were made accordingly.

The 1221 panel members who responded (79.3% response rate) to the invitation were included in the sample. Pearson's chi-square test was used to test for differences in demographics between the respondents and the non-respondents. There was no significant difference between the non-respondents and the respondents concerning gender ($\chi2$ (1, $N = 1539$) = .01, $p > .05$); age ($\chi2$ (4, $N = 1539$) = 4.57, $p > .05$); education ($\chi2$ (8, $N = 1522$) = 12.73, $p > .05$) and income ($\chi2$ (6, $N = 1539$) = 4.06, $p > .05$).

In comparison with the Dutch population [11] gender was almost equally distributed (48% males compared to 49% of the adult Dutch population and 52% females compared to 51% of the adult Dutch population). Respondents younger than 25 years (7% compared to 12% of the Dutch population) and respondents of 65 years and older (5% compared to 17.4% of the Dutch population) are underrepresented in our sample. The other age groups were all slightly overrepresented, namely respondents aged 26 to 35 years (18% compared to 16.5% of the Dutch population), the group of 36 to 50 years (39% compared to 30% of the Dutch population) and the group of 51 to 65 years (31% compared to 24.1% of the Dutch population).

The higher education level of the respondents was also overrepresented in our sample. There were more respondents with a bachelor degree (32% compared to 16% of the Dutch population and a master degree (11% compared to 9% of the Dutch population) and respondents with only primary education or less were underrepresented (1.3% compared to 9% of the Dutch population). Of the respondents, 99.7% owned a computer and/or laptop and 99.2% had access to the internet in their own home. In the Netherlands, 88% of the population has access to a PC and 85% of the population has access to the Internet [12]. The sample of this study is thus not completely comparable to the Dutch population. As social demographics were no part of the hypothesized model to be tested we did not take the somewhat arbitrary step to weigh the results.

5.2 Measures

A questionnaire was designed to examine how people perceive AmI appliances in domestic settings. The questionnaire consisted of two parts. The first part of the questionnaire was dedicated to current possession of ICTs and domestic technologies in the home, past experience with computers, and attitude towards ICTs.

To assess whether respondents with a positive attitude towards current ICTs hold a more positive attitude towards AmI appliances, respondents' attitude towards current ICTs issues was measured with a scale consisting of six positive judgments (scaled 1 to 5 where 1 was totally disagree and 5 was totally agree) following Punie [39]. Punie distinguishes three different attitudes using this scale, namely: tech-phobes, the tech-nuanced and the tech-savvy. A tech-phobe attitude is characterized by a negative attitude towards technological development; a tech-nuanced attitude corresponds with a position between tech-phobe and tech-savvy and a tech-savvy attitude is a positive attitude towards technology.

Some judgments were rephrased to Netherlands-Dutch (Punie's was Belgian-Dutch) and some ICT examples were adjusted to suit current practice in the Netherlands (e.g. telephone was adjusted to internet).

Cronbach's alpha (α) was used as an indication of how well a set of items measures a latent construct. A scale is often regarded as reliable when Cronbach's α is at least .70 [38]. The internal consistency (Cronbach's α) of the ICT-attitude scale was .72.

The second part of the questionnaire started with a short description of what AmI is (i.e. *a vision on the future which includes intelligent appliances that know what you want and automatically can do things for you. These intelligent appliances will also be available for the home'*). After the general introduction of AmI, four specific currently existing AmI appliances, were described in detail to explain their characteristics (i.e. after each description of an AmI application questions followed and then the second application was described, questions followed etc.). The AmI appliances where an intelligent fridge, an intelligent mirror, an intelligent TV and a set of intelligent appliances, labelled intelligent appliances for the home, which consisted of blinds automatically closing, lights automatically turning on and off when entering the door (and leaving the house) and the temperature automatically adjusting to a person when entering a room in the house because the temperature appliance 'knows' what the persons likes. Each application except for the intelligent appliances was accompanied with a photo to give respondents a better idea of the specific application.

Perceived benefits. Perceived benefits of AmI appliances were measured with five items including: more or less perceived enjoyment, making usage more or less easy/complex, having more or less convenience, having more or less personalization, and perceiving more or less utility through usage of the particular AmI application (all scaled 1-5 where 1 was not at all enjoyable and 5 was very enjoyable, etc.). A Cronbach's α of respectively .88 for the intelligent fridge; $\alpha = .89$ for the intelligent mirror; $\alpha = .83$ for the intelligent television and $\alpha = .88$ for the intelligent appliances indicated reliable scales to measure perceived benefits of the four AmI appliances.

Perceived disadvantages. Privacy and loss of control were two measures of the perceived disadvantages of AmI appliances. For each appliance two privacy items and two control items were used to assess how respondents perceive privacy and control aspects of AmI appliances. A high score (5) scale meant that the respondent regarded this aspect of the appliance as very attractive and a low score (1) meant very unattractive. We recoded the scale so that higher values reflect lower perceptions of privacy and control. The items for the intelligent fridge are given as examples (the items for the other three appliances were almost similar, they were only adapted to the specific characteristics of the appliances). The privacy items were *'this intelligent fridge can order foods and give you cooking tips if you give permission to the fridge to keep track of what you are keeping in your fridge'* and *'when and how you use the intelligent fridge is being recorded by an intern system so that the intelligent fridge can better suit your wishes'*. The control items consisted of *'this intelligent fridge can automatically take over a couple of tasks from you such as keeping track of which foods are out of stock'* and *'this intelligent fridge can automatically take over a couple of decisions from you such as ordering foods at the grocery store if you have programmed the fridge to do this'*. The privacy items of the four AmI appliances were summed up to form one overall privacy construct (two items per appliances makes 8 items in total) and the control items (also in total 8 items) were also summed up to form one control construct. The internal consistency of the privacy scale was $\alpha = .88$ and $\alpha = .85$ for the control scale.

Attitude. As a measure of attitude towards the four AmI appliances, respondents rated the use of the four appliances on six five-point bipolar scales. The scale endpoints were defined as good/bad, wise/unwise, beneficial/harmful, pleasant/unpleasant, valuable/worthless and enjoyable/unenjoyable. The internal consistencies of the attitude scales were respectively, $\alpha = .94$ for the intelligent fridge; $\alpha = .95$ for the intelligent mirror; $\alpha = .94$ for the intelligent TV and $\alpha = .95$ for intelligent appliances.

Outcome expectations. Expected outcomes (i.e. "using the ubicomp appliances, how likely are you to _") were measured in a Likert-type scale that ranged from 1 (very unlikely) to 5 (very likely). We used monetary outcomes ($\alpha = .92$), activity outcomes ($\alpha = .89$), social outcomes ($\alpha = .85$), self-reactive outcomes ($\alpha = .89$), novelty outcomes ($\alpha = .80$) and fashion/status outcomes ($\alpha = .86$).

Intention. Three intention measures asked the respondents to rate their intention to use each specific ubicomp appliance if they will become available on a five-point bipolar scale ranging from 'extremely unlikely' to 'extremely likely'. The three intention measures were: 'I intend to use this intelligent fridge if it will be available'; 'I plan to buy this intelligent fridge as soon as it will be available' and 'I will use this intelligent fridge if it will be available'. Cronbach's α was respectively, .95 for the intelligent fridge; .95 for the intelligent mirror; .95 for the intelligent TV and .93 for intelligent appliances.

The questionnaire ended with socio-economic questions (i.e. age, gender, education level, income, household situation, and amount of leisure time during a weekday).

5.3 Data analysis

SPSS v12 was used to analyze the data. Statistical comparisons between groups used Chi-tests for categorical data and Mann-Whitney tests for ordinal data.

Structural Equation Modelling using Amos 6.0 [3] with maximum likelihood estimation was used to test the hypothesized model to predict intention to adopt AmI appliances in domestic settings. As suggested by Holbert and Stephensen [26] the following model fit indices were used: the χ2 estimate with degrees of freedom given that still is the most commonly used means by which to make comparisons across models [27]. Additionally, the standardized root mean squared residual (SRMR) as a second absolute fit statistic [28] in combination with the Tucker-Lewis index (TLI) as incremental index and the root mean squared error of approximation (RMSEA) [10] are reported. Hu and Bentler [28] recommend using a cutoff value close to .95 for TLI in combination with a cutoff value close to .09 for SRMR to evaluate model fit and the RMSEA close to .06 or less. Fit indexes are relative to progress in the field [22]. Although there are rules of thumb for acceptance of model fit (e.g., that TLI should be at least .95), Bollen [9] observed that these cut-offs are arbitrary. A more salient criterion may be simply to compare the fit of one's model to the fit of other, prior models of the same phenomenon.

6 Findings

The questionnaire on AmI appliances was designed to examine the perceptions of future users regarding AmI. It was also designed to get a better understanding of how specific AmI appliances are perceived and what respondents' attitudes are towards these appliances. However, since respondents already have certain attitudes towards today's existing information and communication technologies, we wanted to compare the results of the perceptions of AmI appliances with current attitudes towards information and communication technologies and therefore we also present these findings. Finally, we test which variables are strong predictors for the anticipated adoption of AmI appliances.

6.1 Attitude towards Information and Communication Technologies

Respondents' overall attitudes towards information and communication technologies were measured with a scale consisting of six positive judgments regarding information and communication technology issues (Cronbach's α = .72). Overall, the respondents had positive attitudes towards information and communication technologies. The item *'the disadvantages which some technical appliances can cause just belong to this kind of appliance'* scored the lowest ($M = 3.42$, $SD = .96$) and the item *'I find it good that when I want to know something, I can also get that information via technical appliances'* was the highest ($M = 4.50$, $SD = .70$).

There was no significant difference for gender and education regarding ICT attitudes. Respondents aged 26 to 35 years (mean rank = 664.64) and people older than 65 years (mean rank = 700.16) had a significantly more positive ICT attitude (χ^2 (4, $n = 1221$) = 11.72, $p < .05$) than other age groups. The group of respondents who did not provide answers about their incomes (mean rank = 557.18) and people who have an income of 1.5 times the average (mean rank = 584.65) had a significantly less positive attitude (χ^2 (6, $n = 1221$) = 21.13, $p < .01$) towards information and communication technologies than the other income groups. People who earned three times the average income or more had the most positive ICT attitude (mean rank = 753.79). There was a significant correlation between age and income level ($r = .70$, $n = 1221$, $p < .05$), which indicates that people who are older have a higher income.

Generally, the respondents had a positive attitude towards information and communication technologies. The overall score of the ICT-attitude scale ranging from 6 to 30 is the sum of the six items on a five point scale ranging from 1 = totally disagree to 5 = totally agree. Thus, although the overall attitude towards information and communication technologies was high, based on the mean score of the ICT-attitude scale ($M = 23.88$, $SD = 3.48$), three ICT groups were formed to assess differences in their attitudes towards AmI appliances. The first group (range 6 to 23) had the most negative attitude towards information and communication technologies (labeled the techphobic) and consisted of 42.8% of the sample. The second group (range 24 to 25) was labeled the tech-nuanced group and consisted of 24.2%. The last group, the tech-savvy (range 26 to 30) consisted of 33% of the respondents.

6.2 Perceived Benefits and Disadvantages of Domestic AmI Appliances

To answer the first research question concerning the perceived benefits and disadvantages of AmI appliances, the perceived benefits were measured separately for all four AmI appliances. The perceived benefits of the intelligent fridge and intelligent mirror were regarded as low by the respondents. The mean value *(SD)* for the intelligent fridge ranged from 2.20 *(1.14)* to 3.23 *(.99)* and the mean for the intelligent mirror ranged from 2.35 *(1.18)* to 2.97 *(.90)*. Respondents perceived the intelligent TV, with a mean ranging from 2.34 *(1.04)* to 3.61 *(.98)*, and intelligent appliances, with a mean ranging from 3.00 *(1.16)* to 3.62 *(.89)*, as having slightly greater benefits. See Table 1 for the exact means and standard deviations of the perceived benefits of the four AmI appliances.

Among the three groups with different attitudes towards information and communication technologies, significant differences were found in how they perceive the benefits of AmI appliances. The tech-savvy group perceived all four AmI appliances as having more benefits, followed by the tech-nuanced and the tech-phobes. Consider intelligent appliances as an example. The tech-savvy group (mean rank = 717.45) perceived intelligent appliances as having significantly greater benefits (χ^2 (2, $n = 1221$) = 92.31, $p < .001$) than the tech-nuanced (mean rank = 658.26) and the techphobic groups (502.32).

Table 1. Descriptive statistics and Cronbach's α of perceived benefits and perceived disadvantages of AmI appliances

	M	SD
Perceived benefits intelligent fridge (α = .88)		
Enjoyment	3.03	1.28
Ease	2.20	1.14
Convenience	3.23	.99
Personalization	2.92	1.13
Usefulness	2.40	1.17
Perceived benefits intelligent mirror (α = .89)		
Enjoyment	2.83	1.22
Ease	2.69	1.03
Convenience	2.97	.90
Personalization	2.76	1.10
Usefulness	2.35	1.18
Perceived benefits intelligent TV (α = .83)		
Enjoyment	3.61	.98
Ease	2.82	1.12
Convenience	3.57	.92
Personalization	3.37	1.06
Usefulness	2.34	1.04
Perceived benefits intelligent appliances (α = .88)		
Enjoyment	3.45	.97
Ease	3.00	1.16
Convenience	3.62	.89
Personalization	3.46	.98
Usefulness	3.34	1.16
Perceived disadvantages: Loss of privacy (α = .88)		
P1 intelligent fridge	2.96	1.20
P2 intelligent fridge	3.19	1.16
P1 intelligent mirror	3.23	1.11
P2 intelligent mirror	3.60	1.19
P1 intelligent TV	2.48	1.08
P2 intelligent TV	2.94	1.08
P1 automatic appliances	2.80	1.08
P2 automatic appliances	2.53	1.05
Perceived disadvantages: Loss of control (α = .85)		
C1 intelligent fridge	2.85	1.14
C2 intelligent fridge	3.63	1.19
C1 intelligent mirror	2.90	1.09
C2 intelligent mirror	2.83	1.20
C1 intelligent TV	2.32	.99
C2 intelligent TV	3.63	1.07
C1 automatic appliances	2.74	1.07
C2 automatic appliances	2.63	1.07

The disadvantages of the AmI appliances were in general perceived as varying from not very attractive to neutral to the respondents (see Table 1). With regard to privacy aspects of AmI appliances, the intelligent mirror's sending private information (such as weight and blood pressure) to the doctor was least appealing to the respondents ($M = 3.60$, $SD = 1.19$). Respondents seemed to have fewer privacy concerns with the intelligent TV's keeping a record of programs the user watches and, based on this recorded list, suggesting a list of interesting programs for the user ($M = 2.48$, $SD = 1.08$). A similar response was seen when intelligent appliances keep track of temperatures in the home and adjust the temperature based on the recorded list of previous temperatures ($M = 2.53$, $SD = 1.05$). Respondents did not find it very attractive that AmI appliances could do things for them when this caused a loss of control over tasks typically done by the user. The intelligent fridge ordering food ($M = 3.63$, $SD = 1.19$) and the intelligent TV ordering products ($M = 3.63$, $SD = 1.07$) were found to be the least attractive. The intelligent TV taking over the selection and recording of movies seemed to be a little bit more attractive to the respondents ($M = 2.32$, $SD = .99$).

How privacy and control aspects of AmI appliances were perceived differed significantly among the three ICT groups. The tech-savvy group (mean rank = 506.30) significantly had the fewest problems (χ^2 (2, $n = 1221$) = 77.92, $p < .001$) with the privacy aspects, followed by the tech-nuanced (mean rank = 581.03) and the tech-phobic groups (mean rank = 708.58). The tech-savvy people (mean rank = 502.87) were also significantly more positive χ^2 (2, $n = 1221$) = 76.70, $p < .001$) towards the idea that AmI appliances could take over some control tasks, as compared to the tech-nuanced (mean rank = 591.03) and the tech-phobic (mean rank = 705.59).

6.3 Attitude, Outcome Expectations and Intention to Adopt AmI Appliances

The results for research question two concerning the attitudes and intentions to adopt AmI showed that respondents did not have a pronounced attitude towards AmI appliances. The attitude towards all four AmI appliances varied from a neutral to a slightly positive attitude. People seemed to have a more positive attitude towards the intelligent TV and towards the intelligent appliances than towards the intelligent fridge and intelligent mirror (see Table 2).

Table 2. Descriptive statistics and Cronbach's α of attitudes towards AmI appliances

	M	SD
Attitude intelligent fridge (α = .94)		
Good/bad	3.00	1.09
Wise/unwise	2.89	1.07
Beneficial/harmful	2.88	1.04
Pleasant/unpleasant	2.91	1.22
Valuable/worthless	2.71	1.09
Enjoyable/unenjoyable	3.07	1.36
Attitude intelligent mirror (α = .95)		
Good/bad	2.89	1.10
Wise/unwise	2.91	1.13
Beneficial/harmful	2.74	1.00
Pleasant/unpleasant	2.68	1.19
Valuable/worthless	2.75	1.15
Enjoyable/unenjoyable	2.86	1.32
Attitude intelligent TV (α = .94)		
Good/bad	3.35	.98
Wise/unwise	3.08	.92
Beneficial/harmful	3.06	.96
Pleasant/unpleasant	3.48	1.09
Valuable/worthless	3.15	.94
Enjoyable/unenjoyable	3.58	1.15
Attitude intelligent appliances (α = .95)		
Good/bad	3.45	1.02
Wise/unwise	3.38	1.02
Beneficial/harmful	3.37	1.01
Pleasant/unpleasant	3.60	1.08
Valuable/worthless	3.28	1.02
Enjoyable/unenjoyable	3.51	1.14

The attitude towards AmI appliances differed significantly among the three ICT groups. The tech-savvy group had the most positive attitude towards all four AmI appliances. The intelligent TV is taken as an example. The tech-savvy group (mean rank = 692.17) had a significantly more positive attitude (χ^2 (2, $n = 1221$) = 49.15, p <.001) towards the intelligent TV than the tech-nuanced (mean rank = 638.48) and the tech-phobic (mean rank = 532.96).

Respondents seemed to expect the most from AmI appliances in terms of activity and monetary outcome. These outcome expectancies are more focused on making daily life easier (e.g., "to make your everyday life easier", $M = 3.19$, $SD = 1.11$) and bringing more enjoyment (e.g. "to make daily domestic activities more pleasant" $M = 3.22$ $SD = 1.13$). Social outcomes which focus on the enhancement of social relations or the building of social relations through AmI appliances scored the lowest of all outcome expectations (see Table 3).

For all four AmI appliances (intelligent fridge, intelligent mirror, intelligent TV, and intelligent home appliances), the behavioral intention to adopt the appliances was measured in order to answer research question two. Respondents' intentions to adopt the four appliances were generally low (see Table 4). The intention to adopt the intelligent mirror was the lowest and the intention to adopt the intelligent home appliances (e.g., blinds automatically closing) was the highest.

The intention to adopt AmI appliances differed significantly among the three ICT groups. The tech-savvy group had a higher intention to adopt all four AmI appliances than the tech-nuanced and the tech-phobes. For example, the tech-savvy (mean rank = 685.09) had the highest intention (χ^2 (2, n = 1221) = 45.62, p <.001) to adopt the intelligent mirror compared to the tech-nuanced (mean rank = 637.63) and the tech-phobic (mean rank = 538.89).

6.4 Explaining and Predicting Adoption of AmI Appliances

Prior to the analyses, data were checked for normality; no significant deviation from normality was found (skewness and curtosis Z < 1.96). The variables (e.g., intention,

Table 3. Descriptive statistics and Cronbach's α of outcome expectations of AmI appliances

	M	SD
Activity outcomes (α = .89)		
To make it easier for you	3.38	1.07
Because it offers you more freedom	3.07	1.07
Because it makes the tasks in the home more pleasant	3.25	1.07
To make daily domestic activities more pleasant	3.22	1.13
Because you like to use such appliances	2.85	1.25
To be entertained	2.46	1.20
Monetary outcomes (α = .92)		
To be able to do different things at once	3.07	1.09
To have more control over your daily life	2.88	1.08
Not to have to do everything yourself	2.94	1.12
To make your everyday life easier	3.19	1.11
Because it is convenient that you do not have to carry out certain tasks yourself	3.03	1.14
To save time	3.06	1.21
Social outcomes (α = .85)		
To strengthen my relationship with family and friends	1.98	1.07
To be able to communicate with family and friends	1.95	1.08
To maintain valuable contact with others	2.33	1.17
To belong to a particular group	1.61	.85
To have something to talk about with others	1.74	.95
Self-reactive outcomes (α = .89)		
To have something to do	1.97	1.07
When you are bored	2.02	1.16
To relax	2.66	1.23
When you do not have anything to do	2.10	1.10
To feel less lonely	1.79	.96
As a way to pass time	1.68	.94
Novelty (α = .80)		
Because it is something new	2.28	1.11
To be able to use the internet via the intelligent fridge	1.68	.96
To be able to order products via the intelligent TV	1.86	1.01
To actively monitor your health through the intelligent mirror	2.55	1.29
To discover new possibilities	2.97	1.17
Fashion/Status (α = .86)		
Because these appliances are modern appliances	2.21	1.16
To keep up with the newest technology	2.54	1.19
Because it belongs to your lifestyle	2.10	1.12
Because it increases your status	1.60	.86

Table 4. Descriptive statistics and Cronbach's α of intention to adopt AmI appliances

	M	SD
Intention to adopt intelligent fridge (α = .95)		
I intend to use this intelligent fridge if it is available	2.14	1.25
I plan to buy this intelligent fridge as soon as it is available	1.93	1.08
I will use this intelligent fridge if it is available	1.98	1.14
Intention to adopt intelligent mirror (α = .95)		
I intend to use this intelligent mirror if it is available	1.93	1.08
I plan to buy this intelligent mirror as soon as it is available	1.79	1.01
I will use this intelligent mirror if it is available	1.83	1.06
Intention to adopt intelligent TV (α = .95)		
I intend to use this intelligent TV if it is available	2.76	1.18
I plan to buy this intelligent TV as soon as it is available	2.50	1.13
I will use this intelligent TV if it is available	2.56	1.20
Intention to adopt intelligent appliances (α = .93)		
I intend to use these intelligent appliances if they are available	2.78	1.13
I plan to buy these intelligent appliances as soon as they are available	2.67	1.14
I will use these intelligent appliances if they are available	2.72	1.21

attitude, perceived benefits, and privacy) used for each of the four AmI appliances were summarized to construct 'overall' variables regarding adoption of AmI appliances. In other words, the intention scales of the four AmI appliances were summed up in one overall scale to measure intentions to adopt AmI appliances, the four attitude scales were summed up to measure overall attitudes towards AmI appliances, et cetera.

Measurement model. The initial measurement model generated a poor fit, $\chi^2(1035) = 8472.06$, $\chi^2/df = 8.19$, SRMR = .0649, TLI = .864 , RMSEA = .077 (CI: .075, .078). Items with highly correlated error variances identified by post-hoc modification indices were removed. Although the Cronbach's alpha of the indicators of novelty was above the aspiration level ($\alpha > .70$), the error variances co-varied with various indicators of other constructs and were, therefore, excluded from further analysis. The observed items of monetary outcomes and activity outcomes were loaded on both latent variables. This was also the case for the observed items of social outcomes and self-reactive outcomes. With regard to the content of their items, the four constructs were indeed closely related and were, therefore, reconstructed into two new constructs. The new construct of monetary outcomes and activity outcomes was labeled instrumental outcomes; the combination of the constructs social outcomes and self-reactive outcomes was labeled personal outcomes. This procedure resulted in a reduced number of observed indicators of the latent constructs. The internal consistency of the measures to predict adoption of AmI appliances was above the aspiration level ($\alpha > .70$). The modified measurement model generated a good fit, $\chi^2(209) = 779.32$, $\chi^2/df = 3.73$, SRMR = .026, TLI = .976, RMSEA = .047 (CI: .044, .051).

Structural model. The results obtained from testing the validity of a causal structure of the hypothesized model showed a reasonable fit $\chi^2(222) = 1271.44$, $\chi^2/df = 5.73$, SRMR = .0597, TLI = .959, RMSEA = .062 (CI: .059, .066). Post-hoc modification indices suggested an improved fit by correlating the error terms of personal outcomes and fashion outcomes ($r = .67$, $p < .001$). The respecified model generated a good fit $\chi^2(221) = 930.31$, $\chi^2/df = 4.21$, SRMR = .0355, TLI = .972, RMSEA = .051 (CI: .048, .055). Table 5 summarizes the mean and standard deviation, Cronbach's α, the factor loading (β), and the squared multiple correlation (R^2) of the observed indicators to predict adoption of AmI appliances. The path model with standardized path coefficients is featured in Figure 2.

As shown in Figure 2, there was a significant direct effect of outcome expectations on the intention to adopt AmI appliances. Perceived benefits and perceived disadvantages had a significant direct effect on attitude. The attitude-outcome expectancies path, as well as the outcome expectancies-intention path, appeared to be significant. A correlation was found between perceived benefits and perceived disadvantages of AmI appliances, $r = -.93$, $p < .001$. This indicates that the error terms of the two constructs are very closely related.

Squared multiple correlations (Table 5) showed that the intention to adopt AmI appliances was accounted for 75%, the attitude towards AmI appliances was accounted for 89%, and the outcome expectancies of AmI appliances were accounted for 76%.

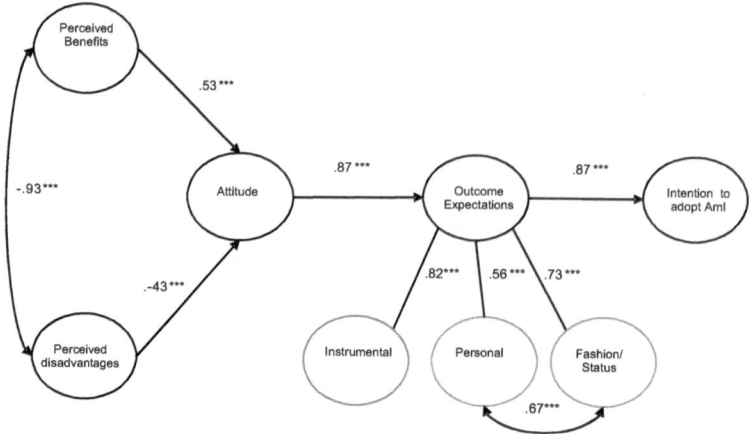

Fig. 2. Standardized path coefficients for the model to predict intentions to adopt AmI appliances

Note. The observed indicators of the latent constructs are not shown (see Table 5).
***$p < .001$. The error terms of the double-headed arrows are correlated.

Table 5. Descriptive statistics, factor loadings, squared multiple correlations and Cronbach's α of the observed indicators to predict intention to adopt AmI appliances

	M	SD	B	R²
Intention (α = .97)				.75
I intend to use this AmI appliance if it is available	2.40	.90	.92	.85
I plan to buy this AmI appliance as soon as it is available	2.22	.85	.98	.95
I will use this AmI appliance if it is available	2.27	.89	.98	.96
Attitude (α = .95)				.89
ATT 1 (good/bad)	3.17	.81	.93	.86
ATT2 (beneficial/harmful)	3.01	.76	.85	.73
ATT3 (pleasant/unpleasant)	3.17	.88	.95	.90
ATT4 (valuable/worthless)	2.97	.81	.92	.85
Perceived benefits (α = .85)				
Enjoyment	3.23	.84	.89	.79
Easy	2.68	.82	.62	.38
Personalization	3.13	.83	.91	.83
Perceived disadvantage (α = .95)				
Privacy	2.97	.83	.94	.88
Control	2.94	.77	.96	.92
Personal outcomes (α = .86)				.32
To have something to talk about with others	1.74	.95	.78	.60
When you do not have anything to do	2.10	1.10	.78	.61
As a way to pass time	1.68	.94	.82	.67
To feel less lonely	1.79	.96	.74	.54
Instrumental outcomes (α = .90)				.68
Because it makes the tasks that you perform in the home more pleasant	3.25	1.07	.87	.76
To make your everyday life easier	3.19	1.11	.84	.71
To make daily domestic activities in the home more pleasant	3.22	1.13	.84	.70
To not have to do everything yourself	2.94	1.12	.80	.65
Fashion/status outcomes (α = .87)				.53
To keep up with the newest technology	2.54	1.19	.83	.70
Because it belongs to your lifestyle	2.10	1.12	.80	.64
Because they are modern appliances	2.31	1.16	.85	.73

7 Discussion

In this study, people's perceptions of AmI appliances in domestic settings and the variables that explain and predict future adoption of these technologies were explored.

The results show that increasing the diversity of formative evaluation methods for AmI contributes to a wider understanding of the acceptance process of AmI technologies. The results of this large-scale, survey-based study indicate that respondents perceived the benefits of the four AmI appliances as varying from low to neutral. Enjoyment and convenience often scored the highest of all the measured perceived benefits for all four AmI appliances (other perceived benefits were ease of use, personalization, and usefulness). Attitudes towards AmI varied from negative to neutral and the intention of prospective users to adopt AmI appliances was low.

The structural equation analysis showed that outcome expectancies of domestic AmI appliances could largely predict the intention to adopt these appliances. This finding indicates that people's expectations about a new technology play a very important role in anticipating the adoption of a new technology. More specifically, instrumental, personal, and fashion outcomes have a large influence on the intention of prospective users to adopt AmI. This finding is interesting from both a theoretical and practical standpoint. For theory development, using outcome expectancies as a predictor for technology acceptance appears to explain people's intention to adopt a new technology to a large extent. From a practical point of view, designers and producers of AmI technologies could focus on these outcomes expectancies, namely instrumental, personal, and fashion outcomes, to better obtain the attention of prospective users.

In this study, the perceived disadvantages of AmI appliances appeared to be the loss of privacy and the loss of control. The findings show that there is statistical evidence for a relationship between perceived benefits and perceived disadvantages of domestic AmI appliances. However, on the basis of the results of this study, the exact nature of this relationship is unclear. It could be that benefits and disadvantages of these appliances act simultaneously, or that one of the constructs has a stronger influence on the other. If the nature of this relationship is known, designers of AmI could keep this in mind when developing these technologies. For example, when it appears that people are willing to accept and use AmI appliances in domestic settings because they derive enough personal benefits from them, then some of the disadvantages of these appliances could be accepted by the majority. Or, if people are not willing to lose their privacy and control and thus perceive the disadvantages of these appliances as being stronger than the benefits, the intention to adopt AmI appliances will probably be lower. Further research could bring more insight into this reciprocal relationship.

Furthermore, the findings suggest that perceived disadvantages, in this case the loss of privacy and loss of control, did have a direct effect on attitudes towards AmI. When the disadvantages with regard to AmI were perceived to be high, the attitude towards these appliances was low and thus more negative. From the start of the development of AmI, loss of privacy and loss of control has been recognized as important concerns for the future success of the adoption of AmI technologies. Even though the specific features of AmI, such as being able to anticipate owner behavior by constantly using data about user behavior and personal routines, make it difficult to exclude all potential privacy and control disadvantages, designers should make the effort to minimize the loss of privacy and control for users from the start of the design process.

The AmI research field is a young domain and there is no common perception among ordinary people about its content and possibilities. The appliances in this study were specifically chosen to include a broad spectrum of domestic AmI appliances and to summarize key technology aspects of AmI. Appliances from other key areas such as health and energy saving might deliver different outcomes in terms of acceptance. However, since these AmI appliances are used more often in previous studies [e.g., 41] we wanted to maintain the continuity of the research field.

In future research, we hope to investigate some of the questions raised by this research study. First, the influence of one's social network is known to be an important factor in technology acceptance. Rogers postulates that diffusion of an innovation happens when this innovation is communicated through certain channels over time in a social system [41]. AmI is still in a research and development phase and not widely available on the market, so our focus was not on social influence. Yet, we do believe that one's social network plays an important role in the acceptance of new technologies, especially, for those technologies meant for private settings such as the home and not for an organizational context where it can obligatory to accept and use new technologies. We are currently expanding the model by including social influence as a variable to investigate the role of social influence in the acceptance process of these technologies.

Second, this study was done in a Western-European country with a high penetration and use of both the mobile phone and the internet [20]. Bell et al. [5] argue that there are cultural differences in technology behavior and that we have to take these into consideration when designing technologies for domestic settings. Even among European countries, differences were found in the use of mobile ambient intelligent services [22]. Therefore, the results cannot automatically be translated to other countries and cultures. Furthermore, the sample was relatively ICT-minded, which could lead to two different conclusions. First, if ICT-minded people are not very positive about ambient intelligent appliances, the population at large would be even less positive. Second, and opposite, less-ICT-minded people would embrace ambient intelligent appliances because they are supposed to be relatively easy to use and can be smoothly integrated into everyday environments. More research is needed to determine how this acceptance process precisely works in these groups. However, it is important to pay attention to this finding to ensure that ambient intelligent appliances will be adopted by everybody and not just by a certain group of people.

Ambient intelligence is cheered and criticized for its possible influential role in people's everyday lives. Obviously, more research is needed to assess the variables and their interrelationships as ambient intelligent appliances become more widespread in societies. Most importantly, variables such as real user experience of AmI should be incorporated into future studies. Overall, this study presents evidence that people's current attitudes and outcome expectations of ambient intelligent appliances are important factors to consider when anticipating the future adoption of ambient intelligent appliances in domestic settings.

References

1. Abowd, G.D., Sterbenz, J.P.G.: Final report on the inter-agency workshop on research issues for smart environments. IEEE Personal Communications 7(5), 36–40 (2000)
2. Ajzen, I., Fishbein, M.: Understanding Attitudes and Predicting Social Behavior. Prentice-Hall, Englewood Cliffs (1980)

3. Arbuckle, J.L.: Amos 6.0 User's Guide. SPSS Inc., Chicago (2005)
4. Baillie, L., Benyon, D.: Investigating ubiquitous computing in the home. The 1st Equator IRC Workshop on Ubiquitous Computing in Domestic Environments. The School of Computer Science and Information Technology, University of Nottingham, http://www.equator.ac.uk/
5. Bell, G., Blythe, M., Sengers, P., Wright, P.: Designing culturally situated technologies for the home. In: CHI 2003 Extended Abstracts on Human Factors in Computing Systems, Ft. Lauderdale, Florida, USA, pp. 1062–1063 (2003)
6. Birnbaum, J.: Pervasive information systems. Communications of the ACM 40(2), 40–41 (1997)
7. Blythe, M.A., Monk, A.F., Doughty, K.: Socially dependable design: the challenge of ageing populations for HCI. Interacting with Computers 17(6), 672–689 (2005)
8. Bohn, J., Coroama, V., Langheinrich, M., Mattern, F., Rohs, M.: Living in a world of smart everyday objects - social, economic, and ethical implications. Journal of Human and Ecological Risk assessment 10(5), 763–786 (2004)
9. Bollen, K.A.: Structural equations with latent variables. Wiley, New York (1989)
10. Browne, M.W., Cudeck, R.: Alternative ways of assessing model fit. In: Bollen, K.A., Long, J.S. (eds.) Testing structural equation models, pp. 136–162. Sage, Thousand Oaks (1993)
11. Central Bureau of Statistics (CBS), 2006a. Population; core figures (May 2007), http://statline.cbs.nl/
12. Central Bureau of Statistics (CBS), 2006b. ICT and media use (July 2007), http://statline.cbs.nl/
13. Chakrapani, C.: Statistics in market research. Arnold Publishers, London (2004)
14. Consolvo, S., Roesller, P., Shelton, B.E., LaMarca, A., Schilit, B., Bly, S., Dourish, P.: Technology for Care Networks and Elders. Pervasive Computing 3(2), 22–29 (2004)
15. Davis, F.D.: User acceptance of information technology: system characteristics, user perceptions and behavioural impacts. International Journal Man-Machine Studies 38(3), 475–487 (1993)
16. Davis, F.D., Bagozzi, R.P., Warshaw, P.R.: User acceptance of computer technology: A comparison of two theoretical models. Management Science 35(8), 982–1003 (1989)
17. Dertouzos, M.L.: The future of computing. Scientific American 281(2), 52–55 (1999)
18. Dourish, P.: Where the action is the foundations of embodied interaction. MIT Press, Cambridge (2001)
19. Ellis, R.: Challenges of work/home boundaries and user perceptions for Ambient Intelligence. In: Cunningham, P., Cunningham, M. (eds.) Eadoption and the Knowledge Economy: Issues, Appliances, Case Studies, pp. 1395–1402. IOS Press, Amsterdam (2004)
20. European Commission: Special Eurobarometer: E-communications household survey (2006)
21. Fishbein, M., Ajzen, I.: Belief, attitude, intention and behavior: an introduction to theory and research. Addison-Wesley, Reading (1975)
22. Forest, F., Arhippainen, L.: Social acceptance of proactive mobile services: observing and anticipating cultural aspects by a Sociology of User Experience method. In: Joint sOc-EUSAI conference, Grenoble, France (2005)
23. Garfield, M.J.: Acceptance of ubiquitous computing. IS Management 22(4), 24–31 (2005)
24. Garson, G.D.: Statnotes. Topics in multivariate analysis (2007)
25. Hilty, L.M., Som, C., Kohler, A.: Assessing the human, social, and environmental risks of pervasive computing. Human and Ecological Risk Assessment 10(5), 853–874 (2004)
26. Holbert, R.L., Stephensen, M.T.: Structural Equation Modeling in the Communication Sciences, 1995-2000. Human Communication Research 28(4), 531–551 (2002)

27. Hoyle, R.H., Panter, A.T.: Writing about structural equation models. In: Hoyle, R.H. (ed.) Structural equation modeling: Comments, issues, and Applications, pp. 158–176. Sage, Thousand Oaks (1995)
28. Hu, L., Bentler, P.M.: Cutoff criteria for fit indexes in covariance structure analysis: Conventional criteria versus new alternatives. Structural Equation Modeling 6(1), 1–55 (1999)
29. Hughes, J., O'Brien, J., Rodden, T., Rouncefield, M., Viller, S.: Patterns of home life: informing design for domestic environments. Personal Technologies 4(1), 25–38 (2000)
30. ISTAG (IST Advisory Group): Scenarios for ambient intelligence in 2010. Institute for Prospective Technological Studies (IPTS), Seville (2001)
31. ISTAG (IST Advisory Group): Scenarios for ambient intelligence in 2010. Institute for Prospective Technological Studies (IPTS), Seville (2001)
32. Langheinrich, M.: Privacy by design - principles of privacy-aware ubiquitous systems. In: Abowd, G.D., Brumitt, B., Shafer, S. (eds.) UbiComp 2001. LNCS, vol. 2201, pp. 273–291. Springer, Heidelberg (2001)
33. Langheinrich, M.: A privacy awareness system for ubiquitous computing environments. In: Borriello, G., Holmquist, L.E. (eds.) UbiComp 2002. LNCS, vol. 2498, pp. 237–245. Springer, Heidelberg (2002)
34. LaRose, R., Eastin, M.S.: A social cognitive theory of Internet uses and gratifications: Toward a new model of media attendance. Journal of Broadcasting and Electronic Media 48(3), 358–377 (2004)
35. McCullough, M.: Digital Ground: Architecture, Pervasive Computing, and Environmental Knowing. MIT Press, Cambridge (2004)
36. Mynatt, E.D., Melenhorst, A.-S., Fisk, A.D., Rogers, W.A.: Aware technologies for ageing in place: understanding user needs and attitudes. Pervasive Computing 3(2), 36–41 (2004)
37. Nguyen, D., Mynatt, E.D.: Privacy Mirrors: Understanding and Shaping Sociotechnical Ubiquitous Computing Systems. Georgia Institute of Technology Technical Report GIT-GVU-02-16 (2002)
38. Nunnally, J.C.: Psychometric Theory, 2nd edn. McGraw-Hill, New York (1978)
39. Punie, Y.: Adoption, use and meaning of media in daily life: continuous limitation or discontinuous liberation (Doctor Thesis). Free University of Brussels, Belgium (2000)
40. Punie, Y.: A social and technological view on Ambient Intelligence in Everyday Life: What bends the trend? European Media, Technology and Everyday Life Research Network (EMTEL2), Sevilla (2003)
41. Rogers, E.M.: Diffusion of innovations, 5th edn. Free Press, NY (2003)
42. Spiekermann, S.: Privacy Enhancing Technologies for RFID in Retail - An Empirical Investigation. In: Krumm, J., Abowd, G.D., Seneviratne, A., Strang, T. (eds.) UbiComp 2007. LNCS, vol. 4717, pp. 56–72. Springer, Heidelberg (2007)
43. Stone, A.: The dark side of pervasive computing. Pervasive Computing 2(1), 4–8 (2003)
44. Venkatesh, A., Stolzoff, N., Shih, E., Mazumdar, S.: The Home of the Future: An Ethnographic Study of New Information Technologies in the Home. In: Gilly, M., Meyers-Levy, J. (eds.) Advances in Consumer Research, vol. XXVIII, pp. 88–96. Association for Consumer Research, Valdosta (2001)
45. Venkatesh, V., Morris, M.G., Davis, G.B., Davis, F.D.: User acceptance of information technology: toward a unified view. MIS Quarterly 27(3), 425–478 (2003)
46. Weiser, M.: The computer for the 21st century. Scientific American 265(3), 94–104 (1991)
47. Wyche, S., Sengers, S., Grinter, R.E.: Historical Analysis: Using the Past to Design the Future. In: Dourish, P., Friday, A. (eds.) UbiComp 2006. LNCS, vol. 4206, pp. 35–51. Springer, Heidelberg (2006)

Adding GPS-Control to Traditional Thermostats: An Exploration of Potential Energy Savings and Design Challenges

Manu Gupta, Stephen S. Intille, and Kent Larson

House_n, Massachusetts Institute of Technology
Cambridge, MA 02142 USA
{manug,intille,kll}@mit.edu

Abstract. Although manual and programmable home thermostats can save energy when used properly, studies have shown that over 40% of U.S. homes may not use energy-saving temperature setbacks when homes are unoccupied. We propose a system for augmenting these thermostats using just-in-time heating and cooling based on travel-to-home distance obtained from location-aware mobile phones. Analyzing GPS travel data from 8 participants (8-12 weeks each) and heating and cooling characteristics from 5 homes, we report results of running computer simulations estimating potential energy savings from such a device. Using a GPS-enabled thermostat might lead to savings of as much as 7% for some households that do not regularly use the temperature setback afforded by manual and programmable thermostats. Significantly, these savings could be obtained without requiring any change in occupant behavior or comfort level, and the technology could be implemented affordably by exploiting the ubiquity of mobile phones. Additional savings may be possible with modest context-sensitive prompting. We report on design considerations identified during a pilot test of a fully-functional implementation of the system.

1 Introduction

With only 5% of the world's population, the U.S. uses 25% of the world's energy [1]. The U.S residential sector is responsible for 21% of the total U.S energy consumption, and heating and cooling accounts for 46% of the total energy consumed in U.S residential buildings. Overall, 9% of total U.S energy consumption is expended on residential heating and cooling [2, 3]. Forty-nine percent of homes in the U.S are unoccupied during the day, and it is estimated that in 53% of U.S homes the temperature (T) is *not* lowered during the daytime when no one is at home in winters (conversely, in 46% the T is not raised in summers) [4]. Even in the 30% of the U.S homes that have programmable thermostats (P-Therms), as many as 44% may not use daytime setbacks to save energy [4]. As **Table 1** shows, as many as 55 million U.S. households – some with manual thermostats (M-Therms) and some with P-Therms – may not change their T settings when no one is home.

Although per capita consumption of energy is much lower in other countries [1], a significant amount of energy is likely being wasted heating and cooling unoccupied environments in many industrialized countries because common thermostats do not

H. Tokuda et al. (Eds.): Pervasive 2009, LNCS 5538, pp. 95–114, 2009.

Table 1. Thermostat usage statistics in the U.S (summarized from [4])

(In millions)	Total homes in the U.S.	Estimated no. of homes using setback when away	Estimated no. of homes not using setback when away
Manual Thermostat	62.16	21.7	40.46
Programmable Thermostat	33.3	18.7	14.60
Total	95.46	40.4	**55.06**

adapt to variable occupancy schedules and because people have difficulty setting and optimizing P-Therms [5]. The challenge, therefore, is to create a system to augment existing thermostats so that regardless of what the home occupants do, the thermostat (1) saves energy, (2) requires non-burdensome user input and no reliance on memory, and (3) doesn't sacrifice comfort, where we define comfort as ensuring that the home is always at a desirable temperature upon return. Additionally, a thermostat needs to be inexpensive to use and install.

We describe a concept for augmenting existing thermostats with a just-in-time heating and cooling mode that is controlled using travel-to-home time computed from location-aware mobile phones. Although existing P-Therms can save substantial amounts of energy when used effectively [6], we show, via a set of simulations using real travel data and home heating and cooling characteristics, that the proposed just-in-time system augmentation might provide energy savings for the substantial number of people *who do not use M-Therms or P-Therms optimally*. The system that we propose does not require users to program occupancy schedules. In fact, *no change in behavior on the part of the home occupants from what they currently do is necessary*. We focus on standalone housing and commuting patterns common in the northern U.S. and leave the question of how these results might generalize to other climates, housing types, and lifestyles for future work.

2 Prior Work

Pervasive computing systems that can infer context clearly offer potential for energy saving. Harris et al. [7], for example, argue that context-aware power management (CAPM) could use multi-modal sensor data to optimally control the standby states of home devices to optimize energy use, reducing so-called vampire power consumption [8]. They conclude that to optimally save energy, in addition to predicting what someone is currently doing, a system should predict what someone is *about* to do. Reliable detection of intentionality to control appliance energy use indoors is a difficult problem that is the subject of ongoing research [9]. Nonetheless, Harle and Hopper [10] showed that even without such prediction, in one office building using location of occupants would have permitted energy expended on lighting and "fast-response" electrical systems to be reduced by 50%.

Although inefficient use of electrical devices can be a substantial source of energy waste in a home or office, others have instead focused on improving home thermostats to lower heating, ventilation and air condition (HVAC) costs. A thermostat balances two competing factors: energy savings and air temperature/humidity comfort levels. There are three common types [11].

Manual thermostats (M-Therms) can be the most energy efficient option. People who set the T very low in winter when they leave the house and then turn up the T when they return achieve maximal energy savings but with significant discomfort upon return to the home. Avoiding that discomfort may be one reason that over 65% of people with M-Therms do not use setbacks when they are away from their homes in winter [4].

Programmable thermostats (P-Therms) automatically regulate the T according to a pre- scheduled program. P-Therms do not adapt to variable occupancy schedules – if schedules change, the user must remember to re-program the system in advance, and reprogramming is often tricky with current interface designs. The lack of responsiveness and difficulty of programming may be one reason that over 43% of people with P-Therms do not use daily setbacks when away in winter [4].

So-called intelligent thermostats have "adaptive recovery control," so that rather than starting and stopping based on timers, they set the T when away to ensure that given typical heating/cooling patterns, the home will reach the comfort T at the right time. These thermostats may also learn the T preferences of the user for different contexts [12] and use occupancy sensors to infer occupancy patterns [13, 14]. Others use light levels to change the T settings in the house [15] or control the air velocity and direction [16]. Some even have persuasive elements, such as informing users about the minimum T settings that can produce the desired comfort level [12, 17]. When these systems imperfectly infer behavior patterns, however, they optimize savings at the expense of comfort, and they typically require complex sensor installations to be retrofit into the home.

Unfortunately, all of these thermostats are often misused. An estimated 25-50% of U.S. households operate the thermostat as an on/off switch rather than a T controller [18]. A common misconception is that the more one changes the T dial, the faster the thermostat will make T change [19, 20]. Also, it has been shown that P-Therms do not save as much energy as predicted [5, 21, 22], most likely because they are difficult to use [5, 23]. Clearly, it is important that the thermostat interface be made as simple as possible.

3   Opportunity

The key idea advocated here is to *augment* current thermostats with the ability to control heating and cooling using travel time, as determined automatically via GPS-enabled mobile phones that will become commonplace.[1] When the thermostat is not being used regularly in setback mode, the thermostat should switch to this "just-in-time" travel-to-home-time mode. In this mode, the thermostat system communicates with the GPS-enabled mobile phones of the residents. Based on the location of the residents as determined by each resident's mobile phone and free geo-location mapping services, travel-to-home time is continuously estimated. The thermostat uses travel time of the home occupants, inside and outside T, and heating/cooling characteristics of the home to dynamically control the thermostat so that energy savings are

---

[1] In this work we use GPS data and the terminology GPS thermostat (GPS-Therm), but phones may also use multiple methods to determine location (e.g., cell tower triangulation or beacons).

maximized without sacrificing comfort. By setting T as a function of the fastest possible return time of the closest resident (and the other factors mentioned above), the system ensures that the home will always be comfortable on return.

The system has the following characteristics: (1) it requires no thermostat programming from the user, (2) it adapts to irregular schedules, (3) it ensures that the user always returns to a comfortable house, (4) it creates opportunities for motivating additional savings using context-sensitive prompting, and (5) and it does not require installation of a complex new sensor system in the home. As we will discuss, in its most straightforward implementation, it does not save more energy than a M-Therm or P-Therm that is regularly used with daily setbacks, but it will save energy overall for the general population because so many people fail to use their thermostats properly. The concept, therefore, is to layer the GPS thermostat (GPS-Therm) capability on top of existing thermostats, so that the GPS system engages only when users are not using a more efficient setback strategy.

This work is inspired by solutions for controlling appliance use in the home or office based on indoor location [9, 10, 24], but the proposed system does not require an extensive sensor or distributed appliance control network to be installed in the home to achieve savings. We make only the following assumptions: (1) that mobile phone adoption trends continue so that in many households everyone who travels alone will have a phone, (2) that within a few years nearly all new phones will have location-finding and Internet data transfer capabilities, and (3) that many homes will have Internet access and home wireless networks. For households where these assumptions hold, we discuss the energy saving potential of the system.

4   Measuring Potential Energy Savings

In this section we describe the results of an exploratory simulation we conducted to better understand the extent to which a GPS-controlled thermostat system might save energy.

4.1   Data Collection

We recruited 8 people living in 4 different homes in the greater metropolitan area of Boston, Massachusetts using mailing lists, flyers, internet advertisement and word of mouth. None of the subjects had any affiliation with the research team. Each person worked outside of his or her home and had a separate car that was used as the main mode of transport. Each house had a heating system that was not shared with any neighboring residences (see **Table 2**). The study was approved by our human subjects review board.

To gather realistic data on travel patterns, between March and June, 2008 a Track-Stick Pro GPS logger [25] was installed on the dashboard of each vehicle of each member of each house. The logger was plugged into the cigarette lighter socket. These loggers were left for up to 3 mo, recording position of the vehicle each minute whenever it was operating. Data were recovered from each logger every 4 weeks.

To measure the heating and cooling properties of the homes, two T and humidity loggers (EL-USB-2, EL-USB-1) [26] and Logit LCV electrical current loggers [27] were installed in each house for 3 d of measurement. One T and humidity logger was placed near the thermostat and another outside of the house on the north facing wall or window, where the sensor was not in contact with direct sunlight. The current logger was attached to the current-carrying wire from the thermostat to the HVAC system to gather information about when the heating system was activated.

During the 3 days that the sensors were installed, each home had its P-Therm reprogrammed by the investigators as follows: set to 50°F (10°C) from 9 AM to 4:30 PM and set to a comfortable T, typically between 67-69°F (19.4-20.5°C), other times.

In addition to the participant homes, data were collected from an unoccupied, newly constructed and well insulated (R-25) 1100 ft² condominium (control house) in December, 2007. A typical day for the type of people who participated in the study would be a person leaving at 7 AM for an 8 AM to 6PM workday with arrival home at 7 PM. This routine was scheduled on the P-Therm for 3 wk, with a comfort T of 69°F (20.5°C) and a setback T of 45°F (7.2°C), the Energy Star recommended settings [28].

Table 2. Participant house details. At the time of the experiments, the cost of natural gas was $1/therm. The cost of oil was $1.20/liter. Hot water heating systems used radiators. All the homes had programmable thermostats.

House	Heating fuel	Heating system	Capacity (Btu/h)	Insulation	Commute travel time	Days of data	Vehicles
1	Gas	Forced air	100k	Low	35 min-50 min	75	1
2	Gas	Hot water	130k	Medium	10 min-15 min	75	2
3	Oil	Hot water	133k	Medium	7 min-10 min	63	2
4	Oil	Hot water	154k	High	20 min-35 min	63	2
Control	Gas	Forced air	100k	Very high	Simulated	90	0

4.2 Evaluation

Software was written to simulate the functionality of manual, programmable, and GPS-controlled thermostats. The simulator, which uses the same algorithm later described when discussing a real-time, fully-functional prototype in Section 6, requires the following for input at each point throughout the day at 1 min intervals: (1) indoor T, (2) outdoor T, (3) latitude/longitude coordinate for each occupant's phone (if available), and (4) heating/cooling tables for the home. Additional information can be provided to the simulator when modeling various conditions (e.g., minimum allowable T in the home, occupant schedules, a T setting for P-Therm and M-Therm, and heating system type).

Heating/cooling tables were created for each house using the T profile data collected over 3 d. A heating table was created for heat gain (i.e. the time it takes for the house to heat up by 1°F (-17.2°C) from each starting T given an outdoor T with the heating system running at full capacity), and a cooling table was created for heat loss. The 3 d of data typically do not span the entire range of outdoor T for winters for the region (0°F (-17.7°C) to 60°F (15.5 °C)). Therefore, values not directly observed were estimated from the 3 d of data and the energy transfer equation for a building [29].

At each point in time throughout the day when a longitude and latitude coordinate is available, to simulate operation of the GPS-Therm, the simulation software sends the location and the occupant's home coordinate to the MapQuest web service [30] and obtains estimated travel-to-home time. The MapQuest web service uses a proprietary algorithm for calculating travel time and distance that appears to use road type, speed limit, and distance, but not local traffic conditions. An outdoor T file was created for the greater Boston area for the entire duration of the study by accessing two online weather archive databases.

Each minute the simulator outputs the target T based on the travel-to-home time, estimated indoor T, and HVAC on/off cycle duration. It also outputs the simulated indoor T and the HVAC cycle state (on/off). Using either the pre-programmed times for leaving for work and returning or the time detected when someone returned home, the system also simulates the operation of the M-Therm and P-Therm.

4.3 Results with Common Travel Patterns for Daily Workers

First we discuss simulator results using the control home and simulated travel patterns where people commute every weekday with average commute times of 15 min, 26 min (Boston's mean commute time) [31], and 90 min. We assume that the comfort T of the home is set to the Energy Star recommended setting of 69°F (20.5°C) [28]. We compare four scenarios. The first is the *baseline*, where the thermostat is set to the comfort T at all times of day. As indicated in Section 1, many people [4] with both P-Therms and M-Therms do not use setbacks at all. The second is *manual setback*, where the T is lowered manually upon leaving the house and raised manually upon returning. The third is *programmable setback*, where schedules are programmed for lowering and raising the setpoint each day based on standard work patterns. In programmable mode, we assume that the system starts heating 30 min before the return time and maintains a target T (setback) of 60°F (15.5°C) during the day when the home is unoccupied. Finally, the last case is the *GPS-thermostat*, where the target T is set as a function of travel-to-home time. All of the savings reported are calculated with respect to the baseline condition.

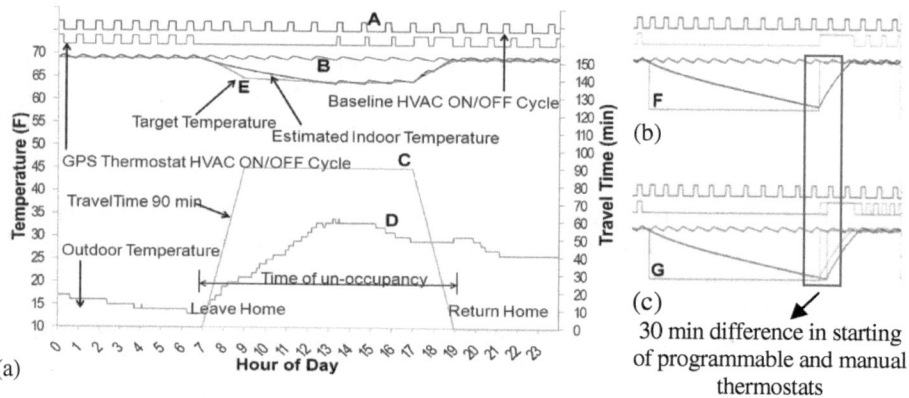

Fig. 1. A typical day with 90 min simulated travel time and simulated results with different thermostat types, as explained in the text

Figure 1(a) shows the simulation results for a 24 h period assuming a long travel time to and from work of 90 min each way. Approximately 1.7 million people in the U.S. commute for 90 min or more each way [26]. Line A shows when the simulation software estimates that the heater will cycle on and off in the baseline condition, with *up* being an on-cycle. Line B is the baseline T, showing how the target T will oscillate around 69°F (20.5°C), assuming the HVAC triggers when the T drops 0.3°F (0.17°C) below the target (at 68.7°F (20.4°C)) and runs until the T exceeds the target by 0.5°F (0.28°C) above the target (or 69.5°F (20.8°C)). Line C shows the 90 min travel time, leaving at 7 AM and returning at 7 PM, assuming an 8:30-5:30 workday. Line D shows the outdoor T fluctuation, which does influence cycle times (a small change can be seen in lines A and B from morning to mid-day in cycle length). Line E is the target T determined based on travel distance. This value is determined using the estimated heating/cooling parameters of the home at various indoor and outdoor T. It drops as the travel time increases and plateaus at the lowest T that will allow the home to heat back up in time to achieve the comfort T given the travel time. In **Figure 1**(b), line F shows the target T for the P-Therm simulation. In **Figure 1**(c), line G shows the target T for the M-Therm simulation.

First we compare the just-in-time GPS-Therm directly to M-Therms and P-Therms. In addition to presenting results for daily savings when the devices are properly used, we present results in terms of "expected energy savings" and "expected monetary savings." Expected savings is equal to the estimated savings multiplied by expected compliance of use of the particular thermostat type. The expected compliance of a M-Therm assumes 35% [4] of users will use manual setbacks when leaving the house. The expected compliance of a P-Therm assumes 56% of users will have it programmed to use setbacks. Although the GPS-Therm requires no action on the part of the user, we assume that 10% of the time the user may forget to take the phone, lack a GPS connection, be out of mobile phone coverage, or discharge the phone battery (resulting in a relatively high estimated compliance of 90%).

Table 3 shows the simulation results for the three thermostat types for different travel times. The savings using the GPS-Therm increase as the travel time increases, whereas the savings from P-Therm and M-Therm are constant and independent of travel time. The expected savings from the GPS-Therm begins to exceed the expected savings from the P-Therm and M-Therm when travel time reaches 60+ min. The simulations show what we expect to be true: that maximal savings can be achieved by simply turning off a heating or cooling system whenever someone leaves home and sacrificing comfort on return. Using travel time provides less substantial savings than P-Therm for people with short commutes and predictable work schedules, because if someone works near home the system does not allow the T to drop very far to ensure that the home can return to comfort quickly enough no matter when a person leaves work for home. Therefore, even though the GPS-thermostat is likely to have a much higher compliance than M-Therms and P-Therms, the simulations suggest advising against generally replacing P-Therms with GPS-Therms. Instead, the travel-time control should activate only in those situations where the system detects that manual or programmatic setbacks will not be in use.

Table 3. Simulation results on daily savings on workdays of different thermostats for different travel times. The data are for the unoccupied, well-insulated control home (see Table 2), assuming an 8:30AM-5:30PM job. The expected savings are adjusted by expected compliance rates.

Type (travel time)	Savings each workday (%)	Savings each workday ($)	Expected compliance (%)	Expected savings (%)	Expected savings ($)
Manual thermostat	24.7	2.16	35	8.65	0.756
Prog. thermostat	21.74	1.9	56.2	12.22	1.068
GPS therm (15 min)	5	0.38	90	4.5	0.342
GPS therm (26 min)	7.6	0.57	90	6.84	0.513
GPS therm (45 min)	11.75	0.88	90	10.57	0.792
GPS therm (60 min)	13.82	1.04	90	12.44	0.936
GPS therm (90 min)	17.05	1.28	90	15.35	1.152

Table 4. The expected savings for the control home when the GPS-Therm mode *augments* M-Therms and P-Therms for a 26 min commute and 90 min commute

Type (commute length)	Compliance (%)	Expected savings per workday (%)	Expected savings per workday ($)
Manual thermostat (26 min)	35	8.65	0.76
Manual therm augmented with GPS therm (26 min)	100	13.59	1.13
Programmable thermostat (26 min)	56.2	12.22	1.07
Programmable therm augmented with GPS therm (26 min)	100	15.55	1.32
Manual thermostat (90 min)	35	8.65	0.76
Manual therm augmented with GPS therm (90 min)	100	19.73	1.59
Programmable thermostat (90 min)	56.2	12.22	1.07
Programmable therm augmented with GPS therm	100	19.69	1.63

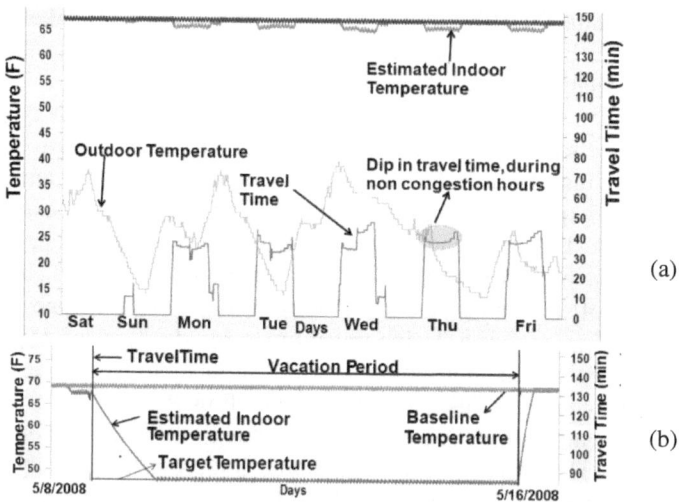

Fig. 2. (a) One week of real travel time data of a participant (House#1) and GPS-Therm simulation. (b) GPS-Therm simulation for a vacation period of participant of House#2

A key benefit of the GPS-Therm concept is that, unlike existing technologies, it adapts to changes in behavior *without requiring any behavior change on the part of the end user.* For the over 50% [4] of households not performing manual setbacks or using programmable thermostats properly, the GPS-Therm could provide a low-burden energy-saving option.

Table 4 shows "GPS add-on" expected savings. Here, expected savings are estimated using the programmable thermostat model for the 56.2% of the programmable thermostat owners who program it, but for the remaining 43.8% who do not program it, we assume that the thermostat defaults to the GPS model. Similarly, we assume that 35% of households that use the M-therms use setbacks but that the remaining 65% would default to the GPS system. For the control home with a 8:30 AM-5:30 PM job and commute of 26 min, the GPS-Therm add-on system would improve the overall expected performance of M-Therms by 4.9% and the overall expected performance of P-Therms by 3.3%. For a commute of 90 min, the savings jump to 11.1% and 7.5% respectively.

4.4   Simulation Using Real Travel Time Data

In the simulation above we assume that the person commuting always leaves and returns at the same time each day. To better evaluate potential savings, we used the real travel data from our 8 participants obtained from the GPS devices in their vehicles. Each house had two participants, so the simulator always used the minimum travel-to-home time of the two. The drive time given by the MapQuest server does not take into account the traffic congestions and delays in commute time during the different hours of the day. Therefore we increased travel times at each hour of the day proportionally to the traffic congestion index for Boston.

Figure 2 shows the travel data and GPS-Therm simulation for a typical week of one participant from House#1. The travel pattern of the participant is fairly regular throughout the work week with a small trip during a weekend. **Figure 3** shows the travelling patterns of the participants in House#2. The simulation algorithm uses min travel time because the system must always be able to condition the environment in time to reach comfort conditions for the closest person. For some homes this puts an upper bound on savings at a short (e.g., 7 min) commute. The figure areas that are

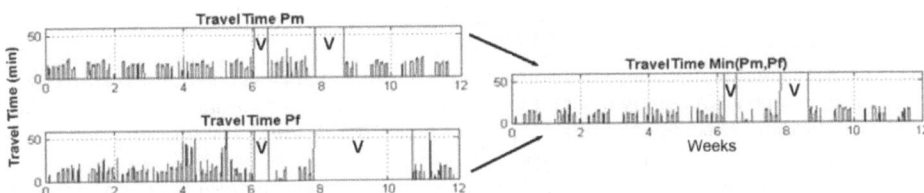

Fig. 3. Travel patterns for entire duration of the study for one household and the minimum combined travel time. V's mark vacation periods.

marked with "V" are the time periods where participants went for a vacation. If only one person in a house went for the vacation, the travelling time of the person staying at the home is considered. Travel time during vacations was manually entered into the dataset in cases where participants left their cars at home. For example, participants in House#2 went to London which is 8 h by airplane. Therefore we manually entered 8 h of travel time during the vacation.

Table 5 shows the simulation results for the entire study duration for all the four participating houses for P-Therms and GPS-Therms. The house details are given in **Table 2**. The baseline for calculating the energy and monetary savings is the cost of maintaining the comfort T throughout the study using no setbacks. [2] For the simulation, the P-Therm for all the participants is scheduled from 7 AM to 6:30 PM every day (and, to be conservative, including weekends). This interval was selected because all of the participants leave and arrive home at about this time. The target T (setback) for the P-Therm was set to 60°F (15.5°C). The effectiveness of the GPS-Therm fully depends on the travel patterns of the individuals and their home heating characteristics. For House#3, the GPS-Therm performs only 0.3% better than the baseline and a P-Therm with daily setback settings would be most effective. House#2 has larger savings because the house occupants had 2 vacations during the study totaling 12 days when the GPS-Therm automatically selects a very low target T that is sustained for the duration of the trips – a situation where the GPS-Therm excels, as shown in **Figure 2**(b). Overall, however, when only considering savings and not return comfort, M-Therms or P-Therms are clearly preferable over the GPS-Therm when they are used properly. The savings for P-Therms are higher than that of M-Therms, because the occupants spent more time in the home than the programmable settings assumed (especially on weekends).

We know, however that compliance rates are low and that use of the GPS system can increase overall *expected* performance without requiring complicated programming or sacrificing comfort if augmented on top of M-Therms or P-Therms that are not being used. **Table 5** (bottom) shows the expected savings that might have been achieved in that case in the larger population for similar homes as the test homes.[3] Savings range from 4.9% to 9.4% for GPS-augmented M-Therm to 7.9% to 12.2% for GPS-augmented P-Therm.

---

[2] Our baseline condition assumes that some of the 64% of U.S. manual thermostat owners who do not set them back regularly do not do so even when leaving for vacation, either due to lack of understanding, concern about plants, pets, or pipes, not desiring to return to a uncomfortable home, or simply forgetting to do so.

[3] The households selected all have dual commuters and therefore the results represent savings that might be achievable for only that type of household. We fully expect, for example, that savings in households with stay-at-home parents and young children might be substantially less, because setbacks are often not appropriate. The GPS-Therm might actually be most convenient in those homes, however, where the occupants have highly variable travel time schedules that are rarely known in advance, and where the occupants may be less willing to tolerate a house that is uncomfortable upon return.

Table 5. Simulated energy and monetary savings for the entire duration of the study for all the participating houses, including energy and monetary savings when using the GPS-Therm to augment manual and programmable thermostats, based on actual commuting patterns and expected compliance rates. CT = comfort temp. LT = lower temp.

	Savings for study duration % ($)			
Thermostat	**House#1 - 75 days** **CT = 67, LT = 65**	**House#2 - 75 days** **CT = 69, LT = 67**	**House#3 - 60 days** **CT = 69, LT = 67**	**House#4 - 60 days** **CT = 69, LT = 67**
Programmable	19.4% ($168)	14.1% ($102.80)	17.1% ($387.50)	13.5% ($244.30)
Manual	18.0% ($165.90)	13.7% ($97.30)	14.6% ($332.90)	12.6% ($225.70)
GPS	2.9% ($25.50)	7.1% ($49.70)	0.3% ($7.50)	0.8% ($15.50)
	Expected savings for study duration % ($)			
Manual defaults to GPS	8.2% ($74.60)	9.4% ($66.30)	5.3% ($121.40)	4.9% ($89.10)
Programmable defaults to GPS	12.2% ($105.60)	11.0% ($79.60)	9.7% ($221.10)	7.9% ($144.10)

4.5 Simulation Using Just-in-Time Questions

Using the GPS data, it is possible to improve the GPS-Therm mode by creating a system that benefits from modest user feedback without requiring the user to proactively remember to change the thermostat or predict schedules far in advance. Suppose when the user is detected to be away from home and not moving (i.e., just arrived at work), the system prompts with a simple question on the phone.

To estimate the savings this small interruption might enable, three additional (winter time) scenarios were simulated for the control house.

- Return at lower T: A user agrees to return at a T slightly lower than his or her comfort T. On the user's return, the home will be at a lower T, but the house will continue to heat up until it reaches the comfort T (see **Figure 4** (a). In this scenario the system will have a lower target T and more energy savings during the day as compared to the T set automatically by the GPS-Therm.
- Specify a time to return home: If the user decides to return back at a specific time, the GPS-Therm will ignore the GPS data and operate like an intelligent thermostat that calculates the lowest possible target T (setback) and heating start time so that when the user returns, the house is at comfort T, resulting in a lower target T and more savings.
- Return at lower T and specify a return time: If the user agrees to return at a lower T and also specifies a return time, maximal savings are achieved (see **Figure 4** (b)).

Occasional questions presented on a mobile phone and *only asked when someone is away from the home* may be an effective way to gather energy-saving information with only modest burden. A properly-programmed P-Therm may achieve similar savings, but remembering to change schedules in advance when someone has a variable schedule may be a challenging task. To evaluate potential savings from a small amount of user input, we simulated expected savings results for two cases for the control house where questions are answered on 2 and 3 workdays on the mobile phone when the system detects that someone has left home. **Table 6** shows the results for the scenarios described above with a travel time each way of 26 min. In short, answering just 2 prompts per week can boost workweek savings by up to 3.6%.

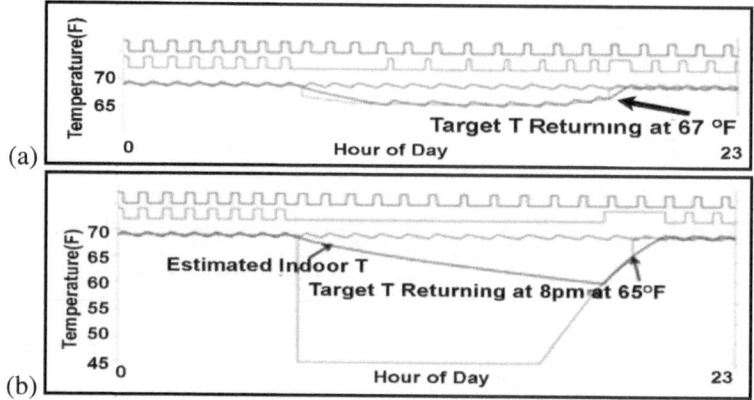

Fig. 4. 24h simulation for workday travel time of 26 min with GPS-Therm where user responded to a prompt to (a) return at lower T and (b) return at a lower T at a specified time

Table 6. Simulated savings for a work week for prompting scenarios with a travel time of 26 min

	Num prompts answered/wk	Expected savings(%)	Expected savings($)
Manual augmented with GPS	0	13.59	1.13
Manual augmented with GPS + Return time	2	16.91	1.37
Manual augmented with GPS + Return time	3	18.62	1.50
Manual augmented with GPS + Lower T	2	14.54	1.20
Manual augmented with GPS + Lower T	3	15.03	1.23
Manual augmented with GPS + Return time + Lower T	2	17.21	1.41
Manual augmented with GPS + Return time + Lower T	3	18.89	1.53
Programmable augmented with GPS	0	15.55	1.32
Programmable augmented with GPS + Return time	2	17.79	1.49
Programmable augmented with GPS + Return time	3	18.92	1.57
Programmable augmented with GPS + Lower T	2	16.19	1.37
Programmable augmented with GPS + Lower T	3	16.52	1.40
Programmable augmented with GPS + Return time + Lower T	2	17.95	1.51
Programmable augmented with GPS + Return time + Lower T	3	19.21	1.61

5 Design Observations from a Real-Time Implementation

To begin to assess the practical feasibility and usability issues that might arise with the proposed GPS-based travel-time mode, a fully-functional prototype was implemented in a participant's house for 2 wk. Due to this study being conducted in warmer months, the system controlled air conditioning rather than heating. We report on some observations from this pilot deployment.

5.1 System Design

The back-end of the GPS-Therm prototype system is a client server model using TCP. The client is the GPS-enabled mobile phone (Motorola 9Qh Global) and the server is a laptop computer that was placed at the participant's house near the location of the thermostat. The server receives the GPS coordinates from the client (via GPRS) once per minute. It contacts the MapQuest web service to get the travel time and distance from the home of the participant. The server then contacts Yahoo weather web service [32] to get the outside T of that area. The algorithm calculates the target T based on the travel time and outside T using the heating/cooling tables measured in the home (using the process described in Section 4.2). Finally the server sends the new target T over a serial connection to the computer-controlled thermostat (RCS TR40 [33]), and in reply the server gets the confirmation and the current room T. The client, in reply to sending the GPS coordinates, receives the travel time, distance, current home T, and energy saving information related to the intervention questions. In some cases, the phone prompts for information, and the responses are sent to the server as well.

The prototype GPS-Therm system is divided into two interfaces: one on the laptop located in the house and the second on the mobile phone. The interface on the phone is minimalist. Nothing is displayed except when the phone detects that the phone user is over two-minutes (drive time) away from home after just having been there. In that circumstance, the phone beeps and a question is displayed on the phone's screen, which remains until the user has a chance to respond – typically on arrival at a desti-nation. The prompt asks the user if he or she is willing return home to a 1-2°F warmer house and a return time (if known). In each case, the interface provides the user with the information on the savings expected when additional data are entered. If arrival at a slightly warmer house is selected, the interface indicates how much time it will take the warmer house to reach the desired comfort T upon return home. The prompts are easily ignored – the user is not forced to answer the questions.

The laptop interface in the prototype system is intended to simulate a replacement thermostat wall interface that would have a small digital display. It provides the user with system status information and full manual control. Unlike most current thermo-stats, it always displays what the system is currently trying to do. As shown in **Figure 5**, the system displays (A) the current home T , (B & C) the current state of the sys-tem, (D) why that is the state, and (E) what the user should do if he or she wants to change the T settings. The interface provides the energy and monetary savings since the system was installed (F). It also provides control to manually change the comfort settings (L). It rewards the participant for making energy savings decisions by

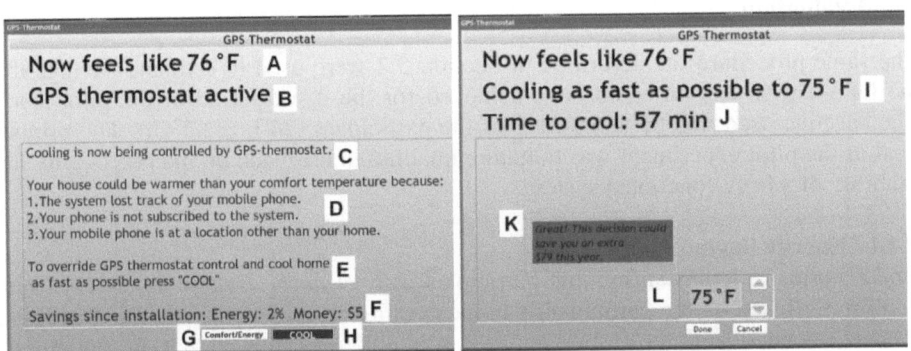

Fig. 5. Screenshots of the laptop thermostat interface (simulating a wall thermostat LCD)

showing monetary tradeoffs in real-time (K) and how the system is going to react to the change (I&J). There is a "COOL" button that can override the GPS control and resume the manual control of the cooling system (H). **Figure 6** (left) shows a typical scenario that may occur when the participant returns to a warmer home and system is not able to track his or her mobile phone.

5.2 Participant

A working professional (realtor and musician) living in a single family house near Boston with a central cooling system was recruited. The study protocol was approved by our IRB and the participant had no affiliation with the researchers.

The participant used his car as the main mode of transport during the 14 d study and lived alone in a stand-alone house with a cooling area of 3000 ft^2 (278.7 m^2). At the time of the experiment, the house was approximately 40 years old, and the cooling system was approximately 23 years old. The cooling capacity of the compressor was 60,000 BTU/h and the whole system was controlled by one M-Therm. The participant had a pet but he mentioned that he left his pet in the basement whenever he was away from the house, and the basement was not included in the area cooled by the air conditioning system. The comfort T of the participant was 74°F (23.3°C). The travel pattern of the participant was irregular because he sometimes worked from home.

5.3 Experimental Setup

A professional electrician installed the computer-controlled thermostat (RCS TR40) in the participant's home. The laptop computer (server) was kept underneath it on a table with the screen clearly visible. The participant had a broadband wireless Internet connection, which was configured so that the client (mobile phone) and server could communicate. The participant was given a GPS-enabled mobile phone (Motorola 9Qh) running the software continuously and was asked to use it as his personal phone (moving his SIM card). The participant was told to recharge the phone every night. A GPS logger was also installed in the participant's car. The remaining procedures were the same as those described in Section 4.3, although adapted for cooling rather than heating.

5.4 Evaluation

The same procedures as described in Section 5.2 were used to estimate savings that the GPS-Therm add-on could have achieved for the 2 wk period of the pilot study. The baseline used was the comfort T of the participant (74°F (23.3°C)). The primary goal of the pilot experiment was to gather qualitative feedback on the practicality and usability of a fully-functional system.

5.4.1 Energy Saving Estimations

Under normal circumstances, this participant said "*I never change my setback temperature.*" Based on the participant's home cooling characteristics and his travel patterns as obtained from the mobile phone, our simulations showed that by running the GPS-Therm prototype for two weeks he therefore saved an additional 3.4% and $2.70. The impact of using the phone's GPS versus a GPS logger in the car was

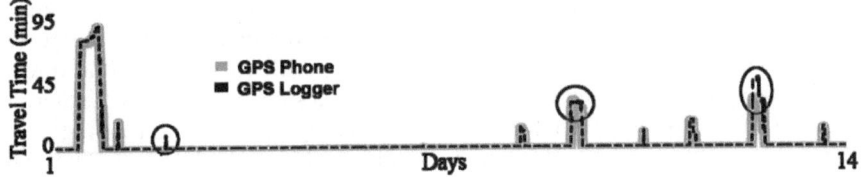

Fig. 6. Travel patterns of the participant for 2 wk. Highlighted portions show the difference between travel data from the phone's built-in GPS and GPS logger in the car, which are modest.

Table 7. Expected savings across similar homes and commute patterns for two weeks

Thermostat	Expected Savings (%)	Expected Savings 2 wks ($)
Manual	1.05	0.84
Manual augmented with GPS	3.3	2.60
Programmable	5.84	4.64
Programmable augmented with GPS	7.3	5.80

minimal. As indicated in **Figure 6**, an additional savings of only 0.2% would have been achieved had the phone GPS worked equivalently to the Trackstick Pro in the car.

As in previous simulations, we used this participant's travel patterns and his home's temperature response characteristics to estimate the expected savings across others with homes and travel patterns like his, assuming the M-Therm and P-Therm compliance rates. **Table 7** shows the expected savings possible for the two week period. Although programmable systems would save the most energy,

5.4.2 Responding to Prompts

During the 14 d study, the participant answered 8 of 24 prompts he received on occasions when he left his home, and he entered some information that led to energy savings in each case. During 3 of his trips, for example, he responded positively to the "return at warmer temperature" and the "specify a return time" questions. Due to the novelty effect, we hesitate to make generalizations about the question-answering. However, we *can* measure the energy saving impact, and each answer he gave saved the participant an average of 0.32%, or $0.25, in energy when the thermostat was in GPS mode. For some people, this amount earned may offset the burden of the interruption.

6 Discussion

The results from our participants using both simulation and the functional system suggest that a GPS/travel-time thermostat mode could save substantial amounts of energy. Here we list issues that may need consideration if such a system were to be implemented at scale.

Issue 1: Aesthetic concerns and impact on design. Most thermostats are wall mounted at eye-level, far from a either an electrical outlet or an Internet outlet. Home occupants may be unwilling to introduce unsightly cables running to the device or invest in costly, messy renovation. To encourage widespread adoption of a thermostat with a GPS mode, the system must (1) be powered by the 24v line running to every thermostat, and (2) connect to the Internet via a wireless link. The extra energy consumed by the continuous wireless link (e.g., 802.11) must be factored into the potential savings. A typical wireless router consumes approximately 0.3KWh/d, which in the Boston area would cost $0.06/d and $22/year, far less than the GPS-mode savings for many people.

Issue 2: Cost. In our prototype, the server was implemented on a laptop, but mobile phones with data connections are sufficient to run the simulations. By leveraging the phones that people will buy anyway, the only new functionality that must be added to a thermostat is a wireless link so the thermostat can be set and read remotely. In homes similar to our participant houses where people do not use setbacks, the technology could pay for itself quickly. On average, winter month savings would range from approximately $45 to $120 in our area. Considering *only* winter months, this leads to estimated payback times from 12 mo for House#2 to 36 mo for House#3. Without knowing how much he could have saved, the participant using the prototype reported that he would pay up to $300 for the device that he tested, an amount adequate to cover the cost of the device.

Issue 3: Unanticipated consequences in behavior. Our participant using the fully-functional system mentioned that his work travelling patterns are irregular and that he did not use setbacks, but that the GPS mode was beneficial because, "I don't have to remember to change my temperature settings before leaving." However, if long term he relied on GPS mode instead of using setbacks for his regularly scheduled trips, he would waste energy. If the system detects someone is relying on GPS mode, the interface should gradually introduce interactive prompting with the goal of encouraging use of the P-therm or gathering information every day that allows the system to operate at the same efficiency. The unintended consequence of someone who otherwise would have adjusted the thermostat instead relying on the GPS mode must be discouraged.

Issue 4: Time away: inference or prompting? Due to the (long) lag time of heating and cooling a home, maximal savings can only be obtained when return time is known or accurately inferred. Otherwise, many people during a normal workweek will work so close to home that the interior T cannot drop/raise sufficiently fast to accrue major savings because return time is so fast. An alternative is to infer typical return-time behavior automatically from prior behavior. Inferring intentionality may be valuable for energy savings [7], but it is also prone to error. An alternative that may lead to a more predictable user experience would be to simply recognize one particular behavior – when someone has stopped traveling -- and then present an easy-to-ignore prompt on the phone. The burden of the interruption could be softened by presenting real-time, tailored information about the savings that the interruption enables. Our simulations suggest that the daily value of answering a question for many homes could be as high as $.25. The participant, "liked the persuasive status messages, because they were showing how much I can save," but even a single extra reward message screen generated the comment that, "the message was adding extra information to an already cluttered day." Is \$.25 savings enough to offset the inconvenience of the interruption? Determining prompting sustainability requires longer-term studies with functional systems.

Issue 5: Explanation of thermostat state and behavior. The GPS-Therm prototype always indicated the current T, target T, and time to reach target T, unlike most thermostats on the market today, to discourage the user from falling into the common thinking that a higher setting leads to faster T change [19]. The participant using the pilot implementation did comment that, "the laptop interface gave me information about what the system was doing and why, which I cannot see on my thermostat." Unfortunately, adding a GPS/travel-time mode creates new special case conditions that can be challenging to convey. For example, if a person leaves his or her phone at work or the phone battery dies, the home will be quite uncomfortable on return; the reason (that the system thinks the occupant is not home) must be conveyed to the user. Handling these special cases must be accomplished without losing the desirable simplicity of the GPS-Therm mode. Here too, longer term studies are warranted.

Issue 6: Incremental prompting frequency tailored based on prior answers. Our pilot interface could put the user in a position where a prompt for savings or return time information must be declined on a regular basis for reasons beyond the user's control, thereby repeatedly creating disconcerting feelings of cognitive dissonance if the person wants to save energy. With respect to the prompts, the participant commented that, "it made me feel good if I make energy saving decisions, and feel bad if I did not." To avoid creating negative feelings, the system could be improved by tailoring prompting rates based on previous frequency of positive responses.

Issue 7: Temperature vs. comfort. Our participants were accustomed to thinking in terms of "temperature" rather than comfort. The participant using the prototype, for example, reported about the minimal phone interface that intentionally did not show temperature that, "I did not find the GPS thermostat phone interface informative [because] it was not showing me the temperature of my house." He further added that the phone should display current T, target T, and energy savings, and the ability to change the T or return time decision at any point of time. However, particularly in summer, the humidity can have an impact on T, as can other factors, such as what one is wearing. Ideally an interface would guide the user to focus on comfort *instead* of T, allowing more fluctuation in T (and therefore more energy savings).

Issue 8: Phone limitations. GPS lock times on the phone ranged from 2-5 min on sunny days with open skies to 15 min on cloudy days in urban areas. Fortunately, the ongoing massive industry investment in improving phone location-based services will only further improve the performance of the GPS-Therm mode. Despite current limitations, however, as mentioned in Section 5.1, the phone still performed adequately to produce savings.

Another consideration for practical deployment is battery life. Using the phone data transfer scheme described in Section 6.1 resulted in a battery life on the MotoQ 9h global of only 10-12 h on a charge. To improve the battery life, we subsequently devised a simple scheme for pilot testing where the GPS switches on every 3 min and remains on for 2 min to get a lock, which enabled 24 h performance.

Other issues. As we ran these experiments, we identified several other areas where our prototype could be improved that we mention briefly: (1) implementing an algorithm which dynamically updates the lookup tables for the T profile of the house based on the current weather conditions like outside T, humidity, wind velocity, and sunshine, (2) implementing algorithms that take into account not only outside T but also forecasted outdoor T, so that a sudden change in weather condition does not create a situation where the GPS-Therm cannot catch up in time to ensure the home is at a comfortable T when the occupants return, (3) detecting "driving" from the GPS data in order to minimize the chance of an ill-timed prompt, (4) modification of the question-prompting to handle multiple participants in the house, including exchange of information between family members as they make decisions that might impact each others comfort, and (5) controlling the hot water heater. We have not considered night setbacks in our simulations, which may also be amenable to this type of interactive control.

7 Conclusion

In summary, in this pilot work we have prototyped, tested and evaluated a GPS-controlled thermostat system. We have shown through simulations that such a system is capable of saving as high as 7% on HVAC energy use in some households. This is less than savings obtained from optimal use of M-Therms or P-Therms, but the GPS mode we propose has the potential to save energy for the more than 50% [4] of the U.S homes that do not change their T settings when there is no one in the house, and the system could be easily and affordably installed in many homes. More work is needed to fully evaluate potential savings and feasibility and usability of the user interface and interactive prompting components and to explore differences in climate, living environments, and lifestyles in other parts of the world.

Acknowledgements

This work was funded by the MIT House_n Consortium. The authors thank the reviewers for helpful comments, Telespial Systems for Trackstick donations, and our participants.

References

[1] CIA, The World Fact Book (cited 2008 08/15/2008),
https://www.cia.gov/library/publications/the-world-factbook/
[2] U.S. DOE Annual Energy Review (cited 2008 08/15/2008),
http://www.eia.doe.gov/fuelelectric.html
[3] U.S. DOE Building Energy Data Book (cited 2008 08/15/2008),
http://buildingsdatabook.eren.doe.gov/docs/1.2.3.pdf
[4] U.S. DOE Residential Energy Consumption Survey (cited 2008 08/15/2008),
http://www.eia.doe.gov/emeu/recs/recs2005/hc2005_tables/detailed_tables2005.html
[5] Nevius, J., Pigg, S.: Programmable Thermostats that Go Berserk? Taking a Social Perspective on Space Heating in Wisconsin. In: Proc. ACEEE Summer Study on Energy Efficiency in Buildings, vol. 8, pp. 233–244 (2000)

[6] Plourde, A.: Programmable Thermostats as Means of Generating Energy Savings: Some Pros and Cons. Canadian Building Energy End-Use Data and Analysis Centre, Technical Report CBEEDAC 2003-RP-01 (2003)

[7] Harris, C., Cahill, V.: Exploiting User Behaviour for Context-Aware Power Management. Proc. Wireless And Mobile Computing, Networking And Comm. 4, 122–130 (2005)

[8] Ross, J.P., Meier, A.: Whole-House Measurements of Standby Power Consumption. In: Proc. 2nd Int'l Conf. on Energy Efficiency in Household App. and Lighting, vol. 108(13) (2000)

[9] Roy, A., Bhaumik, S.K.D., Bhattacharya, A., Basu, K., Cook, D.J., Das, S.K.: Location Aware Resource Management in Smart Homes. In: Proc. IEEE Int'l Conf. on Pervasive Computing and Communications, pp. 481–488 (2003)

[10] Harle, R.K., Hopper, A.: The Potential for Location-Aware Power Management. In: Int'l Conf on Ubiquitous Computing, pp. 302–311 (2008)

[11] Thermostat History (cited 08/15/2008), http://www.prothermostats.com/history.php

[12] Keyson, D.V., de Hoogh, M.P.A.J., Freudenthal, A., Vermeeren, A.P.O.S.: The Intelligent Thermostat: A Mixed-Initiative User Interface. In: Proc. Conf. on Human Factors in Computing Systems, pp. 59–60 (2000)

[13] Fountain, M., Brager, G., Arens, E., Bauman, F., Benton, C.: Comfort Control for Short-Term Occupancy. Energy and Buildings 21, 1–3 (1994)

[14] Mozer, M.C., Vidmar, L., Dodier, R.H.: The Neurothermostat, Predictive Optimal Control of Residental Heating Systems. In: Advances in Neural Information Processing Systems, vol. 9, pp. 953–959. MIT Press, Cambridge (1997)

[15] Titus, E.: Advanced Retrofit: A Pilot Study in Maximum Residential Energy Efficiency. In: Proc. ACEEE Summer Study on Energy Efficiency in Buildings, vol. 1, pp. 239–245 (1996)

[16] Fuji, H., Lutzenhiser, L.: Japanese Residential Air-Conditioning: Natural Cooling and Intelligent Systems. Energy and Buildings 18, 221–233 (1992)

[17] Springer, D., Loisos, G., Rainer, L.: Non-Compressor Cooling Alternatives for Reducing Residential Peak Load. In: Proc. ACEEE Summer Study on Energy Efficiency in Buildings, vol. 1, pp. 319–330 (2000)

[18] The Lawrence Berkeley National Laboratory. Thermostats and Comfort Controls (cited 2008 08/14/2008), http://comfortcontrols.lbl.gov/

[19] Kempton, W.: Two Theories of Home Heat Control. Cog. Science 10, 75–90 (1986)

[20] Gladhart, P., Weihl, J.: The Effects of Low Income Weatherization on Interior Temperature, Occupant Comfort and Household Management Behavior. In: Proc. ACEEE Summer Study on Energy Efficiency in Buildings, vol. 2, pp. 43–52 (1990)

[21] Sachs, H.: Programmable Thermostats. American Council for an Energy Efficient Economy Technical Report (2004)

[22] Analytics, R.: Validating the Impact of Programmable Thermostats: Final Report. RLW Analytics, Inc., Middletown, CT (2007)

[23] Karjalainen, S., Koistinen, O.: User Problems with Individual Temperature Control in Offices. Building and Environment 42, 2880–2887 (2007)

[24] Mozer, M.: The Neural Network House: An Environment that Adapts to its Inhabitants. In: Proc. AAAI Symposium on Intelligent Environments, pp. 110–114 (1998)

[25] TrackStick (cited 2008 08/15/2008), http://www.trackstick.com/

[26] LASCAR Logger (cited 2008 08/15/2008),
http://www.lascarelectronics.com/
[27] LOGiT Current and Voltage Data Logger (cited 2008 08/15/2008),
http://www.supco.com/LOGiT%20Data%20Loggers.htm
[28] U.S. Env. Protection Agency Energy Star Programmable Thermostats Specification (cited 2008 08/15/2008),
http://www.energystar.gov/
index.cfm?c=revisions.thermostats_spec
[29] Mull, T.E.: HVAC Principles and Applications Manual. McGraw-Hill Professional, New York (1997)
[30] MapQuest WebService (cited 2008 08/15/2008),
http://developer.mapquest.com/
[31] U.S. Census Bureau (cited 2008 08/15/2008), http://www.census.gov/
[32] Yahoo Weather WebService (cited 2008 08/15/2008),
http://weather.yahooapis.com/forecastrss?p=ZipCode
[33] Residential Control Systems (cited 2008 08/15/2008),
http://www.resconsys.com/products/stats/index.htm

KidCam: Toward an Effective Technology for the Capture of Children's Moments of Interest

Julie A. Kientz[1] and Gregory D. Abowd[2]

[1] Human Centered Design & Engineering and The Information School
DUB Group
University of Washington
Seattle, Washington USA
jkientz@u.washington.edu
[2] School of Interactive Computing and Health Systems Institute
Georgia Institute of Technology
Atlanta, Georgia USA
abowd@cc.gatech.edu

Abstract. Mobile applications of automated capture present many interesting design challenges, balancing the desire for rich media against ease of use and availability. In particular, capturing rich media of young children has many potential benefits, but remains a difficult challenge due to the many unique constraints of recording children. Motivated by the aim of supporting a parent's need to record the life events of a young child, we have designed KidCam, a prototype rich media capture device. This paper presents the design, implementation, and evaluation of KidCam and its goal of addressing some of the challenges of recording young children. Results from a three-month study with four families show that KidCam addresses some of the challenges of recording rich media of children, but there are still remaining hurdles. We discuss these remaining challenges, potential ideas for how they could be addressed, and emergent uses for KidCam beyond the initial domain for the creation of family memories.

1 Introduction

The capture and access of rich media, such as audio and video, has many potential applications [4,8,9,22] and is one of the three central application themes of pervasive and ubiquitous computing [1]. Our particular interest in this paper is the use of these recordings to support the collection and reflection of children's moments of interest by parents and families. Past capture solutions usually assume a static capture infrastructure, limiting their coverage across spaces, such as all rooms in a large home. Although mobile capture and access applications have been implemented to monitor activities of daily living [20], generate field reports for soldiers on patrol [21], and provide tourist services [6], they typically involve wearable computers, on-body sensing, or both. Despite the potential benefits of these applications, there are many issues associated with these wearable systems, such as weight, fragility, and heat dissipation, which may make them unsuitable for a general population, especially children and

H. Tokuda et al. (Eds.): Pervasive 2009, LNCS 5538, pp. 115–132, 2009.

individuals with special needs [16]. In this paper, we explore how mobile technologies can be designed and developed to aid in capturing rich media experiences in the lives of young children.

The inspiration for a non-wearable, mobile capture and access device came from a need to support parents of young children in documenting important moments of their child's development [17]. For example, many families with children purchase a digital camera, a camcorder, or both for the purposes of recording their child's special moments and development. Parents share these recordings with others, such as grandparents, and archive them for sentimental reflection when the child is older. In addition to collecting keepsakes, many parents may wish to remotely monitor their child while they are asleep or otherwise away from them, and thus often purchase audio or video baby monitors. Parents with questions about their child's development and may wish to record information to share with a pediatrician or a specialist. In addition, there has been increased use of home movies of children to assess how childhood disorders look at young ages [4].

The prevalence of these digital recording devices has grown dramatically over the past decade. Many people own multiple recording devices, including digital cameras, video camcorders, camera phones, web cams, security cameras, and more. However, the presence of these devices does not ensure the capture of important events nor the ability to find and retrieve the relevant media. One of the problems that can arise from owning so many devices is the variety of media types that can be recorded, which can cause people to become overwhelmed with the choice of recording device and storage media. This can prevent both capturing the event and later viewing or sharing the event with others. A further complication is that many traditional capture devices have limited storage and thus do not support continuous recording, which allows people to capture unplanned events.

A potential solution to these issues may be to combine many recording features into one single, semi-mobile, and semi-continuous recording device – a design space that has yet to be explored for this domain. However, how do we design such a device so that it incorporates many of the desired recording features without overwhelming the user? How do we determine if mobile recording is appropriate for this particular domain? To answer these types of questions, we explored the problem by designing such a device with this context of use in mind. The device we designed, called KidCam, was based on our previous research in mobile and continuous capture and design requirements for technology for supporting families [17]. KidCam has the goal of enabling families to record their children's moments of interest through the continuous collection of video using a buffering technique that allows the manual recording of spontaneous segments of videos and remote monitoring and capturing. To determine whether this new design space is appropriate for families in realistic settings, we evaluated the effectiveness of this device through a three-month, long-term deployment with four different families with young children. We found that although we designed KidCam for and with families, there are still some remaining challenges in capture for this domain.

In this paper, we begin with a discussion of related work in mobile capture and access and record-keeping for young children. We then discuss a classification of

recording approaches and how KidCam addresses a gap in existing recording technologies. Next, we discuss the design requirements we used for building a capture
device for recording children and provide an overview of the prototype application.
We then continue with a description and results of the three-month evaluation of
KidCam by four families. Finally, we discuss the implications for the results of our
study and how the field of research in mobile capture and access can be moved toward more effective technologies for recording children's moments of interest.

2 Related Work

In this section, we discuss how our work relates to and expands upon previous work
in relevant areas. This work includes projects related to mobile capture and access
applications along with their associated techniques supporting record-keeping for
young children. We also provide a discussion of a classification for existing recording
technologies and how KidCam fills a particular gap.

2.1 Existing Capture Technologies

In recent literature, there has been a large amount of work in understanding and providing for the mobile capture of rich media such as audio, video, and photographs.
Several studies have looked at how people use camera phones [18,19] and digital
cameras [7] to capture pictures and video segments for personal and social purposes.
These studies aim to understand current practices for how people use the devices they
already own, rather than explore the design of new devices. Other researchers have
designed and built devices that provide automated capture, including one that automatically takes photographs based on scene changes [14] and a proposed system for
automating experience capture for tourists [3]. Although these systems are similar to
what we have built in terms of automating the capture process, they differ in features
and purpose. KidCam supports a variety of capture types, including both still images
and video, and combines everything into one stand-alone apparatus suitable for both
stationary and mobile use.

Other mobile systems have supported more traditional capture and access in that
they support a specific domain. The Personal Audio Loop [10] is an audio-based
wearable system that is used to support near-term memory recall, but does not allow
users to save events for future use. The Soldier Assist System [21] is wearable and
supports the collection of still images, video footage, environmental audio, spoken
audio, and motion information along with automatic indexing into this data for the
purposes of supporting post-patrol debriefings. However, this system is very cumbersome to use, requiring the soldier to wear multiple pieces of hardware that would be
very awkward in civilian settings.

We provide means for recording spontaneous and unplanned events by allowing
for a buffering of video data. Others have explored the use of this type of technique to
help classroom teachers identify the causes of children's behaviors [10] and record
information in informal meeting spaces [12]. Other systems provide automatic triggering using sensors, such as the SenseCam platform [14], which takes still pictures

based on sensor data built into the device (*e.g.,* light sensors, GPS data). StartleCam [13] is a wearable camera that circularly buffers data and automatically archives pictures when the user experiences a startle, which is measured by a significant change in their skin conductivity. The HP Casual Photography project describes another wearable system which constantly record videos and pictures for later viewing [23]. Kid-Cam differs from these by recording not just still pictures, but rich video and audio as well. In addition, all four of these projects require playback or viewing of pictures on a separate device, whereas our system supports the reviewing of videos and pictures on the same device.

In the specific domain of recording young children, the Human Speechome Project [24] uses an extensive recording infrastructure throughout a house to gather linguistic data to help researchers ascertain how children acquire language. While our system could make use of an extensive video recording infrastructure, we aimed to build a device that could be moved from room to room and to places outside the primary home of the child. Though we may sacrifice the amount of footage our system will collect, we believe our system enables capture in more places and will enable parents to collect only the videos they want without as much invasiveness.

2.2 Classification of Existing Capture Technologies for Children

One useful way of classifying existing recording technologies is through two separate dimensions. These dimensions are whether the recording happens continually or whether it is on-demand and whether the devices cover a single, fixed space or are highly mobile. A review of the existing strategies for capturing rich media of children's moments of interest revealed a particular gap in the needs. In particular, existing recording technologies tend to be on the extremes of these two dimensions. While there are benefits at the extremes of each of these dimensions, there are also disadvantages that prevent a desirable and easy-to-use recording system that can capture unplanned moments throughout all the locations a child may need to be. Figure 1 shows a diagram of the capture dimensions and how KidCam fills that gap.

Along the dimension of continuousness is a spectrum of devices that continuously record information without any intervention (*e.g.,* security cameras) at one end and those that only record when explicit user action is taken (*e.g.,* a digital camera) at the other end. The advantage to the continuous recording at the extreme end of the spectrum is that every event is recorded and likely nothing would be missed. The disadvantage to this end of the spectrum is that there are social concerns over privacy and technical problems of storage and searching through many hours of footage to find the appropriate moment of interest. In addition, the quality of this type of recording may be compromised for the sake of storage space or privacy. At the opposite end of the spectrum is the notion of on-demand capture. This has the advantage of being precisely what the user intends to capture and is often of higher quality in terms of captured content. However, the disadvantage to this end of the spectrum is that it is often difficult to capture unplanned moments. The center of this axis is a middle ground where data is constantly recorded but only saved when the user explicitly takes an action. The selective archiving approach described by Hayes *et al.* [10] meets this middle ground.

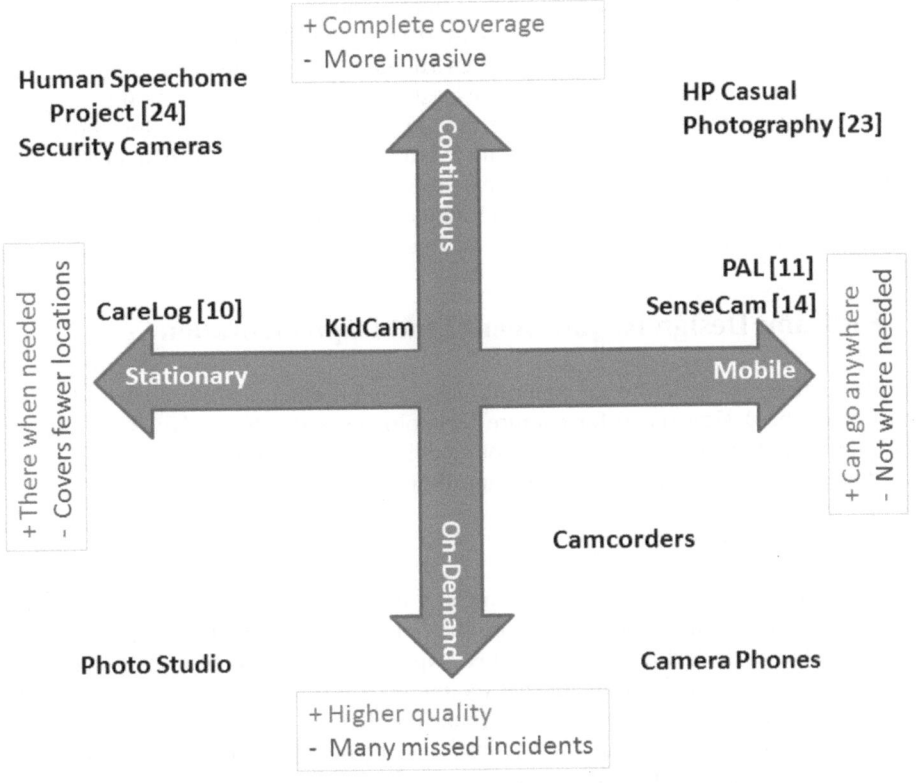

Fig. 1. Classification of existing capture technologies along the dimensions of continuousness and mobility

The second dimension of capture in relation to children's moments of interest is along the axis of mobility, which ranges from stationary (*e.g.*, cameras permanently affixed to a wall) to completely mobile or wearable capture devices (*e.g.*, camera phones). At the stationary end of the spectrum, the advantage is that the capture device is always in a particular space and that space is always covered if something happens. The disadvantage to this is that the device only covers this space, and if events of interest occur outside of the covered location, the event will be missed. Because children are often mobile and have many different areas where events occur, it would be difficult to provide coverage for all potential areas of interest. At the other end of the spectrum would be a completely mobile system that could be anywhere the child may be, such as a wearable device. The advantage to this is that the device can capture events anywhere. The primary disadvantage to the mobile approach is that it is possible the device would not be in the correct location or orientation when the moment of interest occurs. In addition, mobile capture devices tend to be fairly low quality due to their need for compactness and have limited storage capabilities. At the center of this spectrum is a device that is portable enough to be taken anywhere, but can still be placed in a fixed location while in a room.

Because of the disadvantages of each of the far ends of the two dimensions, we designed KidCam to meet at the halfway point of each. KidCam uses a selective archiving technique that constantly buffers when running, and parents can choose to save videos from a 20 minute buffer stream of past recorded events. For mobility, the device is lightweight enough to be easily moved from location to location, but comes with a built-in stand that can be set in a fixed location, thus is considered semi-mobile. One interesting note is that there is a second gap in the classification, which is in the area of semi-mobile but fully continuous. This could describe a technology that is constantly recording, but can be easily moved from one location to the other.

3 KidCam: Design Requirements and Implementation

In this section, we describe the specific design requirements we used based on our analysis of the design space for capture technologies and a formative study on technologies to support recording data on young children. We also provide an overview of how these results influenced the KidCam prototype implementation.

3.1 Design Requirements

The requirements for KidCam came through an in-depth formative study we conducted on how new parents might wish to record their children's developmental progress [17]. In this study, we interviewed new parents, experienced parents, secondary caregivers, and medical professionals on the design requirements for how designers might develop technology to support record-keeping for young children. From this work, we determined several important considerations and challenges for recording technologies for children.

- Families enjoy taking digital pictures and videos of their children to preserve sentimental memories and share with family and friends. Digital photos allow them to take multiple pictures to try to get the "best shot." Videos allow them to record their child's voice and actions and are often used during special occasions, like birthday parties and holidays.
- Parents do not need or want to continuously record every move. Rather, they care most about recording interesting moments, such as accomplishing a significant milestone, such as saying their first words, or a sentimental purpose, such as reading a bedtime story together. Very young children also sleep for a significant amount of the day, so continuous recording may not be necessary. Thus, shorter, filtered video segments are optimal.
- Children are in several locations throughout the day. Although they may have bedrooms with cribs, they are often in the living room, the kitchen, a playroom, or parents' bedrooms. They may also visit daycare centers or the homes of childcare providers where they also spend a significant amount of time. Thus, capture systems should be mobile enough to work in many places.
- It is difficult to predict when moments of interest may happen. A parent may prompt a child to do various activities, but many times a child will spontaneously act when a parent does not expect it. Any recording technologies should be able to capture these unplanned moments.
- Continuous, always-on video recording may be more invasive than a mobile device where parents control the recording. In addition, a mobile solution

would alleviate having to instrument an entire house with cameras, which can be impractical for everyday use and violate home aesthetics.

In designing KidCam, we aimed to build a technology that would meet each of these considerations and enable parents and caregivers to record throughout the home and other places where the child may go, such as daycare or the home of another caregiver. We followed an iterative design process to ensure that KidCam was meeting the needs of the users for which we were designing, including some low-fidelity prototyping and short-term evaluations on the interface design.

3.2 Overview of Prototype Implementation

The final design of KidCam uses the Sony VAIO™ UX running Windows XP, an ultra-mobile PC (see Figure 2), though any model of ultra-mobile PC would work. The VAIO has two built-in cameras (one on the front and one on the back), a microphone, a touch screen interface, a mini-qwerty keyboard, Bluetooth and 802.11 wireless communications, and 30 GB of storage space. We then wrote a software application using C# that provides a user interface and supports all of the capturing and reviewing of the audio-visual data.

The user interface is themed as a child monitor and recording device, which child-friendly graphics and colors and large widgets for touch-screen interaction. The basic functionality enables the recording of video, audio, and still pictures using either the front or the back camera, as well as reviewing multimedia data based on different annotations that are provided either during or after capture. A commercially available mobile RAM® mount stand was added to the system to allow people to situate the device and camera to whichever angle they need in a variety of environments. When attached to the mount the entire unit stands about 9 inches (23 cm) high. The device can be easily removed from the stand for hand-held recording and viewing. The battery life of the device enables it to run for approximately 1.5 to 2 hours while unplugged. We recommended that parents leave the device plugged in while it is situated in the stand. Overall, the device is completely mobile when detached from the base, and measures approximately 6 inches (15 cm) wide, 4 inches (10 cm) high, and 1.5 inches (4 cm) thick, and weighs 1.1 pounds (0.5 kg). When attached to the base, it is slightly less mobile weighing approximately 3.75 pounds (1.7 kg).

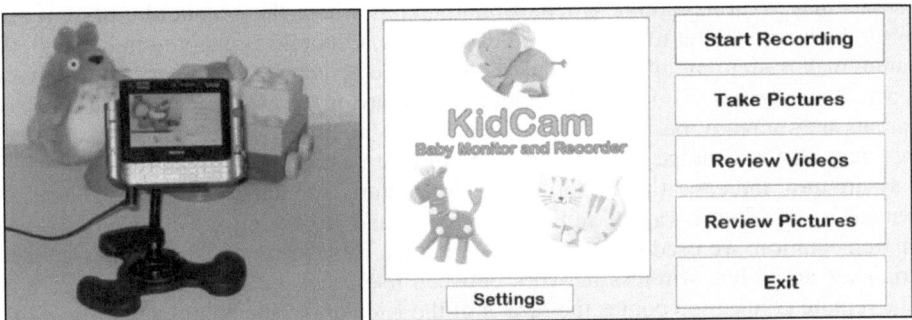

Fig. 2. View of KidCam prototype on a Sony VAIO (left) and a screen shot showing the main menu of KidCam's interface (right)

3.3 Continuous Video Buffering for Saving the Past and Future Events

To archive videos, we wanted to allow for continuous recording during an event and have users specifically choose to save videos either during, before, or after an event occurs. To accomplish recording prior to an event, users can set the recorder to save video for a specified number of minutes in the future. For example, parents may witness their children spontaneously take their first steps and wish to go back and record those moments, or at the child's first birthday party, the parent may set KidCam to record from the beginning of opening presents until they are finished. Thus, we implemented a video buffering system similar to that which a digital video recorder uses. The concept of our design was similar to that of the notion of selective archiving [10], which allows for saving only the past events of interest. When the user chooses to save a video file (see Figure 3), she specifies how far in the past and how far into the future to save the video using a range slider widget. The device then copies a segment from the buffer to a video file that corresponds to the beginning and ending of the desired video segment. While the device is buffering multimedia data, the interface shows a live preview of the video so it can be easily positioned to the desired angle while in the stand or used like a handheld video camcorder.

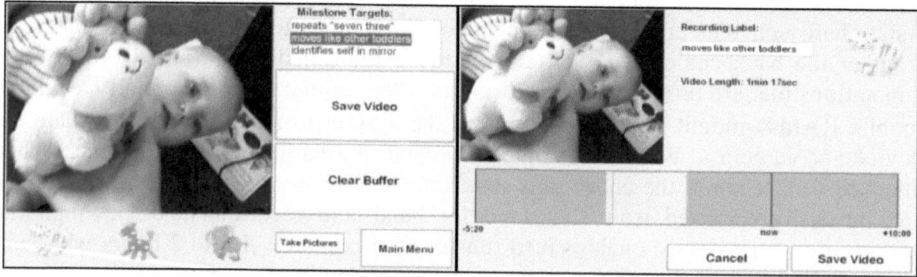

Fig. 3. Interfaces for previewing current capture of KidCam (top) and saving videos using sliders to indicate the start and stop points to archive (bottom)

3.4 Remote Monitoring and Remote Capture

Parents may want to capture video or photographs where they cannot be present or it would be inconvenient to be present. For example, if a child is napping in her crib, the parent may wish to monitor from the kitchen while he is making dinner. Alternatively, a parent may wish to monitor his child at daycare from his desk at work. Because parents may already be using a baby monitor to monitor their child while sleeping, this may also encourage them to have the device near their child more often, which may in turn increase the opportunities to capture spontaneous events. Thus, we wanted to develop a way of remotely viewing and triggering the KidCam. For our implementation, we used the Nokia n800 Internet Tablet™ to create a remote connection over an ad-hoc wireless network between itself and the KidCam (see Figure 4). The remote connection copies the screen of the KidCam to the Nokia and provides for remote interaction through the touch screen of the Nokia Internet Tablet. The live audio-visual feed from the KidCam can be remotely accessed on the internet tablet, though at a reduced video frame rate.

Fig. 4. KidCam's interface can be remotely viewed and controlled via another device (*i.e.*, a Nokia Internet Tablet) that can replicate its screen and audio

3.5 Media Reviewing Interface

Users of KidCam may also need to review videos in a variety of locations. For example, if parents have recorded videos of their child playing with toys a relative has given her, they may want to show it to that relative while at their house for a visit. Thus, we have implemented a file viewing interface for the device that enables quick reviewing of videos and pictures (see Figure 5). The media file reviewing interface is divided into a screen for reviewing videos and a screen for reviewing still pictures. The video review screen allows the user to sort the files by date and time, length of video, or name. In the list, when the user clicks on the video file, it will play the video. The still picture review screen is similar to the video review screen, but shows thumbnails instead of text labels for easy viewing. For both the video and the picture reviewing interfaces, the user can choose to delete videos and pictures they no longer want to keep. They may also choose to "export" videos, which will copy them to an export directory for later synchronization with a home computer.

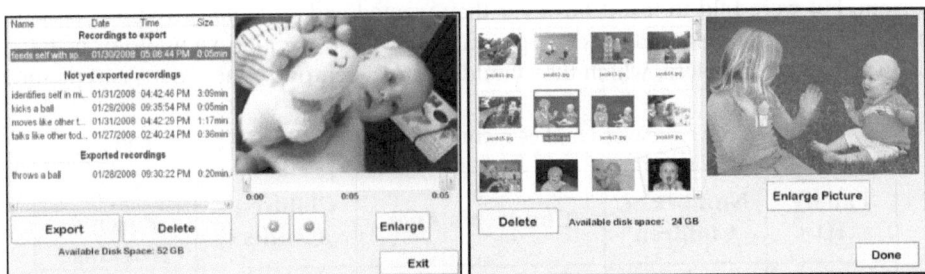

Fig. 5. KidCam interfaces for reviewing videos (left) and still pictures of their child (right). The video and the pictures can both be enlarged to take up the full screen.

4   Evaluation Study of KidCam

In this section, we describe the design of and results from the KidCam study, in which we provided the device to four families to use over a three-month period to capture their children's experiences.

4.1 Study Design and Participants

The iterative design process showed that KidCam was functional and could successfully capture data. However, we wanted to evaluate the effectiveness of KidCam in a manner that was ecologically valid and to determine if it was actually useful to families. Thus, we deployed the device with four families for approximately 3 months. The goal of the study was to evaluate the use of KidCam in realistic settings with real families and determine how often parents would run the device and how many photos and videos they would capture and review using the device. We also wanted to expose any problems with the design and refine our design requirements to develop future versions of the KidCam system.

Families in this study were recruited as part of a larger study investigating technology solutions to support record-keeping of young children [15] and were given KidCam in addition to a software application for tracking their children's progress. We recruited 4 families who were all clients of a single pediatrician's office in a suburb of Atlanta, Georgia. All families were of similar demographics and socio-economic statuses, which was college-educated, middle class, married, and with two parents in their early to mid 30s. Three of the families were American-born, and one of the families (Family 2) had a father who was born in South America. All families were computer literate and were familiar with digital photos and digital cameras. The number of children in each family was either 1 or 2, and all children in the study were aged 5 and under. Table 1 shows the composition of each of the four families we recruited for our study. The families in the study were compensated monetarily for their participation in the study as a means of recruitment and retention over the 3-month period, but it was not in any way contingent on using the system.

For the study, each family was given a KidCam recording device and a remote viewing device (*i.e.,* a Nokia internet tablet). They were instructed on how to use the system, but were told to use it however they wanted and were not given any specific instructions. They also had a developmental milestone tracking technology installed on their home computers, which they could use to synchronize the videos and pictures

Table 1. Overview of families recruited to evaluate the KidCam system

Family ID	Number of Children	Children's Ages	Children's Genders	Both Parents Working?
1	2	12 months, 4 years	M, M	No
2	1	9 months	M	Yes
3	2	9 months, 3 years	M, F	No
4	2	15 months, 5 years	M, F	Yes

taken with KidCam to their home computers. Synchronizing KidCam with the home computer also updated a list of three upcoming milestones for which parents should be watching. We logged all interactions with the KidCam device to its local hard drive, as well as all photos and videos recorded (unless the parents explicitly deleted them). Prior to being given the device, we interviewed parents about their current techniques for capturing and reviewing their children's moments of interest. We also met with them once halfway through the three month time period to download the logs, photos, and videos and conduct a mid-study interview on their use of the device and suggestions for improvement. We conducted a similar meeting at the end of the three-month period, where we also collected the device and conducted a final interview, which probed on their use of KidCam, suggestions for improvement, and a description of their ideal recording device. Following the study, each interview was transcribed for further analysis.

4.2 Study Results

In this section, we provide data on the general usage of KidCam by each of the families as determined by the logs. We then present the general perceptions of KidCam from the families as reported during the mid-study and final interviews.

4.2.1 Overall Use

The videos that parents recorded were appropriate for what might be needed for preserving child memories or aiding a pediatrician in analyzing whether a child has achieved a developmental milestone. They tended to be shorter segments of the child's development or fun experiences they wanted to save for later, with the average length per video being 3 minutes, 43 seconds. Family 2 recorded longer videos in general, because they used the device to record the child's family events, such as the dinner for his first birthday party. For photos taken with the device, the quality was not as high as what parents were used to with digital cameras, which is why parents reported that they did not use that feature often.

Overall, use of KidCam was lower than we expected. Parents reported in their interviews that they did not have a practice for setting up KidCam regularly and thus did not have it running unless they explicitly remembered to get it out and set it up, which is not often in busy families with small children. In total, KidCam was only running an average of 12 hours and 27 minutes across 16.5 days during the entire

Table 2. General usage of KidCam for each of the 4 families in the study

Family ID	# of videos	Average video length	# of pictures	Total running hours	Number of Days of Use
1	10	0 min., 54 sec.	1	8 hrs, 10 min	11
2	16	7 min., 13 sec.	10	3 hrs, 16 min	20
3	3	5 min., 19 sec.	3	15 hrs, 25 min	15
4	9	1 min., 28 sec.	2	22 hrs, 58 min	20
Average	9.5	3 min., 43 sec.	4	12 hrs, 27 min	16.5

three-month period. In addition, families averaged 9.5 videos and 4 photos per family at the end of the study. This does not include videos and photos that were recorded but deleted, which for some families, was fairly often due to their varying uses of the device explained below. Table 2 shows the individual usage for each of the families. Despite the lower than expected use, when parents did use the device, they liked the buffering capabilities and the ability to record unplanned events. We discuss more about how we can improve frequency and utility in Section 5.

4.2.2 User Perceptions

In general, the families in our study liked the concept of KidCam, despite their low use of it. They appreciated the functionality and ability to capture unplanned moments while KidCam was in the room and turned on. The size of the device was appropriate, and the stand made it easy to situation to view the room. Most parents reported said they just forgot to turn on the device when they were playing with the child or forgot to move the camera when the child moved to a different space. Parents reported that they rarely used the baby monitor function because it did not provide much functionality beyond their already existing baby monitors. Many described it as being more complicated to use than their existing monitors. In addition, parents typically only used their baby monitors while the child was sleeping and did not think their child would do anything interesting while they were sleeping.

> **Mother, Family 3:** *"[A baby monitor] is only for when he's sleeping and he does nothing we want to record when he's sleeping... the reason why it was nice to have KidCam is because it would capture things awake. But like I was saying, nine times out of ten he was over here and it's so far away that it doesn't get him..."*

Parents became discouraged when they did remember to start KidCam and try to play with their child to get them to do interesting actions, but the child was not in the mood to perform. They also commented that KidCam itself became an attractive toy, and when they would interact with it, the child became interested in the device itself and want to play with it or watch themselves on the camera's playback screen.

> **Father, Family 4:** *"He'll do something and it will be so fascinating, and then when you try to get the video to record it, the process of going to get it, or whatever, set it up. Then he'll be distracted by it, and it's like, "Oh, let me look at the toy." Rather than do the trick."*

Family 1 used KidCam as a way to record and analyze their child's activities. The mother reported wanting to record and save her child's progress, but only if she was unsure about whether the child was able to do something. That way, she could go back and play what he had just done to see if he had actually performed the skill correctly. She mentioned that if she knew her child could do something, she did not feel the need to record it.

> **Mother, Family 1:** *"If I knew it, then I wasn't going to record it. But if I wasn't sure, then I got the video thing out and I went through the list of what I wasn't sure of. Like I really wasn't sure about, for example, that throwing the ball. And even when I got it on video I replayed it a couple of times to make sure..."*

Besides just recording videos for review later, Family 2 used KidCam as a way to immediately play back what had happened when they were not in the room. However, the mother said she did not feel the need to save these videos because they were only for entertainment as opposed to trying to capture a specific moment for later.

Mother, Family 2: *"And the video is always more fun because I can just turn it on. And what was nice with the video, I think I did last week, is I had to come downstairs and make his breakfast, so I put the little KidCam in there while he was playing in his crib. And it was kind of neat to go back and see what he does when I'm not around."*

For the families with older children, several parents mentioned that the children liked to perform for the camera and watch themselves on the playback screen. The mother of Family 2 reported that this generated a significant number of videos that they did not necessarily want to keep, so they deleted them after the child was finished.

The pictures that families took tended to be of their child doing cute things or making silly faces they wanted to save and maybe share with family later. Videos were taken of a specific special event, such as Family 2 and their child's first birthday party, parents playing with their child trying to get them to do specific actions, or children doing silly things such as singing a song, dancing, or making a mess with cake. Video and picture content seemed to depend on the family, as Family 1, which had a history of developmental delay, took more videos of developmental activities such as stacking rings and jumping, while Families 2, 3, and 4 focused more on sentimental activities like playing and family time. Figure 6 shows an example of a video and a picture taken by Family 1 and Family 2 respectively.

Finally, despite being satisfactory in initial usability studies, the prototype KidCam device was a bit cumbersome for everyday use. Because the UMPC was a Windows XP machine, parents had to wait for it to boot before beginning to use the device and then wait for the application to load and the video to begin buffering, which could take several minutes. This often happened when parents forgot to plug in the device and charge the battery. In addition, touch screens are not necessarily optimal for quickly taking pictures. The baby monitor required starting up a second device and

Fig. 6. Still from video taken by Family 1 of their child stacking rings (left) and photo taken by Family 2 of their child having breakfast (right)

application and took even longer to get started, which is the main reason cited for not using that feature. As a side note, these results highlight the usefulness of real-world deployments, as they did not surface until parents tried to use the device in real situations.

5 Discussion, Implications, and Future Improvements

Through the implementation and evaluation of a prototype that attempts to balance two dimensions of recording, continuousness and mobility, we have uncovered additional challenges and potential solutions for the difficult problem of recording children's moments of interest. In this section, we discuss the implications for this design space and how future improvements to KidCam can address these problems.

Quality of recorded media. There were many considerations and tradeoffs we needed to make in designing and implementing this device for recording young children. The requirement of semi-mobility meant that we had to sacrifice some features that would be needed for various aspects of child recording. For example, the requirement to capture video using a single, self-contained mobile device meant we had to sacrifice high-resolution video and recording from multiple angles simultaneously. This lower quality, single view video and low resolution photographs may affect what the media can be used for in the future. For example, parents desire higher quality photographs for printing or keeping in scrapbooks and the videos are not high enough quality for any sort of automated analysis or tagging using computer vision techniques. The quality on smaller off-the-shelf devices may improve with time, and could be improved now by making a custom hardware device.

Need for semi-automated continuous capture. The second dimension of the design was to provide for semi-continuous recording through the use of buffering and selective archiving. While many parents appreciated this approach to recording, the device was not quite continuous enough. There was a separate step in turning on the recording device to begin buffering, which we contribute to be the biggest hurdle to high frequency of use. Redesigning the device to run continuously without the extra step of booting it and turning it on could significantly increase the frequency at which the video buffer is available for recording. Thus, in the classification described in Section 2.2, we believe that KidCam should be moved toward the continuous end of the continuous *vs.* on-demand dimension.

Need for quick interactions. While parents liked the idea of being able to go back and choose the time and precise times for the archived video, in practice it just took too long to find the specific start and end points for a particular video, when parents were already likely busy and wanting to interact with their child rather than review portions of the video to decide what to save. The small screen size and touch screen interface of the ultra-mobile PC were also not ideal for quick interactions. Thus, we suggest keeping this feature for those who want it, but also providing the ability to do a "quick save" by just tapping a button which will save a small video clip with a preset start and end point around the point of capture. We could accomplish this by using a physical button on the device, a remote control (similar to how the CareLog system [10] initiated the recording of videos), or through voice commands.

Semi-automated trigger of media archival. The use of KidCam required parents to manually trigger when events of interest occurred that they would like to save for later. In practice, this was somewhat problematic, as a parent interacting with a child may not have the time to go to the device to manually trigger the recording. We suggest that future technology might employ the use of wearable or environmental sensors or computer vision to detect the presence of individuals in the scene to occasionally archive pictures and videos while moments of interest are occurring. This could be through communication with accelerometers embedded in different child's toys or special markers stitched onto children's clothing. This adds an additional task for parents to decide on rules for when automated capture should occur. These rules can be designed to occur by default, or could be programmed by parents using an end-user programming environment such as CAMP [25].

Better integration into existing lifestyles. We intended for KidCam to be used as a baby monitor to encourage more frequent use. However, the device was not as easy to set up as commercial baby monitors. Furthermore, parents did not typically associate baby monitors as something they used while the child was awake. Thus, we believe that KidCam needs to rely less on parental interaction and designers should find additional ways to build recording into the family's life. For example, the camera might be integrated into a favorite stuffed animal of the child or capture from existing video baby monitors that the family already owns. In addition, parents reported just forgetting to turn on the device when they were playing with their child. Recording technologies could be tied to an explicit activity that parents do anyway, such as initiating video buffering when the light switch in the child's bedroom is turned on. In addition, video buffering could be automatically activated during times of day when the children are most active. Finally, the device could be made more child-friendly so that older children can also participate in the recording process.

Physical design of the capture device. In addition to the need for physical buttons for triggering the capture of the media, we learned several other valuable lessons about the physical design of the device. The size and weight of the current device were about right for a semi-mobile device. However, one aspect to consider may be durability and cost. Parents thought KidCam was fairly expensive and fragile and thus wanted to keep it out of reach of their children. However, the device had a nice screen and was colorful, so naturally attracted children to want to play with it. Thus, KidCam should be redesigned to be durable enough to withstand a child playing with it, similar to commercially available digital cameras built specifically for children. In addition, steps that can be taken to reduce the cost of the device may lead parents to take more risks with the device. It can also increase the range of families that can have access to the recording device, as the technology used for the prototype was fairly expensive and may be beyond many families' budgets for technology purchases.

Issues of privacy. We expected there would be concerns over privacy from the semi-continuously recording capability provided by KidCam. Because the device is portable, parents can take it to a variety of places with other people present who may not consent or be comfortable with to continuous recording, even if the buffer deletes video that is older than 20 minutes. For example, a parent may take their child to play group at another person's house and want to record her interactions with the other children. However, in our study, we found that parents only used KidCam at home and did not take it to other places. In addition, because they were in complete control of the position of the device and what was in the field of view and could see what was being recorded, they did not feel a sense of invasion. This also suggests that the ideal placement of KidCam on the continuousness dimension could be more toward the fully continuous end of the spectrum and more automated recording techniques may be appropriate for this domain. In addition, it indicates that a child-specific recording device can be thought of as an appliance for the home, rather than a truly mobile or wearable device.

Uses of a semi-mobile, semi-continuous capture device in other domains. Though we designed KidCam to be used by families of young children, there are aspects that will be useful for a variety of applications. Through interviews with families and further analysis of applications, we uncovered several emergent uses of KidCam that are quite different from our original design requirements. For example, people may want to record family or friends in a variety of places, such as at holiday dinners or parties. Teachers may want to record different activities throughout a school for training newer teachers on how to improve their teaching skills [2]. Other uses might be to support recording for traditional capture and access in more than one location. The traditional capture and access model in ubiquitous computing was to instrument a space, such as a classroom, meeting room, or operating room. This mobile architecture allows the model to stay with a single person, such as a teacher, a meeting manager, or a physician. However, changes in the context of use will require the design to adapt certain capture behaviors, such as to record longer videos in meetings or classroom lectures.

6 Conclusion

The capture of rich media for young children remains a difficult but interesting challenge. In this paper, we explored the design and use of a mobile capture device, called KidCam, which allows families to capture video and photographs of children for generating sentimental keepsakes and monitoring activities. KidCam was designed to fill a void in capture technologies along the combined dimensions of mobility and continuous recording. However, a three-month deployment study with four families showed there were still problems. Despite the lower than expected usage of the device, parents identified interesting uses of KidCam and helped reveal additional design guidelines for the space of mobile capture and access. The study also underscores the importance of conducting real-world deployments when evaluating pervasive computing applications, as data from our formative evaluations did not predict some of the problems uncovered by parents.

The main contribution of this work was to explore the design space of mobile capture and access for a specific domain beyond existing work. A classification scheme of existing devices along two dimensions – continuousness and mobility – showed that KidCam addressed a gap by being semi-mobile and semi-continuous. We identified that this technology still suffered from some of the same disadvantages as other existing technologies within this classification, such as the camera not being turned on at the appropriate times and manual capturing interfering with family activities. We suggest that further design exploration is needed to make KidCam more continuous, rather than semi-continuous, through the use of automated recording techniques and semi-automatic capture through the use of sensors or through more pervasive actions of parents. The study of KidCam showed that this space, although full of challenges, remains a high-need domain for technology researchers to explore.

Acknowledgements

We would like to thank the families who participated in our study for their time and cooperation. We also thank Stefan Puchner, Yi Han, Tracy Westeyn, Shwetak Patel, and Gillian Hayes for their assistance in this effort. This work was supported by the National Science Foundation under Grant No. 0745579.

References

1. Abowd, G.D., Mynatt, E.D.: Charting past, present, and future research in ubiquitous computing. ACM ToCHI 7(1), 29–58 (2000)
2. Allen, D., Ryan, K.: Microteaching. Addison-Wesley Publishing Company, Inc., Reading (1969)
3. Ashbrook, D., Lyons, K., Clawson, J.: Capturing Experiences Anytime, Anywhere. IEEE Pervasive Computing 5(2), 8–11 (2006)
4. Baranek, G.T., et al.: Object play in infants with autism: methodological issues in retrospective video analysis. American Journal of Occupational Therapy 59(1), 20–30 (2005)
5. Brotherton, J.A., Abowd, G.D.: Lessons learned from eClass: Assessing automated capture and access in the classroom. ACM ToCHI 11(2), 121–155 (2004)
6. Fels, S., et al.: Building a context-aware mobile assistant for exhibition tours. In: The First Kyoto Meeting on Social Interaction and Communityware (1998)
7. Frohlich, D., Fennell, J.: Sound, paper and memorabilia: resources for a simpler digital photography. Personal and Ubiquitous Computing 11(2), 107–116 (2007)
8. Geyer, W., Richter, H., Abowd, G.D.: Towards a Smarter Meeting Record–Capture and Access of Meetings Revisited. Multimedia Tools Appl. 27(3), 393–410 (2005)
9. Hansen, T.R., Bardram, J.E.: ActiveTheatre: a Collaborative, Event-based Capture and Access System for the Operating Theatre. In: Beigl, M., Intille, S.S., Rekimoto, J., Tokuda, H. (eds.) UbiComp 2005. LNCS, vol. 3660, pp. 375–392. Springer, Heidelberg (2005)
10. Hayes, G.R., Gardere, L.M., Abowd, G.D., Truong, K.N.: CareLog: a selective archiving tool for behavior management in schools. In: Proc.of CHI 2008, Florence, Italy, April 5-10, pp. 685–694 (2008)
11. Hayes, G.R., Patel, S.N., Truong, K.N., Iachello, G., Kientz, J.A., Farmer, R., Abowd, G.D.: The Personal Audio Loop: Designing a Ubiquitous Audio-Based Memory Aid. In: Dunlop, M.D. (ed.) Mobile HCI 2004. LNCS, vol. 3160, pp. 168–179. Springer, Heidelberg (2004)
12. Hayes, G.R., Poole, E.S., Iachello, G., Patel, S.N., Grimes, A., Abowd, G.D., Truong, K.N.: Physical, Social, and Experiential Knowledge in Pervasive Computing Environments. IEEE Pervasive Computing 6(4), 56–63 (2007)

13. Healey, J., Picard, R.W.: StartleCam: A Cybernetic Wearable Camera. In: Proc. of ISWC 1998, October 19-20, p. 42 (1998)
14. Hodges, S., Williams, L., Berry, E., Izadi, S., Srinivasan, J., Butler, A., Smyth, G., Kapur, N., Wood, K.: SenseCam: A Retrospective Memory Aid. In: Dourish, P., Friday, A. (eds.) UbiComp 2006. LNCS, vol. 4206, pp. 177–193. Springer, Heidelberg (2006)
15. Kientz, J.A., Arriaga, R.I., Abowd, G.D.: Baby Steps: Evaluation of a System to Support Record-Keeping for Parents of Young Children. In: Proc. of CHI 2009 (2009)
16. Kientz, J.A., Hayes, G.R., Westeyn, T.L., Starner, T., Abowd, G.D.: Pervasive computing and autism: Assisting caregivers of children with special needs. IEEE Pervasive Computing 6(1), 28–35 (2007)
17. Kientz, J.A., Arriaga, R.I., Chetty, M., Hayes, G.R., Richardson, J., Patel, S.N., Abowd, G.D.: Grow and Know: understanding record-keeping needs for tracking the development of young children. In: Proc. of CHI 2007, pp. 1351–1360 (2007)
18. Kindberg, T., Spasojevic, M., Fleck, R., Sellen, A.: The Ubiquitous Camera: An In-Depth Study of Camera Phone Use. IEEE Pervasive Computing 4(2), 42–50
19. Kindberg, T., Spasojevic, M., Fleck, R., Sellen, A.: I saw this and thought of you: some social uses of camera phones. In: Ext. Abs. of CHI 2005, pp. 1545–1548 (2005)
20. Lester, J., Choudhury, T., Kern, N., Borriello, G., Hannaford, B.: A hybrid discriminative/generative approach for modeling human activities. In: Proc. of IJCAI 2005, pp. 766–772 (2005)
21. Minnen, D., Westeyn, T., Presti, P., Ashbrook, D., Starner, T.: Recognizing soldier activities in the field. In: Proc. of ISWC 2007. pp. 236–241 (2007)
22. Pedersen, E.R., McCall, K., Moran, T.P., Halasz, F.G.: Tivoli: an electronic whiteboard for informal workgroup meetings. In: Proc. of CHI 1993, pp. 391–398 (1993)
23. Pilu, M.: On the use of attention clues for an autonomous wearable camera. HP Tech. Report HPL-2002-195R1 (2003)
24. Roy, D.: The Human Speechome Project. In: Proc. of the 28th Annual Cognitive Science Conference (2006)
25. Truong, K.N., Huang, E.M., Abowd, G.D.: CAMP: A magnetic poetry interface for end-user programming of capture applications for the home. In: Davies, N., Mynatt, E.D., Siio, I. (eds.) UbiComp 2004. LNCS, vol. 3205, pp. 143–160. Springer, Heidelberg (2004)

Mobile Device Interaction with Force Sensing

James Scott, Lorna M. Brown*, and Mike Molloy

Microsoft Research, Cambridge, UK

Abstract. We propose a new type of input for mobile devices by sensing forces applied by users to device casings. Deformation of the devices is not necessary for such "force gestures" to be detectable. Our prototype implementation augments an ultra-mobile PC (UMPC) to detect twisting and bending forces. We describe examples of interactions using these forces, employing twisting to perform application switching (alt-tab) and interpreting bending as page-down/up. We present a user study exploring users' abilities to reliably apply twisting and bending forces to various degrees, and draw implications from this study for future force-based interfaces.

Keywords: Force, sensors, mobile devices, interaction.

1 Introduction

Many pervasive computing applications rely on mobile devices. To give three examples, such devices can be used in the control of intelligent environments, they can be used to locate users and objects and provide location-aware applications, and they can be used for communications to support applications requiring pervasive connectivity. Thus, improving users' experience of interacting with mobile devices is a key ingredient in pervasive computing.

For mobile devices such as tablet or slate PCs, ultra-mobile PCs (UMPCs), PDAs, or new smart phones such as the iPhone, much of the surface of the device is devoted to the screen. The trend for increasing screen sizes is continuing, since larger screens facilitate better information presentation. However, since users also want their devices to be small, less and less space is available for physical controls such as numeric or alphanumeric keys, buttons, jog dials, etc. While these devices typically have touch-sensitive screens, dedicating a portion of the screen to an input interface reduces the available screen area for information presentation.

In contrast, the sensing of physical forces made by the hands grasping a mobile device can provide an input mechanism for devices without taking up screen space or requiring external controls mounted on the surface area of on the device. By sensing forces such as bending or twisting that users apply to the device casing itself, input can be provided to the device's operating system or applications. The body of the device does not need to actually bend for forces to be detectable, so this sensing mechanism is compatible with today's rigid devices.

---

* Now at Vodafone.

H. Tokuda et al. (Eds.): Pervasive 2009, LNCS 5538, pp. 133–150, 2009.

In this paper, we describe the concept of force sensing as a user input. We detail our prototype implementation based on a UMPC using thin load sensors. We present examples of device interactions where force sensing may prove useful, namely to provide functionality normally associated with shortcut keys such as alt-tab and page-down/up on mobile devices without keyboards. We then describe our user study into the capability of users to control the forces applied to devices, allowing us to evaluate the usefulness of force sensing as an input mechanism.

2   Related Work

In the past decade there has been a move to consider new interaction techniques for mobile devices beyond the use of physical or virtual on-off buttons, including work using accelerometers, e.g. Rekimoto [1] and Hinckley et al. [2], or inertial sensing [3].

The Gummi concept and prototype [4] propose a vision for a fully flexible mobile computer. The elastic deformation of such a device can be used to interact with it, e.g. to zoom in and out by bending it like a lens. Our work differs in that users do not actually bend the device significantly when they apply forces to it, allowing for simpler integration with current device designs which incorporate many rigid components.

Previous work on force sensing for rigid devices uses direct pressure (see Figure 1, top left), whereas our work explores several different types of forces such as bending and twisting (Figure 1, bottom left and right). Researchers at Xerox PARC used pressure sensors [5] both as a physical "scroll bar" so that squeezing 2/3 of the way along the top edge would navigate 2/3 of the way along a document, and to detect which side a user was gripping a device to inform "handedness-aware applications". Direct pressure input has also been explored in the context of a touchpad [6] or touch screen [7]. Jeong et al. [8] also used direct pressure applied to a force sensor mounted on a 3D mouse to control movement speed in virtual environments.

A number of input devices have been proposed which use force sensing in various ways but do not have any computing or display electronics. The Haptic media controller by Maclean et al. [9] uses orthogonal force sensing (using strain gauges) to sense forces applied to different faces of a thumb wheel in addition to rotation of the wheel to provide richer interactions. The haptic rotary knob by Snibbe et al. [10], which can sense the rotational forces applied by a user. TWEND [11] and Bookisheet [12] are flexible input peripherals which can sense the way they are bent.

Another related field of work is that of Tangible User Interfaces [13], which if defined widely can include any type of physical UI such as force sensing. In Fishkin's terminology [14], the current focus of our work is on "fully embodied" interaction in that the output is shown on the device itself, in contrast to the input-only systems above. However, our ideas could in future also be applied to augment passive objects while retaining their rigidity, e.g. to provide computer inputs for the control of intelligent environments.

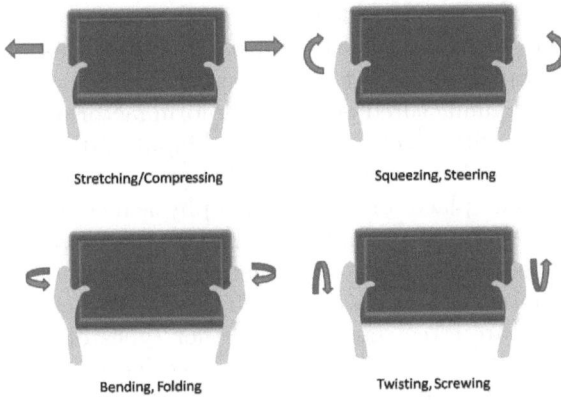

Fig. 1. Four force gestures that can be sensed

3   Sensing User-Applied Forces

Physical forces that users apply to device casings can be sensed and used as input for applications or the operating system running on such devices, with graphical, audible, haptic or other types of feedback provided to the user. Such forces can be detected with a variety of sensors including load sensors which sense direct mechanical compression between parts of the casing, strain gauges which sense the elastic stretching of the case due to the force applied, or pressure sensors sensing gaseous or liquid pressure in a channel in the device.

Users can apply various types of force, with four examples shown in Figure 1. Each force can be applied in two directions (the way indicated by the arrows and the opposite directions), and can be applied to a variable degree. Forces can be applied along different axes (e.g. vertically as well as horizontally) and to different parts of a device. Thus, force sensing can potentially be a rich source of input information and is not limited to the simple presence or absence of force. In this work, we focus on two of the forces shown, namely bending and twisting.

Unlike with flexible technologies such as Gummi, detecting force does not require that the device must actually bend significantly, or be articulated around a joint. The device can be essentially rigid, with only very small elastic deformations due to the applied pressure. This is crucial as it allows force sensing to be deployed in many types of device which rely on current-day components such as LCD screens or circuit boards which are rigid, and avoids the need to use less mature or more expensive flexible technologies in a device.

3.1   Qualities of Force Sensing Input

Force sensing has some intrinsic qualities that make it an interesting alternative to other forms of input. With force sensing, the user interacts with the casing of the device, turning an otherwise passive component that just holds the device together into an active input surface. Forces applied to the casing are mechanically transmitted through it and parts attached to it. Therefore, unlike for physical controls such as keys or dials, force sensors do not necessarily need to be located at the external surface of the device in the part of the device that the user holds, and device cases can be made with fewer holes for physical switches, which can facilitate more robust, more easily manufactured, and smaller form factor devices. Force sensing shares this advantage with accelerometer-based input and similar sensors.

Another advantage of force sensing over other physical controls is the ability to apply an input in many places or ways. When physical controls are present on a device, their positions lead to an implied orientation and grip for the device, while with force sensing, the device can be easily made more symmetric, allowing a device to avoid having an intrinsic "right way up" which may be useful in some scenarios.

We can also compare force sensing with other types of input such as touch or multi-touch input on the screen, the use of accelerometers or other physical sensors, etc. Compared to shaking or tilting the device or using a touch screen, force sensing allows the screen to be kept at the most natural viewing angle, unobscured by a finger or stylus. During force sensing input the device can be essentially stationary, thus inputs can be made subtly which may be useful in some situations (e.g. during a meeting).

One potential disadvantage of force sensing is its ability to be triggered accidentally, e.g. while the device is being carried or handled. This is in common with many other input mechanisms for mobile devices, such as accelerometer-based, touch-based or even button-based. With force sensing, it may be necessary to use a "hold" button or "keylock" key-combination for users to indicate that input (including force sensing input) should be ignored.

Another disadvantage of force sensing may be in the need for using two hands, which is the mode of operation of our prototype based on a UMPC (see below). One can envision one-handed use of force sensing, e.g. using another object/surface as an anchor, or using a device small enough to both support and apply forces to in a single hand. However, the desirability and usability of this for input remain questions for future work; advantages such as the ability to keep the screen at the correct viewing angle and unobscured may be lost.

A further issue with force sensing is in the effects of repeated shearing and bending forces that the user applies to the device over time. While this may result in higher mechanical demands on the casing of the device, there is also an advantage to the device being "aware" of its physical environment, so that it can warn the user (e.g. audibly) if excess forces are applied, either during force-based user interface actions or otherwise (e.g. placing heavy books on the device).

Although this section compares force sensing to other forms of input, it is important to note that an either-or decision is not implied, as many input mechanisms can be built in to the same device. Mobile computers are general-purpose tools with many applications, and thus the inputs required are complex. By employing multiple input mechanisms with separate functionalities, we can avoid overloading any one input mechanism. The best choice of input mechanism at any given time will depend on the application, the user, the usage scenario, the environment/context of use, and so on. Thus, we present force sensing as a mechanism to add to the richness of input on a device rather than a way of replacing another input.

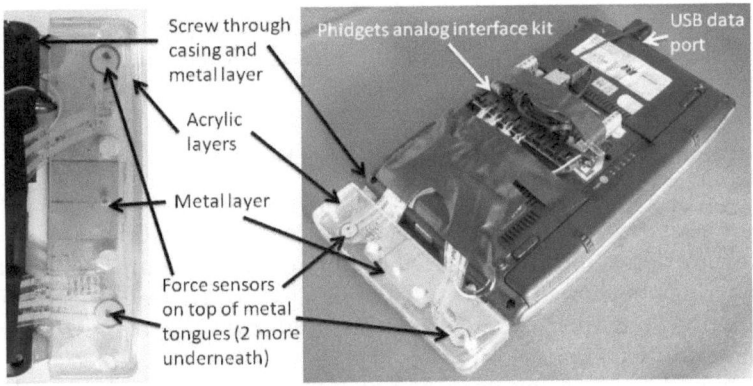

Fig. 2. Force sensing prototype using augmented UMPC

4   Prototype Force Sensing Hardware

We built a prototype force sensing device by augmenting a Samsung Q1 UMPC with a custom additional casing incorporating four force sensors, as shown in Figure 2. We used an additional casing rather than an integrated solution for feasibility reasons; UMPCs are tightly packed with components and have precisely fitted cases, making it difficult to incorporate force sensing hardware internally. However, in an device with force sensing designed in from the beginning, an additional casing would be unnecessary. We pilot-tested both strain gauges (which detect tension forces) and pressure sensors (which detect compression forces) and found the latter to be easier to obtain useful data from and easier to integrate in the mechanical construction of a first prototype.

The additional casing comprises an exterior acrylic section made of layers cut with a laser cutter, screwed together and hand-filed to round the corners. This tightly surrounds a central hand-cut metal layer (magnesium alloy) which is rigidly attached to the body of the UMPC at the same screw points that hold the top and bottom half of the UMPC casing together.

For the force sensors we use four FlexiForce 0-25 lb load sensors from TekScan Inc. These are small (the sensing area is 10 mm wide and can be made smaller) and thin (0.2 mm) making them simple to incorporate into the prototype, and in future work to incorporate into a redesigned device casing. The force sensors are placed tightly between "tongues" on the metal layer and the acrylic layers (using blobs of epoxy to make sure of a good fit), two on each side of the metal layer, at the top and bottom. The placement of the force sensors and shape of the tongues and casing are dictated by the forces that we intended to sense; this prototype was built to sense "twist" and "bend" gestures (see Figure 2).

The Flexiforce sensors lower their electrical resistance from over 30 MΩ when no pressure is applied to around 200 kΩ when squeezed hard between a thumb and finger. We apply a voltage to one contact of each Flexiforce sensor and measure the voltage at the other, with a 1 MΩ pull-down resistor. These voltages are measured in the prototype using analog to digital converters in a Phidgets

Analog Interface Kit which is attached to the back of the UMPC (as shown in Figure 2) and software on the UMPC queries the force values through a USB interface.

4.1 Determining Force Gestures Using Sensors

With load sensors we can detect only compression forces. We could, of course, preload a sensor so we can detect release from compression, or even load a sensor to halfway to detect both compression and release of compression. However, initial experiments showed this to be unreliable since friction causes the "zero" point to move each time the device is manipulated. Therefore, the prototype design senses only compression.

When building the casing described above we had intended to sense each of the four gestures (twisting and bending in two directions each) using the sum of inputs from two force sensors. Twisting in each direction should compress two diagonally opposite sensors, while bending should compress either the two front sensors or the two back sensors. The initial gesture magnitude calculations were therefore done according to the following rules:

```
TwistClockwise = BottomBack + TopFront
TwistAnticlockwise = BottomFront + TopBack
BendTowards = TopBack + BottomBack
BendAway = TopFront + BottomFront
```

where `TwistClockwise` etc are the gesture magnitudes and `TopFront` etc indicate the raw sensor values (which rise with increased pressure). However, this did not work as well as expected. After some experimentation, we found that pressure was concentrated at the edges of the metal tongues and not centred on the tongues where we had placed the force sensors. We derived a more optimal positioning whereby the top tongue's two sensors were placed at the side edge of the tongue, to best detect the bending gestures, while the bottom tongue's two sensors were placed at the top and bottom edges of the tongue, to best detect twist gestures. Thus, the second term in each of the equations above was removed. However, while this greatly improved the true positive rate, it did not eliminate false positives, i.e. the sensors also sometimes trigger when the wrong gesture is applied, e.g. the `BottomFront` sensor indicating `TwistAnticlockwise` would sometimes trigger for `BendAway`. We therefore further modified the rules, such that when two sensors trigger, the gesture detected corresponds to the one compatible with these two sensors, and the other gesture is "damped" by subtraction. This results in reliable performance in practise.

```
TwistClockwise = BottomBack - TopBack
TwistAnticlockwise = BottomFront - TopFront
BendTowards = TopBack - BottomFront
BendAway = TopFront - BottomBack
```

4.2 Interaction Using Force Sensing

Force sensing is a general input mechanism providing a number of scalar values that can be mapped onto any desired functionality. To illustrate the potential

 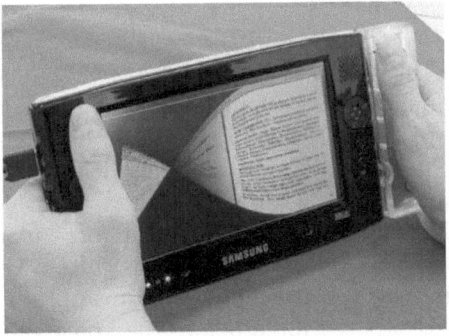

Fig. 3. Two examples of force-based interactions: (left) bending for page turning and (right) twisting for application-switching, with visual feedback

usefulness of force sensing, we built demo software showing how it can be used for one class of interactions, that of replacing "shortcut keys".

Our motivation for considering this example compelling is as follows. Mobile devices such as UMPCs are capable of running applications that desktop PCs run, but have reduced I/O capabilities. One of the implications of this is that the convenient shortcut keys that users may normally employ with a full keyboard may be unusable on a UMPC since they keys are not present or difficult to use quickly due to their size or placement. Force-based interactions are an interesting alternative, taking advantage of the fact that UMPCs are designed to be gripped by the hands during use.

Page-Down and Alt-Tab Force-Based Interactions. We implemented two force-based shortcut key interactions on the UMPC, as shown in Figure 3. The bend action is used to indicate "page down", and the twist action is used to change foreground application (i.e. "alt-tab" under Windows). With both gestures we provide an accompanying visual feedback which mimics a real-life interaction with the application window or document in question. For page-down, we provide visual feedback as if the user was flicking through a book (by bending the book and allowing pages to flick across). Thus, right hand page appears to flip up and over to the left hand side. For the inverse action, page-up, we use the inverse force, i.e. the user bends the device as if to see the backs of their hands. For alt-tab, we twist the window vertically from one side of the screen to the other, revealing the virtual reverse side of the window, which is the next window. The windows are kept in a persistent order, unlike alt-tab which puts least-recently-used first. This is so a user can go forwards or backwards from the window they are on in a consistent fashion. Users receive differing visual feedback with the twist direction matching the direction of the force applied.

For prototyping purposes we implemented these gestures as mock-ups rather than integrating them into a running system. They are implemented using C# with .NET 3.5 and the Windows Presentation Foundation (WPF) library and we use Windows Vista as the UMPC's operating system. A single 3D polygon mesh is used for alt-tab, while two meshes are used for the page-down visualisation (the current page and the next page). We have since integrated the alt-tab gesture with Windows Vista so that twisting causes a screenshot of the current and next applications to be captured and the animation to be run with those images.

For both interactions, we have currently implemented the interaction by changing a single page/application at a time, and the user can control how far along the animation the system goes by applying a harder or softer force. If the user applies sufficient force to move to the end of the animation (when the new page/application is fully revealed) this is committed and as the force returns to zero the new page/application will continue to be visible. If the user releases the force before they have made the animation reach the end, then the animation goes back to the beginning as the user releases the force, without any persistent change, i.e. the same page/application is in view. This allows users to apply small forces to see the potential effects (e.g. seeing which window would appear when twisting that way) before committing with a larger force.

The UMPCs were used for a demo event at Microsoft where hundreds of novice users tried it. Many were able to use the device with no explicit instruction (by watching it being used by someone else, or reading instructions on a poster), and nearly all users were able to operate it with a few seconds of coaching.

Other Interactions. We do not claim that the mappings chosen are in any way "best", merely that they illustrate the potential of force sensing as an input. Other mappings could, of course, be used instead, e.g. twisting could be used to indicate "cut" and bending "paste".

We did choose the mappings based on the ease of providing visual feedback that has strong physical analogies with the applied forces, as we expected (and found during demos) that this made force-based interaction intuitive to discover through observation, and easy to remember and to use. Such physical analogies are popular choices in past work, e.g. Harrison's page turning gesture [5]. This analogous feedback could be extended to other interaction mappings, e.g. twist for "cut" could involve the cut text twisting in on itself, and bend for "paste" could involve the the document visually bending and splitting to make space for the pasted text to appear at the cursor. Again, we expect that the visual mapping will assist users in remembering which force performs which UI action.

Another modification to be explored is that, instead of different force levels being used to move through the animation of a single change, different force levels can indicate levels of change or rates of change. For example, the bend force could cause pages to flip over at a slow rate if applied weakly (such that one page could be flipped at a time), or at a quick rate if a strong force was applied, giving the effect of rapidly flipping through a book.

Aside from visual feedback, the current prototype provides a click sound when enough force is applied to reach the end of the animation and lock in the new view. Richer audio cues can be incorporated in future, e.g. cues which also match the physical action taken such as a crumpling sound or page-flicking sound. Haptic feedback can also be used, particularly since the user is known to be gripping the device, furthering the analogy with physical movements.

While these are all exciting possibilities, before looking further into them, in this paper we wish to address more basic questions: how repeatably and accurately can a number of users apply forces such as bending and twisting to devices. A study of these issues is the subject of the rest of this paper.

5   User Study

We conducted a quantitative user study to assess the capabilities of force sensing as an input method using our prototype. We chose this form of study so as to gain understanding of the fundamental capabilities of users to apply forces in a controlled fashion to mobile devices, thus informing the design of force sensing interfaces. The study aimed to assess the number of distinct levels of force that a user could reliably "hit", and the speed at which they could do this, for both twisting and bending forces. The methodology was based on that used by Ramos et al. [15]. A screenshot of the software used in the study is shown in Figure 4.

5.1   Participants

Twenty participants were recruited from within our research lab for a between groups study with two conditions: in the Bend condition users used the bend gesture to interact with the device, while in the Twist condition they used the twist gesture. Ten participants (5 male, 5 female) were assigned to the Twist condition and ten (6 male, 4 female) to the Bend condition. 80% of the participants were aged 25-35, the remaining participants divided between the following age groups: 20-24 (1 participant), 36-40 (1 participant) and 50-55 (2 participants). All but one of the participants were right handed. Around half of the participants were researchers and the others came from a range of job roles, including IT support, human resources and marketing. Participants were each given a box of chocolates worth around 5 GBP in thanks for their time.

While conducting the tests, participants were provided with an office chair with adjustable arms and a desk and could choose to sit in any comfortable

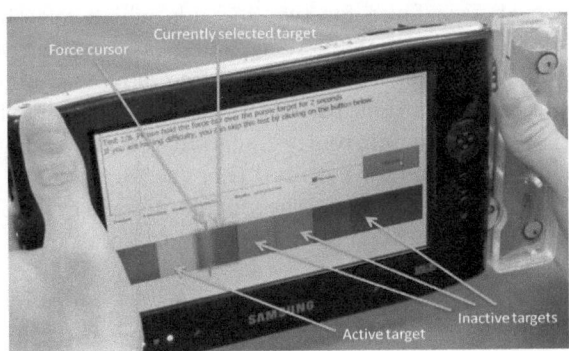

Fig. 4. Screenshot of software for user trials

way. Some participants leaned on the desk, some sat back in the chair, and some changed positions during the trial.

5.2   Methodology

Each user trial consisted of a familiarization phase, a training phase, and an experimental phase. After the first and second phases, device calibration was undertaken allowing the user to choose their preferred maximum force values to allow for differences in individual strength.

In the familiarization phase, we explained the operation of the device, demonstrated the force gesture being studied, and asked the users to experiment with applying forces for a period lasting a minimum of two minutes. Whilst applying forces, users received feedback in the form of a visual "force cursor", representing the magnitude of the force applied, moving along a horizontal bar. At zero force, the cursor was in the middle of the bar, and the user moved the cursor in either direction by twisting/bending one way or the other. The mapping from twisting/bending to left/right was decided by observing the most natural mapping in pilot tests; this was not a source of confusion to the participants after the training phase.

During the calibration that followed, users configured the device with chosen maximum forces by applying a comfortable but hard force in each direction a minimum of three times. We considered the maximum force a user to be comfortable with to be an individual personalization setting (akin to mouse sensitivity).

Both the training and test phases involved the same overall structure of blocks of tests, with the training phase simply allowing the user to become accustomed to the type of tests being applied. This method was based on the discrete task variation of the Fitts' Law task [16]. A block of tests involved a set number of targets on the screen uniformly filling the space between zero force and the maximum configured force in each direction (i.e. for "two target" tests there were actually two targets in each direction). Figure 4 illustrates a test with four targets on each side (i.e. targets with the same width but varying "amplitudes" in the normal Fitts' Law terminology). For each test, the user was first made to keep the device at zero force for two seconds which left the cursor at the home position in the centre of the screen. Then, a single target was coloured purple and the user was asked to move the force cursor as quickly as possible into the coloured target, and then to hold the force cursor inside the target for 2 seconds. To aid the user, the currently-aimed-at target was highlighted with a yellow glow. After the target was acquired (i.e. the force cursor was kept inside the target for 2 continuous seconds), the purple marking disappeared and the user was instructed on screen to return the force to zero before the next target appeared.

We deliberately chose not to include a button for users to click on a target as soon as they enter it, for three main reasons. First, because the trial was conducted using a two-handed grip, adding a button would add a significant new demand on one of the hands applying forces. Second, with force sensing users can change the position and grip style of their hands, so it was not clear where a button would be placed. Third, we wanted to explore the potential for button-free interaction with force sensing.

In a given block of tests, users had to acquire each target precisely once, though the targets were presented in random order during a block. During a block of tests the targets were presented immediately one after the other (with a minimum 2-second zero-force period in between). After each block of tests was completed, the user was offered the chance to take a break, and had to press a touch screen button to continue to the next block.

During the training phase there were four blocks with 2, 4, 6 and then 8 targets in each direction, allowing the user to become accustomed to the system through progressively harder tasks. After the training phase, the user was given the opportunity to recalibrate the force maximums.

During the main experiment phase, the number of targets in a given block was varied between 2, 3, 4, 5, 6, 7 and 8 targets in each direction, with blocks for each number of targets appearing three times, once during each of three cycles. During the cycles, the order of number of targets was randomized. In total, each participant was asked to acquire 210 targets during the experiment phase, covering a wide range of widths and amplitudes. The total duration of the study was around 50 minutes per participant.

In case of major difficulty in acquiring a target, after 10 seconds a "skip" touch screen button became available for the user along with on-screen instructions that they could skip the target if they were finding it too difficult.

6   Results from the User Study

In all the times presented below, the final two seconds are not included, i.e. the time is reported until the target is entered by the force cursor at the beginning of the two continuous seconds required for the test to be complete.

One participant's data has been excluded from the Bend condition as the prototype broke during the trial and it had to be aborted. The prototype was subsequently repaired before further trials. Therefore the data reported herein is only for the remaining nine participants.

Analysis of the data for the Twist condition showed outlier behaviour for one participant (average times more than two standard deviations worse than the mean for the majority of targets). Therefore, this data has been excluded to allow analysis of the behavior of the non-outlying participants. At the same time, we must conclude that some users may have difficulty using a force sensing system and may be more comfortable with another input mechanism.

In this section we present the raw results and statistical analysis, discussing their implications in the next section. We do not present or analyse data from the training phase except where specially noted. While we offered users the ability to skip targets, in a total of 3780 experiment-phase targets presented to the 18 non-excluded participants, only 10 were skipped (0.3%).

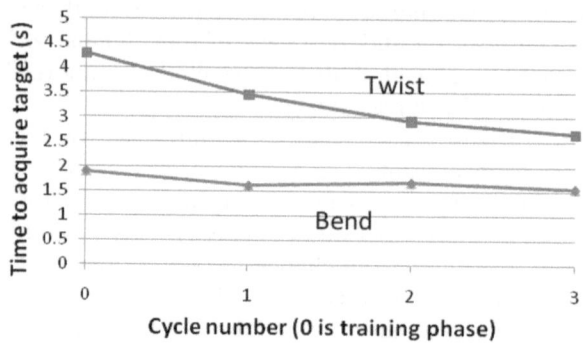

Fig. 5. Effect of participant experience on target acquisition time

6.1   Learning Effects and Effects of Gesture Type

Figure 5 shows the effect of the cycle on the average target acquisition time (cycle 0 is the training phase). Two within groups ANOVAs (analyses of variance) were carried out, one for each gesture type, to investigate whether there was any significant learning effect. For the Bend gesture, the ANOVA showed no significant difference between cycles. The ANOVA for the twist gesture, on the other hand, showed a significant effect of cycle ($F(3,24)=7.09$, $p<0.01$). Post-hoc Tukey tests showed that there was a significant difference ($p<0.01$) between the training cycle and cycles 2 and 3.

Excluding the training phase, the mean target acquisition time (across all targets) was 1.6 seconds for Bend and 3 seconds for Twist. A between groups ANOVA showed that the average time was significantly faster for the bend gesture $F(1,16)=14.34$, $p<0.01$).

6.2   Effect of Number of Targets/Target Width

Figure 6 shows the effect of number of targets on the average target acquisition time, with error bars indicating the standard deviation for variation between participants. Since targets in a given block of tests filled the whole force bar, the number of targets is inversely proportional to the width of each target. Note that in the figures and discussion, we describe the number of targets in each direction, i.e. there are twice as many targets on the screen.

The graph shows that, for both conditions, the average target acquisition time increased as the number of targets increased (and width decreased). Two within-group ANOVAs were carried out, one for each gesture. These showed significant effects of number of targets on performance for both gestures (for Bend, $F(6,160) = 55.11$, $p<0.01$ and for Twist, $F(6,160) = 23.06$, $p<0.01$). Post hoc Tukey tests showed significant ($p<0.05$) differences between numbers of targets as follows.

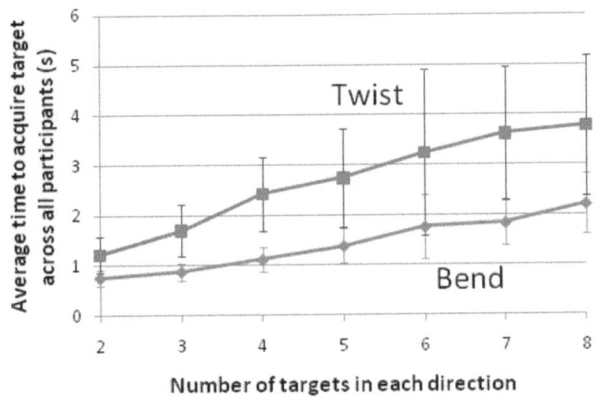

Fig. 6. Effect of different numbers of targets on average target acquisition time

For Bend:
2 was significantly faster than 4, 5, 6, 7 and 8
3 was significantly faster than 5, 6, 7 and 8
4 and 5 were significantly faster than 6, 7 and 8
6 and 7 were significantly faster than 8

For Twist:
2 was significantly faster than 4, 5, 6, 7 and 8
3 was significantly faster than 5, 6, 7 and 8
4 and 5 were significantly faster than 7 and 8

6.3 Effects of Target Position

Figure 7 plots the effect of target position, direction, and number of targets in each direction on target acquisition time. For the Bend condition, overall, the time taken to acquire a target decreased as the distance increased (although it can be seen that this effect is less when fewer targets are present). It can be observed that the most difficult target to acquire was the closest target to the left of the zero force position. This is borne out by the analysis. Within-groups ANOVAs were performed for each number of targets and showed that, for every number of targets except 3 and 4 targets, there was a significant effect of target position ($p<0.05$). Post hoc Tukey tests showed that this was due almost entirely to the first target to the left being significantly slower to acquire than almost all other targets.

A similar pattern is observed for the Twist condition. The first target to the left is again the hardest to acquire. The analysis confirms this: individual ANOVAs for each number of targets showed a significant effect of target position ($p<0.05$) for all numbers of targets except when there were only 2 targets. Post hoc Tukey tests again revealed that, apart from a few other isolated cases, this result was due to the first target to the left taking significantly longer to acquire

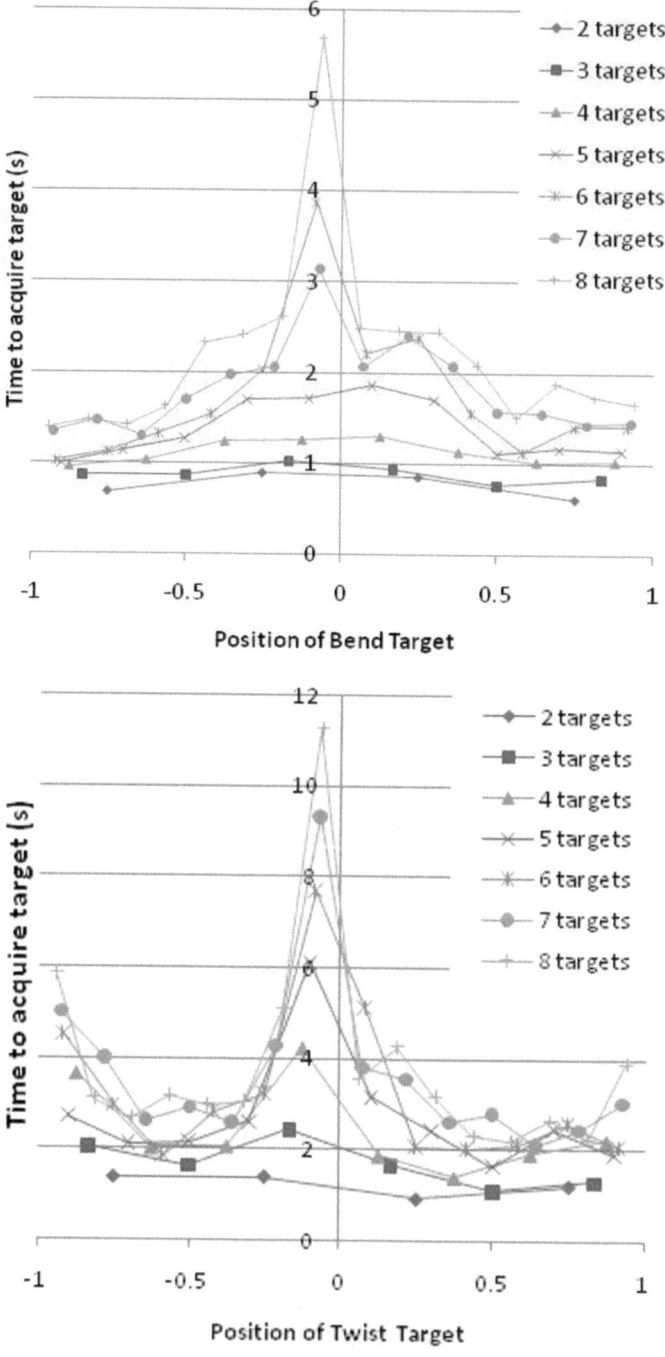

Fig. 7. Target acquisition times for different numbers of targets and target positions. Note different y axis scales.

than almost all other targets. It should be noted that "left" for twist and bend are two very different gestures.

We also observe an increase in the time taken for the furthest target to the left in the Twist condition, especially when 5–8 targets are present. However, ANOVA tests found no significant difference between these points and any other positions. Nonetheless it is interesting to note that there is a slight difference between bend and twist for the highest-force targets. Bending remains at the same level of ease, while twisting hard enough to reach the furthest targets seems more difficult.

7   Discussion

Our data shows that target acquisition is significantly faster in the Bend condition than in the Twist condition. The Bend gesture was learned faster, and also performed better throughout the experiment. There are a number of possible causes of this difference. Despite designing for stiffness, our prototype deforms noticeably during bending but much less so for twisting, which may make it easier for users to apply force in the direction that our force sensors are aligned with, i.e. the deformation acts as a guide. This deformation also meant that users may have avoided setting maximum bend force too high for fear of breaking the prototype, thus making the "far" targets simpler to reach. Twisting performance improved significantly during the experiment, so it is possible that with everyday use a user might reach equivalent proficiency in both gestures.

7.1   Fitts' Law

As Fitts' Law is applicable to many input devices and its implications affect the design of user interfaces, we tested its applicability for force sensing. The Fitts' Law model (Shannon formulation [16]) is expressed as follows, where ID is the index of difficulty, A is the amplitude or distance to the target and W is the target width (inversely proportional to the number of targets).

$$ID = log_2(A/W + 1)$$

If Fitts' Law were to apply to force-based interfaces, we would expect that the target acquisition time would increase as the number of targets increased, and this was borne out by our results. However, Fitts' Law would also cause us to expect that target acquisition time would increase as the distance to target increased. Our data shows that this is not the case: acquisition time does not increase as distance to target increases, and for the case of the closest-left target for both gestures, it actually significantly decreases for further away targets. MacKenzie [16] noted that force-based devices (e.g. isometric joysticks) undergo negligible limb motion compared to a device like a mouse, and that Fitts' Law may be a poor fit for modeling the performance of such devices; our results confirm that this is true for twisting and bending forces.

7.2   Reaction Time, Movement Time and Jitter

To further understand our results, we looked at a breakdown of target acquisition time into several stages: reaction time before any movement occurred, movement time until the target was first entered, and jitter time during which the target was left and re-entered any number of times before the final entry (when the force level was held in-target for 2 seconds continuously). Reaction times were generally low (0.6s for bend, 0.8s for twist) and consistent (they do not change based on size or position of target). The exceptions to this are for the lowest-force targets for which reaction times are slightly higher, which can be explained by the user attempting to apply a very small force, and therefore applying an undetectably small force to start with. This reinforces the conclusion that small forces are harder to apply than large forces.

Movement time to first entering a target generally shows a minor increase as the distance to the target increases. The exception is for the far-left case in many-target cases of the Twist condition, for which the average time spikes higher. By observing participants we could see that this was due to participants sometimes being unable to reach these targets despite applying hard force, and later finding that they needed to be more precise about the angle at which the force was applied. This may be eliminated with practise by the user or with a further refined implementation which is more forgiving with regards to the angle that the force is applied.

Reaction time plus movement time typically accounts for up to 1 second of acquisition time for bend targets, and up to 1.5 seconds for twist targets. Therefore, by examining Figure 7 we can see that jitter time accounts for much of the acquisition time with 5 targets or more. Unlike other input systems such as a mouse which requires zero effort to hold in one place, the user must apply active effort to hold the force in one place for 2 seconds, and errors in keeping a constant force cause jitter.

We can speculate as to a number of sources of the high jitter time when users apply low forces for both gestures. The prototype device experiences some elastic deformation which interferes with sensing at small force levels, however, the fact that the same result is found for two different force gestures suggests that it is not a problem with the particular sensor mounting used. Some degree of elasticity is present in all devices. Another explanation is that, given the users are already applying a force with their hands to support the weight of the UMPC, it may be difficult for users to modulate the forces they apply by small amounts while keeping the UMPC balanced in their hands, and it is easier to apply larger forces of similar or greater magnitude than the supporting forces.

7.3   Implications for Design of Force-Based Interfaces

A number of implications for the design of interfaces using force sensing can be drawn out of these results. The bend gesture performs significantly better than the twist gesture and, therefore, can be used when more targets are required. Target acquisition time increases as the number of targets increases, so there is a tradeoff between speed and number of targets. Since acquisition time increases for targets with low force, user interfaces should use avoid the use of such targets, or use wider targets at low forces.

To avoid jitter time overhead in force sensing based systems, we could explore adding a "select" mechanism which might be a button, another force gesture (e.g. a squeezing gesture), or something else. Alternatively, as with our example use of force sensing for alt-tab and page-down/up, we can make the interactions threshold-based, therefore eliminating any jitter as the force can immediately go back to zero after the threshold is reached.

8   Conclusions

Force sensing can be used for user inputs through applying physical forces such as twist and bend to mobile devices. We described a prototype implementation using pressure sensors added to a UMPC, and example uses of force sensing to perform shortcut key functionality that is otherwise missing from mobile PCs, for application switching (alt-tab) and page turning (page down/up). These interactions benefit from visual feedback which is related to the physical forces the user applies, therefore making them easier to learn and use.

We presented a user study into human abilities to apply bending and twisting forces at certain levels. We found that users performed bending quicker than twisting, that up to 4-5 separate levels of force were applyable by users without excess "jitter" in holding the force at that level, and that low levels of force were more difficult for users to apply than higher levels of force.

This work opens up a rich set of further research into force sensing as an interaction mechanism. One area of exploration is how best to map force inputs to user interface actions, for shortcut key replacement or otherwise. Alongside this it will be useful to investigate how best to integrate visual, audio and haptic feedback for force-based interaction. Future work can also explore other types of force sensor and how best to integrate force sensing into mass-produced mobile devices. Finally, it will be interesting to see how force sensing can be used in combination with other input types (e.g. touch-based or position/orientation-based) in order to take best advantage of each mechanism.

References

1. Rekimoto, J.: Tilting operations for small screen interfaces. In: Proceedings of UIST 1996. ACM Press, New York (1996)
2. Hinckley, K., Pierce, J., Sinclair, M., Horvitz, E.: Sensing techniques for mobile interaction. In: Proceedings of UIST 2000 (2000)
3. Williamson, J., Murray-Smith, R., Hughes, S.: Shoogle: Excitatory multimodal interaction on mobile devices. In: Proceedings of CHI 2007. ACM Press, New York (2007)
4. Schwesig, C., Poupyrev, I., Mori, E.: Gummi: a bendable computer. In: Proceedings of CHI 2004. ACM Press, New York (2004)

5. Harrison, B.L., Fishkin, K.P., Gujar, A., Mochon, C., Want, R.: Squeeze me, hold me, tilt me! an exploration of manipulative user interfaces. In: Proceedings of CHI 1998. ACM Press, New York (1998)
6. Rekimoto, J., Schwesig, C.: Presenseii: bi-directional touch and pressure sensing interactions with tactile feedback. In: Extended Abstracts of CHI 2006. ACM Press, New York (2006)
7. Mizobuchi, S., Terasaki, S., Keski-Jaskari, T., Nousiainen, J., Ryynanen, M., Silfverberg, M.: Making an impression: force-controlled pen input for handheld devices. In: Extended Abstracts of CHI 2005. ACM Press, New York (2005)
8. Jeong, D.H., Jeon, Y.H., Kim, J.K., Sim, S., Song, C.G.: Force-based velocity control technique in immersive v. e. In: Proceedings of GRAPHITE 2004. ACM Press, New York (2004)
9. MacLean, K.E., Shaver, M.J., Pai, D.K.: Handheld haptics: A usb media controller with force sensing. In: Proceedings of HAPTICS 2002. IEEE, Los Alamitos (2002)
10. Snibbe, S.S., Shaw, R., Roderick, J.: Haptic techniques for media control. In: Proceedings of UIST 2001. ACM Press, New York (2001)
11. Herkenrath, G., Karrer, T., Borchers, J.: Twend: Twisting and bending as new interaction gesture in mobile devices. In: Extended Abstracts of CHI 2008. ACM Press, New York (2008)
12. Watanabe, J., Mochizuki, A., Horry, Y.: Bookisheet: Bendable device for browsing content using the metaphor of leafing through the pages. In: Proceedings of UbiComp 2008 (2008)
13. Ishii, H., Ullmer, B.: Tangible bits: Towards seamless interfaces between people, bits and atoms. In: Proceedings of CHI 1997. ACM Press, New York (1997)
14. Fishkin, K.: A taxonomy for and analysis of tangible interfaces. Personal and Ubiquitous Computing 8(5), 347–358 (2004)
15. Ramos, G., Boulos, M., Balakrishnan, R.: Pressure widgets. In: Proceedings of CHI 2004. ACM Press, New York (2004)
16. MacKenzie, I.S.: Movement time prediction in human-computer interfaces. In: Readings in human-computer interaction, 2nd edn. Kaufmann, San Francisco

Inferring Identity Using Accelerometers in Television Remote Controls

Keng-hao Chang[1], Jeffrey Hightower[2], and Branislav Kveton[3]

[1] University of California, Berkeley, California, USA
[2] Intel Research, Seattle, Washington, USA
[3] Intel Research, Santa Clara, California, USA

Abstract. We show that accelerometers embedded in a television remote control can be used to distinguish household members based on the unique way each person wields the remote. This personalization capability can be applied to enhance digital video recorders with show recommendations per family-member instead of per device or as an enabling technology for targeted advertising. Based on five 1-3 week data sets collected from real homes, using 372 features including key press codes, key press timing, and 3-axis acceleration parameters including dominant frequency, energy, mean, and variance, we show household member identification accuracy of 70-92% with a Max-Margin Markov Network (M^3N) classifier.

1 Introduction

Personalizing the television watching experience has become a hot topic as service providers, content creators, and consumer electronics manufacturers all search for ways to expand their user-base, provide exciting and relevant programming, increase the effectiveness of advertising [1], incorporate digital home technologies like interactive TV [2], and distinguish their devices' features and usability. Most personalized capabilities and services are not possible, however, without first knowing who is watching TV. The work presented in this paper addresses this challenge of distinguishing between television watchers in a household.

Our new method of distinguishing TV viewers applies the lightweight biometric of analyzing people's hand motions and button press sequences on remote controls. This method is effective yet simple enough to be invisible and embedded pervasively. Users can simply grasp the remote control as needed and watch TV without any effort to explicitly login or identify themselves. Our system observes people's hand motion in the background and analyzes whether it matches existing signatures.

2 Related Work

There are other research and commercial efforts to develop ways to detect and identify TV viewers. Some existing approaches ask TV viewers to validate their

H. Tokuda et al. (Eds.): Pervasive 2009, LNCS 5538, pp. 151–167, 2009.

identity explicitly. Digital video recorders such as TiVo™ ask users to enter a user-name and password with an on-screen remote-driven virtual keyboard in order to access some personalized services. Orca Interactive (www.orcainteractive.com) uses a custom remote control device to read users' fingerprints. These logins and cryptographic-grade biometrics have high accuracy (in particular, a low false-positive rate) compared to the sensor-based approaches like the one presented in this paper. They are also quite secure and thus useful for authorizing sensitive transactions like purchases or subscriptions. But logins and intentional actions are cumbersome to perform repeatedly and their prompts almost certainly interfere with natural TV watching behavior. Another approach that is similar to our work uses computer vision for face detection and recognition. Hwang and colleagues' work is a good example of this approach [3]. Similar to our approach, facial recognition can be used to identify people in a "background" fashion, without explicit user input. However, a sense of privacy intrusion can come along with an embedded camera staring at every activity that happens in the livingroom, bedroom, or wherever the TV is positioned.

Outside the television domain, the work most similar to our contribution is that of Hodges and Pollack who showed that users manipulating everyday kitchen appliances (coffee making materials in their experiments) could be distinguished with approximately 77% accuracy based on their patterns of usage [4]. They applied decision trees for their classification, as did we in our initial work before we improved our results using a higher-level sequence information with a Max-Margin Markov Network (M^3N) classifier.

There is also quite a bit of work on combining machine learning with physical sensors to infer what activities someone is engaged in. This activity recognition research is worth mentioning in that it involves similar components to our work—namely, machine learning plus sensors like accelerometers—but it does not specifically focus on determining identity. Three specific examples are Bao and Intille's work using body-worn accelerometers to recognize physical activities [5], Philipose and colleagues' work where participants wear an RFID bracelet to sense which objects they are interacting with to infer their Activity of Daily Living based on web-mined models [6,7], and Consolvo and colleague's Ubifit persuasive fitness technology [8,9].

3   Feasibility Study

To consider the feasibility of the entire project, we first conducted a small study to understand how people use remote controls to watch TV. We recruited five of our lab colleagues videotaped them watching TV and channel surfing. We found some interesting patterns in the video recordings, which not only made us feel more comfortable to proceed in the research, but also inspired some of our feature selection approach.

Remote Control Orientation. One participant, shown in Fig. 1b, did not hold the remote control horizontally while switching channels. Another participant did not aim the remote control at the TV when switching channels.

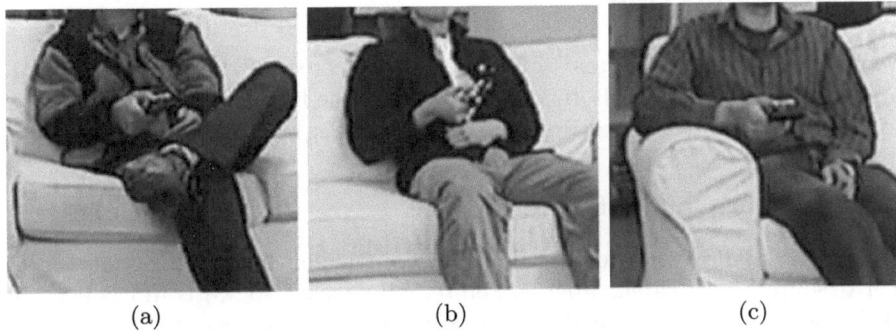

(a)   (b)   (c)

Fig. 1. Snapshots of different hand motion patterns as captured in our plausibility study. In comparison to (a), the participant in (b) holds the remote with different orientation, and the participant in (c) leans his arm on the sofa, which stabilizes his movements.

Physical Support. Some participants put their hands on the sofa arm, as shown in Fig. 1c, which stabilized their hands and induced less vibration on the remote control. Some participants usually put their arm on their lap. Another participant regularly left the remote control directly on the coffee table and switched channels without holding it.

Shaking while Surfing. One participants tended to shake the remote control in a seemingly unique way while surfing. Specifically, between each button press, he wiggled the remote while deciding whether to switch to the next channel.

Based on these high-level observations, we hypothesized that hand motion pattern might be distinguishable if we look at acceleration features *before*, *during*, and *after* each button press. Features before the press roughly capture the hand motion when the remote control was picked up or held between surf actions; features during the press capture distinctiveness in the orientation of remote as well as the dynamics of actually pressing the buttons; features after a press capture how the remote falls back to the arm, couch, lap, or table. We use these characterizations simply as a principled way to create features from the data stream for use by the machine learning algorithms. As such, it is not important that these descriptions are precise or exactly capture how all people use remotes.

In addition to our motion pattern observations, the ethnographic literature, specifically the work of Langan, revealed that females tend to switch to a planned channel by channel numbers while males tend to surf channels more using channel up and down buttons [10]. Though this work predates innovations like on-screen program guides and digital video recorders, which may alter or nullify some of the potential gender differences, it nonetheless led to our second hypothesis that capturing data about which keys were pressed and in which sequence may be valuable. Again, the veracity of these ethnographic claims is probably not critical since we use them simply as principles to justify including button press and button timing information as features for the machine learning algorithms.

4   Data Collection

We created the hardware and software needed to record the acceleration forces imposed on remote controls and also to capture the button presses. We used this capability to conduct a real-world data collection study in five households.

4.1   Hardware and Software

We wanted to have no dependency on a particular brand of remote or type of TV source (e.g. cable, fiber, satellite, broadcast) so we could collect data on the devices already owned by our participants. Thus, we designed our data collection hardware to easily integrate into a variety of TV setups. The hardware and software components are described below and shown in Fig. 2.

Accelerometer Module. Our 3-axis accelerometer module can be attached and wired into the power source of any remote control. In our deployments, we would purchase the same model remote control used by the household and modify it to attach our accelerometer module. At the conclusion of the study we would return the household's original remote. The accelerometer module continuously measures and transmits all the acceleration forces imposed on the remote control. The module hardware is a custom 3-axis accelerometer board connected to a Telos sensor mote [11], which acts as a relay to transmit the data to the logging laptop. The module is enclosed in a custom plastic case. We optimized the module to use as little power as possible and

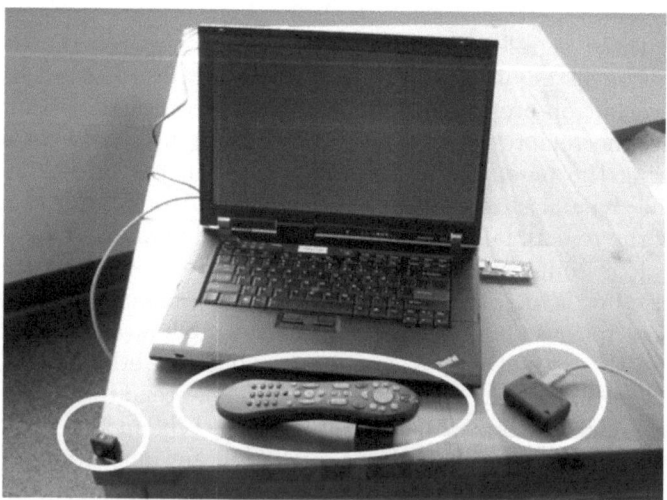

Fig. 2. Data Collection Components: top, a logging laptop; bottom from left to right, a video camera pointed at the room to gather ground truth about who was watching TV, a remote control with our accelerometer module attached, a universal infrared code receiver

found that on most remote controls it would last 2-3 days while continuously recording data.

Infrared Receiver. We use the Tira-2.1 multi-protocol infrared receiver made by HomeElectronics. In our deployments, we placed this receiver by the TV to capture the infrared signals caused by button presses on the remote control. Each infrared code is timestamped and logged by its unique ASCII code string. Infrared remotes will transmit the same signal several times (typically three times for a "normal" button press) to make sure the TV receives the signal, or continuously if the participant keeps pressing a button.

Logging Laptop Computer. A laptop computer receives and logs the acceleration and button press data streams. In our deployment it would be placed next to or behind the participants' TV operating with its lid closed. Acceleration data is wirelessly transmitted to the laptop through another paired Telos mote plugged into the laptop's USB port. The infrared code stream is received through a direct USB connection to the Tira infrared receiver. All data is timestamped with a precision of 100ns.

Video Camera. The last component is a video camera pointed at the room where the TV is located. The logging laptop automatically starts capturing a video clip whenever it receives an infrared button press and stops encoding the clip after 10 seconds without any additional button presses. We use these video clips in our experiments to hand-label ground truth about who was watching TV. To give participants control over their privacy, before returning the data collection system to the researchers, household members were given a way to access and review the video files in rapid playback to delete any video clips they did not wish to share with us. We omitted this data from our experiments.

4.2   Data Collection Study

We conducted a real-world data collection study in five households in Seattle metropolitan area of the United States. The households were recruited through one of their members working with us as colleagues. Doing completely outside recruiting seemed unnecessary for this study because all members in each household except one were not familiar with the project. Furthermore, we did not see a significant potential for bias even in the one member of each house who is our colleague since manipulating a TV remote is a simple physical activity. Everyone in the households already knew how to use a remote control (except one child who was very young and has not yet learned to use a remote) and there is no additional learning curve added by our technology, therefore there is no potential bias where someone who is technically trained might be able to learn our technology more quickly.

The composition of the five households are different: the first household is a four person family with two parents, a pre-teen, and a teen; the second is a couple; the third is a couple with a child who is too young to use remote controls; the fourth is a house with two graduate student roommates; the fifth is a large house with four graduate student roommates.

We asked each household to simply watch TV as they usually do while having the data collection system installed for one to three weeks. The system collected data 24 hours a day. Since we knew our sensor module mounted on participants' remote controls would last around 3 days when wired into a typical remote, we gave each household several extra sets of batteries and instructions to replace the batteries in their remote "every other day or whenever they thought about it." These informal instructions were sufficient in that we only saw one dropout in the data across all the households due to a battery dying. Even in this case we probably did not miss any data because the participant told us that she immediately replaced the batteries when she realized they were drained, which is not surprising because her remote would not work without fresh batteries since it shared power with our module.

We wanted to collect three weeks of data from each household, but the amount of time we collected data in each house varied between one and three weeks due to participants' summer vacation schedules, limited data collection hardware (we built two complete data collection rigs), and one mother who stopped participating after two weeks when she decided that the family had watched enough television for the summer and needed to spend more time on other pursuits.

5   Experimental Method

By iteratively adding features and analyzing their performance, we settled on a two-level classification technique: **button-press-level classification** and **session-level classification**, the former containing *motion-features* and *button-features* and the latter using motion-features and *inter-button-features*. Features for button-press-level classification occur before, during, and after a single button press. In this classifier, an inference about who is using the remote control is computed with each button press. At the higher level, session-level classification aggregates a sequence of button-press-level classifications and also has additional features that describe the longer sequence of button presses (e.g. the histogram). In this classifier, an inference about who is using the remote control is computed at the end of each session window. We will evaluate classification accuracy at both levels. Our feature extraction routines are implemented in MATLAB.

5.1   Button-Press-Level Classification Features

Classification at the button-press-level makes use of motion-features and button-features. Motion-features are computed from the accelerometer data. Twelve different time-windows (three types with four lengths of each type) are first located in the data stream around the current button press at time t. The three window types are *preceding, centered,* and *succeeding* capturing hand motion before, during, and after a button press, respectively. The preceding window has the right end point located at $t - 0.5$ seconds, the midpoint of the centered window is anchored at t seconds, and the start point of the succeeding window is positioned at $t + 0.5$ seconds. The four window lengths are $\{0.5, 1, 2, 4\}$ seconds.

The same set of features is computed for each of the 12 windows. The reason for having 12 windows is that the window size and type may influence the value of a extracted feature and, since we do not a priori know the best choice, we generate a variety and and let the classification algorithms decide the utility of the features by assigning them importance weights.

For each of the twelve windows we compute the (1) energy, (2) dominant frequency, (3) magnitude of the fundamental frequency, (4) mean, (5) variance, (6) maximum, (7) minimum, (8) median (9) range, and (10) correlation coefficient. The first nine features are extracted for each of the x, y, and z axes of the accelerometer. Energy, describing the total amount of hand motion, is calculated by the squared sum of the results of Fast Fourier Transform (FFT) with the DC component excluded. The fundamental frequency is defined as the frequency with the highest magnitude from the result of FFT (again, with DC removed), which provides information about shaking. The mean in three axes serves as an indicator of the remote control's orientation. In addition, the correlation coefficient is extracted from each of the x-y, y-z, x-z axis pairs, calculated by $(\Sigma a_i b_i - \overline{a}\overline{b})/((n-1)\sigma_a\sigma_b)$ where a and b are sequences of n measurements with mean $\overline{a}$ and $\overline{b}$ and standard deviation σ_a and σ_b.

Button-features used by the button-press-level classifier include (1) the infrared code of the button press signal, (2) the number of times the code was sequentially transmitted, (3) the approximate duration of the key press, and (4) a time-of-day to let the classification algorithms take into account habits of when particular people in a household watch TV in a day. In Section 4, we mentioned that a button signal may repeat if a participant keeps pressing the button. Therefore, we merge multiple sequential button presses into one classification step to create the the second and third features, which serve as an approximation of the button press duration.

5.2   Session-Level Classification Features

Patterns or frequency in a series of motions and button presses may also benefit user identification. On first glance, however, there is a "chicken and egg" problem: we want to extract features from consecutive button presses to identify a person, but we do not know whether a given sequence of button presses were made by the same person. Fortunately, a little domain knowledge gives us an effective heuristic: if there is continuous acceleration imposed on a remote control, then the same person is holding the remote during this period of time. While no heuristic is ever completely correct, this one turns out to be both reasonable and effective in practice. In examining the video clips we captured to hand-code the ground truth, we never saw someone operate the remote and then hand it directly to someone else to operate. Therefore we look for periods where the remote control is stationary, specifically where the energy of acceleration drops to near 0 for for more than s seconds, and label this point as a session boundary and possible transition between users. Based on video analysis, we set $s = 8$ and the found that the heuristic effectively segmented the data such that over 98% of the sessions were indeed occupied by the same person. Using these heuristic-derived session boundaries, we can now calculate the following features session-level features:

Motion-Features. The same set of motion-features described in Section 5.1 are also calculated over the entire session. This captures hand motion in a "macro" view, spanning across multiple button presses.

Inter-Button-Features. We generate features about all the button presses in a session including: (1) the number of button presses, the (2) mean and (3) variance of the intervals in between button presses, and (4) a histogram (count of appearance) of button presses in the session. The first three features indicate the frequency of pressing button behaviors, and the histogram shows the habit which buttons are used more often than the others.

6   Results

We use WEKA [12], a popular suite of machine learning software from the University of Waikato, to test the performance of several machine learning methods including Naive Bayes Classifier, C4.5 Decision Tree, Random Forest, and Linear Support Vector Machine (SVM). Our evaluations use ten-fold cross validation. To bring more realism to our results, cross validation is done over large contiguous blocks of time. For example, dividing the data at the individual button-press level to evaluate button-press-level classification would artificially boost accuracy, so we instead divide 5 days worth of data into 10 half-day blocks.

6.1   Button-Press-Level Classification

Table 1 shows the results for identifying users at the granularity of single button presses. We found few accuracy differences between the various machine learning algorithms on these data sets so we report Random Forest results for all data sets. Table 1 also shows statistics about each data set including the number of participants in the household, the total number of button presses recorded, the distribution among participants, and the baseline. Since some members of a household will watch more TV than the others, the baseline is the accuracy that would result if an oracle knew which person in the household pressed the most buttons on the remote control and always reported them as the answer. This baseline oracle would achieve decent accuracy, but a poor F-measure. In general, our classification accuracy is 12% better than the baseline (a 35% relative improvement) with classification at button-press granularity.

6.2   Session-Level Classification

The accuracy of the button press-level result was encouraging, so we believed that accuracy could be improved further by incorporating temporal relationships between button presses. Our first attempt was to simply smooth over sequential button-press-level estimates to try and exploit the theory that two button press

Table 1. Accuracy of button-press-level identity prediction using a Random Forest learning algorithm

	Statistics				Results	
Household	#Participants	#Presses	Distribution	Baseline	F-Measure	Accuracy
1	4	8756	0.63/0.33/ 0.04/0.01	62.8%	0.80/0.52/ 0.22/0.87	70.96%
2	2	695	0.82/0.18	81.6%	0.93/0.55	87.48%
3	3	122	0.25/0.75/ N/A	74.6%	0.68/0.89/ N/A	83.61%
4	2	834	0.72/0.28	72.4%	0.90/0.65	84.53%
5	4	1240	0.36/0.15/ 0.07/0.42	42.0%	0.69/0.39/ 0.74/0.70	66.45%

events that happen closely in time are more likely to be made by the same person. This simple smoothing approach did not boost accuracy, however, because when the classifier made a mistake, the confidence measure for the wrong decision was still high making filtering or majority voting work poorly. Instead we adopted a more principled way to improve accuracy using the previously described session-level features.

At the session level, we trained two machine learning classifiers: Linear SVMs and Max-Margin Markov Networks (M^3Ns) [13]. The features consist of the session-level features described in Section 5.2. The SVMs predict using a linear combination of features. The M^3Ns extend the SVMs such that they capture the temporal relationship between consecutive sessions. More specifically, for M^3Ns we model the fact that the same user usually uses the remote control several session in a row. Note that at the time we report these results WEKA does not support the M^3N graphical model so we implemented this algorithm ourselves. As shown in Table 2, this more complex approach leads to better results than SVMs—Linear SVMs are on average 11% better than the baseline (a 30% relative improvement) and the M^3Ns are on average 17% better (a 46% relative improvement), which is 6% better than the SVMs. Accuracy for the third household does not show a large improvement because the training set is extremely small relative to the other 4 households. Insufficient training data always impacts accuracy in any supervised machine learning. Even with a tiny training set, the accuracy is still no worse than the baseline oracle and improves slightly using the M^3N graphical model.

Though it is tempting to do so, the values in Table 1 and Table 2 are not directly comparable because partitioning button presses into sessions changes the nature of the problem—specifically, the "baseline oracle" in the session case knows which person in the household watches the most total TV instead of which person pressed the most buttons on the remote. With this difference in mind, we can conclude that temporal modeling and session-level classification does indeed offer an improvement over button-level classification because the former shows a 17% improvement over its baseline while the latter has an improvement of only

Table 2. Accuracy of session-level identity prediction. Household three's accuracy does not show a large improvement due to insufficient session training data

	Statistics			Accuracy	
Household	#Participants	#Sessions	Baseline	SVM	M^{3}N
1	4	458	53.9%	61.79%	69.87%
2	2	124	76.6%	90.32%	91.94%
3	3	28*	75.0%	75.00%	78.57%
4	2	90	65.6%	81.11%	88.89%
5	4	340	44.1%	61.78%	72.06%

12% over its baseline (the relative improvement is even greater at 46% versus 35%). Session-level classification is also more realistic because it more closely matches the ways people actually use and share remote controls.

6.3   Feature Evaluation

The Linear SVMs trained in section 6.2 provide a way to evaluate the importance of features because an SVM assigns importance weights to its features for class prediction. The prediction is made by weighted linear combination of features, i.e. $y = \arg\max_c \sum_k w_{ck} x_k + b_c$. We can think of the weight w_{ck} as a vote assigned to a particular feature x_k. The feature values themselves are normalized in their variance to a value between 0 and 1.

The first analysis is to look at how many features are actually important to the classification. The classifiers were given 372 different features as input, but, as the rank-order weight plot in Figure 3 shows, only about 10-20 features have high weight after feature selection. These features contain most of the classification power for that particular participant.

The rank-order weight analysis in Figure 3, however, does not reveal the features' variance, i.e. whether the set of highly weighted features is the same or different across participants. To illuminate this issue, Tables 3, 4, 5, and 6 break down the ten most indicative features for each participant for households 1, 2, 4 and 5, respectively (household 3 is excluded excluded due to its insufficient data). Each feature has in parentheses its level (Session or Button), followed by a dash, followed by its feature type (Hand Motion or Button Press Feature). For category B-M, the feature is abbreviated as feature_windowType_windowLength_axis, e.g. fundamental frequency extracted in center window of length 2 in y-axis as fundFreq_center_2_y. Similarly, for category S-M, the feature is abbreviated as feature_axis, e.g. maximum in the session window for the y-axis is max_y. In addition, buttons and their codes are hashed into integers to be uniquely identified. Finally, there is a special category B/S-B because, for classification at session granularity, aggregating individual button presses in a session actually generates a button press histogram that spans features in both session and button level.

Looking across households we can see that the highly weighted features are definitely not identical but there are some similarities and frequently occurring

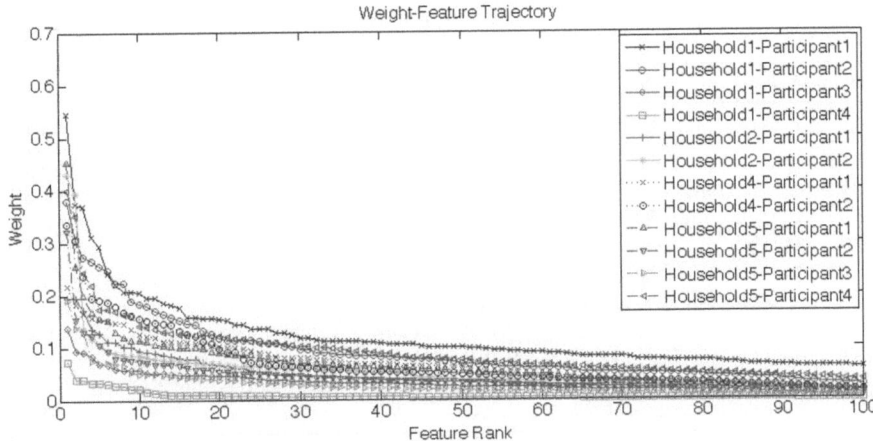

Fig. 3. Weights of the Linear SVM's features for different participants plotted in rank order shows that, although there are 372 features input to the classification algorithms, a small set of them are selected because they contain most of the classification power for a particular person

Table 3. Top features for predicting participant identify in Household 1

Participant 1			Participant 2		
Category	Feature	Weight	Category	Feature	Weight
(B/S-B)	button code/histogram 16	0.545	(B/S-B)	button code/histogram 25	0.379
(B/S-B)	button code/histogram 15	0.370	(S-M)	correlation_xy	0.307
(B/S-B)	button code/histogram 26	0.370	(B/S-B)	button code/histogram 18	0.274
(B-M)	fundFreq_center_2_y	0.312	(B-M)	fundFreq_center_0.5_z	0.266
(B-M)	fundFreq_center_1_y	0.294	(B/S-B)	button code/histogram 24	0.255
(B-M)	range_suceeding_1_z	0.242	(B/S-B)	button code/histogram 22	0.248
(B-M)	fundFreq_center_2_x	0.221	(B-M)	energy_center_4_y	0.224
(B/S-B)	button code/histogram 6	0.207	(B-M)	magnitudeFundFreq_center_4_y	0.224
(B-M)	magnitudeFundFreq_center_4_z	0.207	(B-M)	correlation_center_1_xz	0.190
(S-M)	correlation_xz	0.205	(B-M)	energy_center_2_y	0.183

Participant 3			Participant 4		
Category	Feature	Weight	Category	Feature	Weight
(B-M)	fundFreq_center_4_z	0.138	(B-M)	correlation_center_0.5_xy	0.074
(B/S-B)	button code/histogram 5	0.094	(B/S-B)	button code/histogram 18	0.040
(B-M)	max_preceding_4_y	0.091	(S-M)	fundFreq_z	0.040
(B-M)	correlation_succeeding_1_xz	0.083	(B/S-B)	button code/histogram 22	0.034
(B/S-B)	button code/histogram 27	0.070	(B/S-B)	button code/histogram 6	0.033
(B-M)	max_preceding_2_y	0.067	(S-M)	max_y	0.031
(B/S-B)	button code/histogram 13	0.059	(B/S-B)	button code/histogram 5	0.028
(B-M)	magnitudeFundFreq_center_2_z	0.058	(B/S-B)	button code/histogram 28	0.027
(B-M)	correlation_succeeding_2_xz	0.057	(B/S-B)	button code/histogram 7	0.022
(B-M)	fundFreq_center_0.5_y	0.057	(S-M)	range_y	0.022

Table 4. Top features for predicting participant identify in Household 2

Participant 1			Participant 2		
Category	Feature	Weight	Category	Feature	Weight
(B/S-B)	button code/histogram 123	0.196	(B/S-B)	button code/histogram 143	0.430
(S-M)	correlation_xy	0.195	(B/S-B)	button code/histogram 142	0.393
(B/S-B)	button code/histogram 115	0.172	(B-M)	correlation_succeeding_1_xy	0.136
(S-B)	number of presses	0.136	(B-M)	fundFreq_center_y_2	0.117
(B/S-B)	button code/histogram 119	0.128	(S-M)	min_x	0.106
(B/S-B)	button code/histogram 121	0.111	(S-M)	energy_x	0.093
(S-M)	correlation_yz	0.111	(B-M)	range_succeeding_2_y	0.089
(S-M)	range_x	0.102	(B-M)	max_succeeding_4_x	0.085
(B-M)	correlation_succeeding_2_xz	0.102	(B-M)	max_succeeding_2_x	0.085
(B-M)	correlation_center_1_xy	0.093	(B-M)	var_succeeding_2_y	0.085

Table 5. Top features for predicting participant identify in Household 4

Participant 1			Participant 2		
Category	Feature	Weight	Category	Feature	Weight
(B-M)	fundFreq_center_0.5_x	0.217	(B-M)	fundFreq_center_0.5_z	0.335
(B-M)	energy_center_4_z	0.184	(B/S-B)	button code/histogram 118	0.306
(B/S-B)	button code/histogram 181	0.165	(B/S-B)	button code/histogram 174	0.237
(B-M)	energy_center_0.5_x	0.157	(B/S-B)	button code/histogram 212	0.195
(S-M)	correlation_xz	0.153	(B-M)	magnitudeFundFreq_center_2_z	0.192
(B-M)	energy_center_0.5_z	0.151	(B-M)	energy_center_0.5_y	0.188
(S-M)	var_x	0.147	(B-M)	energy_center_2_y	0.179
(S-M)	energy_z	0.145	(S-M)	min_z	0.168
(S-M)	range_z	0.136	(B-M)	magnitudeFundFreq_center_1_z	0.162
(S-B)	number of presses	0.125	(B-M)	energy_center_1_y	0.156

features. For the button-features, the code of the button press signal appears the most, which indicates the habit of pressing certain button is a good indicator of who is using the remote. For example, one TiVo[TM] user in the household may avoid commercials with the skip-forward-30-seconds button while another always presses the fast-forward arrow. Unsurprisingly, the session-level inter-button-press histogram feature also shows up, indicating that the count and sequence of button presses is also a good discriminator of users. The frequency of pressing buttons are also distinguishing in some cases. Motion-features at both the button-press-level and session-level are selected. In particular, the fundamental frequency, magnitude, and energy are reported several times, meaning the shaking remote behavior is a distinctive pattern in some cases. The maximum, minimum, and mean features are listed, showing the orientation of a remote control can be somewhat indicative. In addition, windowing acceleration with different types and lengths also helps. In general this analysis gives us reassurance about the results since the selected features seem to match our intuitive

Table 6. Top features for predicting participant identify in Household 5

Participant 1			Participant 2		
Category	Feature	Weight	Category	Feature	Weight
(B/S-B)	button code/histogram 226	0.453	(B/S-B)	button code/histogram 308	0.321
(S-M)	correlation_xz	0.255	(B/S-B)	button code/histogram 227	0.153
(S-B)	number of presses	0.200	(B-M)	magnitudeFundFreq_center_2_y	0.129
(B-M)	var_succeeding_4_x	0.167	(S-M)	correlation	0.123
(S-M)	variation_x	0.155	(B/S-B)	button code/histogram 159	0.105
(B-M)	range_succeeding_4_x	0.151	(B-M)	range_center_0.5_x	0.094
(S-M)	range_x	0.130	(B-M)	range_center_0.5_y	0.080
(B-M)	fundFreq_center_1_z	0.120	(S-M)	correlation_xy	0.078
(B/S-B)	button code/histogram 148	0.114	(S-M)	med_z	0.076
(S-M)	energy_z	0.113	(B-M)	range_center_0.5_z	0.073

Participant 3			Participant 4		
Category	Feature	Weight	Category	Feature	Weight
(B/S-B)	button code/histogram 308	0.191	(B/S-B)	button code/histogram 152	0.399
(B-B)	button press duration	0.137	(B/S-B)	button code/histogram 151	0.352
(B-B)	button signal repetition	0.132	(S-M)	energy_y	0.244
(B-M)	magnitudeFundFreq_center_1_z	0.076	(S-M)	max_z	0.219
(S-B)	number of presses	0.072	(B-M)	var_preceeding_1_y	0.174
(B-M)	range_center_0.5_x	0.071	(B-M)	energy_center_4_x	0.173
(B-M)	magnitudeFundFreq_center_1_y	0.069	(S-M)	var_z	0.169
(B-M)	magnitudeFundFreq_center_4_z	0.065	(B-M)	fundFreq_center_1_x	0.164
(B/S-B)	button code/histogram 227	0.053	(S-M)	range_z	0.161
(B-M)	range_center_2_x	0.050	(S-M)	mean_z	0.142

ideas about which features would be useful. Although it does not seem to be the case that a particular subset of features is universally useful for all households, from the results shown we can conclude that households members do have sufficiently different behavior combinations such that machine learning methods are able to find a unique feature subset and infer identity.

6.4   Button-Features versus Motion-Features

Tables 3, 4, 5, and 6 reveal a similar number of highly weighted button-press and hand-motion features, which raises additional interesting questions: Do hand-motion or button-press features contribute more to the overall accuracy? Are button-press features alone sufficient to identify users? How does one type of feature complement the other? To answer these questions we ran the experiments again with only hand motion features and with only button press features. The results are shown in Tables 7 and 8.

Table 7 suggests that for button-press-level classification, button-features alone work as well or better than if motion-features are included. In fact, motion-features seem to drag down the overall accuracy when combined with button-features. Using motion-features alone performs similarly to using all features. If this were the end of the story, the conclusion would be to abandon the accelerometer hardware

Table 7. Accuracy comparison with subsets of features of button-press-level identity prediction

Household	Accuracy with all features	Accuracy with only motion-features	Accuracy with only button-features
1	70.96%	69.82% (↓)	77.58% (↑)
2	87.48%	85.90% (↓)	96.26% (↑)
3	83.61%	96.72% (↑)	87.70% (↑)
4	84.53%	85.61% (↑)	83.21% (↓)
5	66.45%	64.84% (↓)	74.11% (↑)

Table 8. Accuracy comparison with subsets of features of session-level identity prediction

Household	Accuracy with only all features		Accuracy with only motion-features		Accuracy with only button-features	
	SVM	M3N	SVM	M3N	SVM	M3N
1	61.79%	69.87%	58.52%(↓)	60.70%(↓)	57.86%(↓)	57.21%(↓)
2	90.32%	91.94%	86.29%(↓)	87.90%(↓)	90.32%(-)	92.74%(↑)
3	75.00%	78.57%	78.57%(↑)	78.57%(-)	75.00%(-)	78.57%(-)
4	81.11%	88.89%	77.78%(↓)	81.11%(↓)	63.33%(↓)	63.33%(↓)
5	61.18%	72.06%	54.12%(↓)	57.94%(↓)	43.24%(↓)	44.41%(↓)

and simply use a button press logger as the sole input to the classifier. However, Table 8 shows that there is a different trend in session-level classification—we may not want to give up the accelerometer quite yet. At the session level, only by using both types of features can the system achieve top accuracy. In fact, using button-features alone results in the lowest accuracy, sometimes by a significant margin (except for household 2 where it merely holds even).

Why is this trend different from the one in Table 7? We made a hypothesis that in button-press-level classification there might be more momentary deviation in motion-features. Hence, they worsen the overall accuracy when combined with button-features. In contrast, in session-level classification motion-features are calculated in the span over several button presses, which smoothes out the momentarily noise happened in the button-press level. The motion-features are therefore more informative and contribute the overall accuracy in session-level classification. We can also look at this from another perspective. The session-level motion-features may reveal patterns of moving the remote while pressing a series of buttons, which is less prone to having high variance.

7   Future Work

A potential confounding factor in our results, though not one we believe to be significant, is that our accelerometer module did change the shape of participants' remote controls. Though we always attached the sensor module in a place where it did not interfere with any of the normal hand positions, it may still have changed our participants behavior in some way. Toward this end it would be beneficial in future studies to shrink the acceleration module significantly and embed it inside the void space in the remote's plastic so the overall form of the remote is unchanged.

Though we are pleased with how well acceleration and button presses seem able to identify users, there are many other features, sensors, and sources of information that we would like to add to try and improve the accuracy. For example, hand shape, detected using pressure sensors or capacitive field sensors, may be a very good indicator of who is using the remote. Program guide information is another potential source of input as different people in a household may be attracted to different TV shows or categories of TV show. To test these new sensors and ideas, we plan to follow this work with a longer study of least 8 families over more than 1 month. In this new deployment we will also evaluate an application that makes use of this new personalization capability, specifically a digital video recorder that can recommend TV programs to each household member instead of the control case where recommendations are provided based on the behavior of the entire household.

Television users are probably not willing to go through a training phase where they repeatedly tell the system who is using the remote with each button press. Therefore, our approach would be much more practical if we could apply semi-supervised machine learning to the problem. Specifically, if we could automatically cluster sequences of similar button presses and sessions we could reduce the burden to the point where the user must only be infrequently prompted to verify their identity to provide a training label for the machine learning. The prompts would gradually decrease as the models improved. Even better would be a completely unsupervised learning technique where, given the number of users in a household, the system clusters and learns the models completely on its own, perhaps learning the users' names out-of-band by observing a login name (e.g. when the user is making an online purchase) that correlates with a particular cluster. Semi- and un-supervised learning are only possible if there are sufficient distinguishing features in the data. In the future, we plan to test the feasibility of these approaches on this type of data as well as study the tradeoff between the labeling effort by users and the learning curve of the system.

8   Conclusion

We have built and evaluated the technologies to test the hypothesis that accelerometers embedded in a television remote control can distinguish household members based on the unique way each person wields the remote. Based on real TV watching data collected from five households with 2-4 people of various demographics in each household for 1-3 weeks we achieved user identification accuracy of 70-92% by including both button press-level and session-level features in a Max-Margin Markov Network (M^3N) classifier. We analyzed the feature selection finding that only 10-20 of the 372 features hold most of the distinguishing power for a given household and participant, but the actual features vary somewhat by participant. We also found that button press features alone (without motion data) work well in a simple classifier that triggers with each button press, but to get the greater accuracy from inferring over longer sessions, both button press and motion features are desirable.

Though more accuracy is always better, we believe our results are already sufficient to enable useful TV personalization applications such as improved targeted advertising and digital video recorders that provide program recommendations per user instead of per device. Additional sensors, such pressure sensors or capacitive field sensor to detect users' hand shapes, may boost accuracy even further and a semi-supervised learning approach would make the system more deployable. Ultimately, combining our approach with an existing heavyweight mechanism such as login-password or secure biometrics could result in a complete TV personalization system that is natural and invisible for everyday personalization enhancements while supporting infrequent but authentication-critical situations like financial transactions.

References

1. von Rimscha, M.B., Rademacher, P., Thomas, N., Siegert, G.: The future of TV commercials. In: Annual Meeting of the International Communication Association, San Francisco, CA (May 23, 2007)
2. Lu, K.Y.: Interaction design principles for interactive television. Master's thesis, Georgia Institute of Technology (May 2005)
3. Hwang, M.C., Ha, L.T., Kim, S.K., Ko, S.J.: Real-time person identification system for intelligent digital TV. In: Consumer Electronics, ICCE 2007, Las Vegas, NV, pp. 1–2 (2007)
4. Hodges, M.R., Pollack, M.E.: An 'Object-use fingerprint': The use of electronic sensors for human identification. In: Krumm, J., Abowd, G.D., Seneviratne, A., Strang, T. (eds.) UbiComp 2007. LNCS, vol. 4717, pp. 289–303. Springer, Heidelberg (2007)
5. Bao, L., Intille, S.S.: Activity recognition from user-annotated acceleration data. In: Ferscha, A., Mattern, F. (eds.) Pervasive 2004. LNCS, vol. 3001, pp. 1–17. Springer, Heidelberg (2004)
6. Patterson, D.J., Fox, D., Kautz, H., Philipose, M.: Fine-grained activity recognition by aggregating abstract object usage. In: ISWC 2005: Proceedings of the Ninth IEEE International Symposium on Wearable Computers, pp. 44–51. IEEE Computer Society, Washington (2005)
7. Philipose, M., Fishkin, K.P., Perkowitz, M., Patterson, D.J., Fox, D., Kautz, H., Hahnel, D.: Inferring activities from interactions with objects. IEEE Pervasive Computing 3(4), 50–57 (2004)
8. Consolvo, S., McDonald, D.W., Toscos, T., Chen, M.Y., Froehlich, J., Harrison, B., Klasnja, P., LaMarca, A., LeGrand, L., Libby, R., Smith, I., Landay, J.A.: Activity sensing in the wild: a field trial of ubifit garden. In: CHI 2008: Proceeding of the twenty-sixth annual SIGCHI conference on Human factors in computing systems, pp. 1797–1806. ACM, New York (2008)

9. Consolvo, S., Klasnja, P., McDonald, D.W., Avrahami, D., Froehlich, J., LeGrand, L., Libby, R., Mosher, K., Landay, J.A.: Flowers or a robot army?: encouraging awareness & activity with personal, mobile displays. In: UbiComp 2008: Proceedings of the 10th international conference on Ubiquitous computing, pp. 54–63. ACM, New York (2008)

10. Langan, C.R.: A case-study of how people within the same household differ in their use of television. web document (April 1997),
    http://www.aber.ac.uk/media/Students/crl9501.html

11. Polastre, J., Szewczyk, R., Culler, D.: Telos: Enabling ultra-low power wireless research. In: Information Processing in Sensor Networks, 2005. IPSN 2005, pp. 364–369 (April 2005)

12. Holmes, G., Donkin, A., Witten, I.H.: Weka: A machine learning workbench. In: Second Australia and New Zealand Conference on Intelligent Information Systems, Brisbane, Australia (2007)

13. Taskar, B., Guestrin, C., Koller, D.: Max-margin Markov networks. In: Advances in Neural Information Processing Systems 16 (2004)

The Effectiveness of Haptic Cues
as an Assistive Technology for Human Memory

Stacey Kuznetsov, Anind K. Dey, and Scott E. Hudson

Human Computer Interaction Institute, Carnegie Mellon University,
5000 Forbes Avenue, Pittsburgh, PA 15213
{stace,anind,scott.hudson}@cs.cmu.edu

Abstract. Many people experience difficulty recalling and recognizing information during everyday tasks. Prior assistive technology has leveraged audio and video cues, but this approach is often disruptive and inappropriate in socially-sensitive situations. Our work explores vibro-tactile feedback as an alternative that unobtrusively aids human memory. We conducted several user studies comparing within-participant performance on memory tasks without haptic cues (control) and tasks augmented with tactile stimuli (intervention). Our studies employed a bracelet prototype that emits vibratory pulses, which are uniquely mapped to audio and visual information. Results show interaction between performance on control and intervention conditions. Poor performers on unaided tasks improve recognition by more than 20% ($p<0.05$) when haptic cues are employed. Thus, we suggest vibro-tactile feedback as an effective memory aid for users with impaired memory, and offer several design recommendations for integrating haptic cues into wearable devices.

Keywords: Wearable computing, haptic interfaces, memory cues.

1 Introduction

Memory recall and recognition continue to pose a challenge for a variety of people during routine activity. Numerous wearable devices aid human memory retrieval with context-aware audio and video cues (e.g., [1, 2, 3]). However, these cues are disruptive in environments that require acute visual or audio focus, and are often inappropriate in socially-sensitive situations. Haptic feedback is a discreet and unobtrusive alternative - it can be conveyed by a bracelet or anklet, minimizing audio and visual disruptions. Since interaction with haptic feedback as well as the device itself may be entirely concealed, tactile cues can be leveraged to assist users suffering from amnesia, Alzheimer's disease, dementia and other functional memory impairments, without drawing social attention to their condition.

We hypothesize that tactile cues can aid recall and recognition of visual or audio information. We present 4 studies which employ a wearable device that maps distinct haptic pulses to new concepts. When these concepts are re-encountered at a later time, corresponding cues are replayed. We find that high-performers on non-haptic tasks perform neutral or worse when assisted by cues. However, low-performers significantly improve on recognition tasks when haptic cues are employed.

H. Tokuda et al. (Eds.): Pervasive 2009, LNCS 5538, pp. 168–175, 2009.

2 Prior Work

Prior research has explored cross-modal priming for visual and haptic stimuli, where information was presented in one modality (e.g., haptically) and tested in another (e.g., visually). Ballesteros et al. have shown that people remember haptically presented stimuli after *repetitive* tactile exploration. Reliable haptic priming has been shown for adults, as well as healthy elderly and Alzheimer's patients [4, 5]. Piateski et al. compared human recognition of vibro-tactile patterns applied to the hand and the torso, and concluded that torso-recognition is superior [6]. Since people are adept at retrieving haptic memories, we propose to explore multimodal priming where retrieval of video and audio concepts is assisted with haptic cues.

Several user interfaces have leveraged haptic stimuli to direct attention and aid memory [7, 8]. Young et al. employed haptic cues ("taps") to orient users' visual attention to different quadrants on a screen [9], implying the possibility of cross-modal links between haptic and visual attention. Wall, Brewster and Kildal have employed haptic feedback as memory "beacons", enabling visually impaired users to mark points of interest on tactile displays and navigate by recognizing the 'beacon' pulses [10]. The Multimodal Collaboration Environment for Inclusion of Visually Impaired Children (MICOLE) project employed 'tactons' in memory games where visually impaired users were presented with vibro-tactile pulses and, at a later time, asked to identify which pulses they have experienced before [11]. While prior work leveraged haptic feedback in the context of attention, navigation or memory games for visually impaired users, we hope to explore the effectiveness of vibro-tactile cues as a memory aid for people who experience memory difficulties during everyday tasks.

3 Wearable Haptic Feedback Device

We envision a context-aware wearable device that augments human memory with haptic cues. To this effect, a prototype of the wearable device has been implemented using a small vibratory motor. This motor is integrated into a bracelet (Fig. 1), powered by an external Atmel AVR microcontroller [12] board, although the final instantiation of the device would be controlled wirelessly. Distinct pulse signatures are produced by varying the motor speed, pulse length, and frequency. Each pulse is encoded by analogue signals, ranging from 0-255, with perceptible motor speeds starting at 120 (effectively, no pulse below the analogue signal of 120).

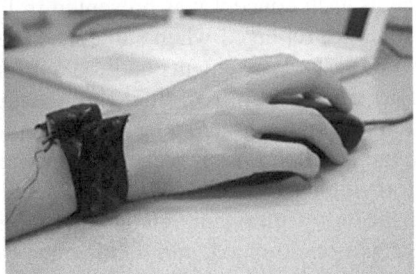

Fig. 1. Vibratory motor integrated into a bracelet to serve as a haptic feedback device

Table 1. Example pulse encodings, durations and descriptions

Encoding (100 ms segments)	Description	Duration (ms)
120,0,120	Short pulse, pause, short pulse	300
120,140,160,180,200,220,240	Continuously increasing long pulse	700
120,120,120,0,0,0,255,255,255	Soft pulse, pause, strong pulse	900

4 Methodology

We developed several user studies to validate our hypothesis that tactile cues can aid recall and recognition of visual and audio information. Our studies focused on 4 different challenges for human memory: blending (combining multiple distinct concepts into one) [13], auditory recognition, visual recognition, and free recall. Blending occurs when people confuse ("blend") aspects of different objects or locations into one seemingly familiar concept, most notably, during witness accounts in court proceedings. Poor auditory memory poses a challenge for a wide range of common tasks such as foreign language learning or following vocal instructions. Similarly, visual recognition is crucial, for example, in remembering faces or geographic locations. Finally, free recall ability influences retrieval of itemized information such as tasks from a to-do list, important dates, events, or phone numbers.

In order to evaluate the effects of a wearable haptic feedback device on the memory challenges outlined above, we conducted four user studies, each lasting for approximately 15 minutes. Each study consisted of three parts:

(I) *Training Phase*: Participants were presented with several distinct visual or audio stimuli, one at a time for five seconds each.

(II) *Pause*: Participants were given a distraction task of reading and rating three comic strips. This pause ensured that our experiments were not obscured by variance in short-term or working memory capacity.

(III) *Test Phase*: In this final phase, participants were asked to recall the concepts they were presented in phase I. Answers and response times were recorded.

Participants were asked to wear the haptic feedback device on their dominant hand. Participants in each study performed two versions of the tasks: control (without any haptic feedback), and intervention (with haptic feedback). To account for learning effects, the order of the control and intervention tasks were counterbalanced within and across participants. In the intervention tasks, a unique haptic pulse was played twice as each concept was presented in Phase I. During recognition tasks in Phase III, pulses were replayed when subjects were asked to recognize corresponding concepts. In recall tasks, participants had the option to replay the pulses multiple times.

Participants were recruited through the CMU Center for Behavioral Decision Research and compensated $10 upon completion of the study. Subjects (mostly graduate students) were not prescreened for memory impairments. The majority participated in all four studies, although several subjects skipped some studies due to time constraints and the non-Chinese language prerequisite for the audio study. Although the task order was randomized, and haptic cues were not reused between tasks, an interference effect is nonetheless possible among different haptic stimuli.

Haptic Cues and Blending
(24 participants, 15 male, 9 female, ages 21-65)
To determine whether haptic cues reduce blending, we implemented a shape-color recognition test. During phase I, subjects were shown 7 shapes of varying color, filled or hollow. The test phase presented 7 multiple-choice questions that showed three images - one familiar shape and two never-before-seen shape-color combinations. Participants were asked to identify which image they had seen before. During non-control tasks, haptic pulses associated with the familiar shapes were played.

Haptic Cues and Auditory Recognition
(20 participants: 12 male, 8 female, ages 21-65)
Non-Mandarin speaking participants listened to 5 Mandarin phrases, each repeated twice, while viewing the English translation. During the test phase, a Mandarin phrase was played while participants were shown an English translation and asked if it corresponded to the phrase. Each phrase was tested twice – once with a real translation and once with a translation that corresponded to another phrase in the set (10 total pairs).

Haptic Cues and Visual Recognition
(22 participants: 16 male, 6 female, ages 21-65)
Participants were shown 5 black and white portraits from the AT&T face database [14], and a name associated with each face. In phase III, subjects were shown a portrait-name pair, and asked if the name matched the face. Each picture was tested twice – once with the correct name, and once with a name belonging to a different face in the set (10 pairs total). For non-control tasks, haptic pulses corresponding to faces were played for both correctly and incorrectly matched name tasks.

Haptic Cues and Free Recall
(24 participants: 16 male, 8 female, ages 21-65)
We implemented a free recall test based on the Ebbinghaus nonsense syllable experiment [15]. In the control phase, participants were shown 7 nonsense syllables for 4 seconds each, for example "hik" or "lup". During the test phase, participants were asked to type in the syllables they could remember. In the test phase of the intervention study, subjects had the option to replay cues from the training phase.

5   Results

Tactile feedback led to a marginally significant 11% improvement in audio phrase recognition (p = 0.0753). There was no strong effect for face or shape recognition. Furthermore, haptic cues hurt free recall by an average of 20% (p = 0.0094).

Although haptic feedback did not have a strong positive main effect, our data shows significant interaction between subjects' baseline performance and the effects of the intervention for the recognition tasks. To examine these effects, we label participants who perform above average on unaided control tasks as *high performers* and subjects who perform below average as *low performers*. Low performers were significantly better at all three recognition tasks when haptic cues were employed, while high performers did not show improvement or performed worse.

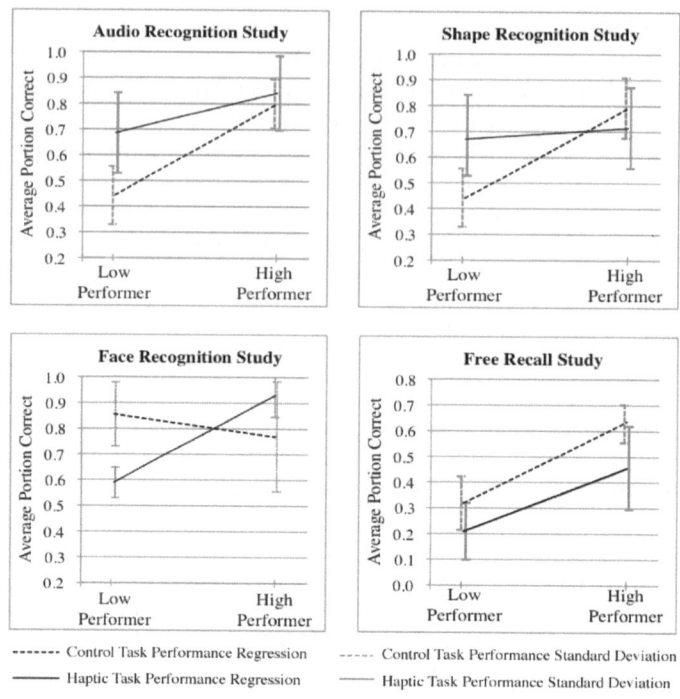

Fig. 2. Performance on shape, face, and audio recognition tasks with haptic cues interacts with performance on the control tasks (p<0.001, p<0.001, p=0.0239 respectively). Free recall shows no significant interaction, as both types of users performed consistently worse. with cues.

A 2-way Anova regression analysis was used to determine the interaction between control-level performance and performance on haptically-aided tasks. For each study, participants were classified as high or low performers based on the portion of control (non-haptic) memory tasks they completed correctly. Task condition (control or intervention) and performer type (high or low) were treated as independent factors. Significant interaction effects were found for all three recognition tasks (Fig. 2).

Fig. 3 juxtaposes the effects of haptic cues on low performers for each of the four memory tasks. Low performers showed an average improvement of 27% (p<0.001), 23% (0<0.0064), and 24% (p=0.0062) in face, shape and audio tasks respectively. Conversely, the average accuracy of high performers dropped by 16%, (p=0.018) and 18% (p=0.02) on the face-recognition and free recall tasks when haptic cues were employed. There was no significant difference in the audio and shape recognition studies for high performers, although the averages dropped by 4% and 10%.6 Discussion

High performers often performed worse on tasks that included haptic cues. Anecdotally, several participants felt that the haptic cues were 'distracting'. It follows that some subjects were unable to concentrate on the memory tasks while attending to the haptic cues. In part, this cognitive overload stems from the crude nature of our prototype. The vibratory motor did not allow for fine-grain control of speed or frequency, making some pulses 'too intense'. Furthermore, haptic feedback did not adapt to

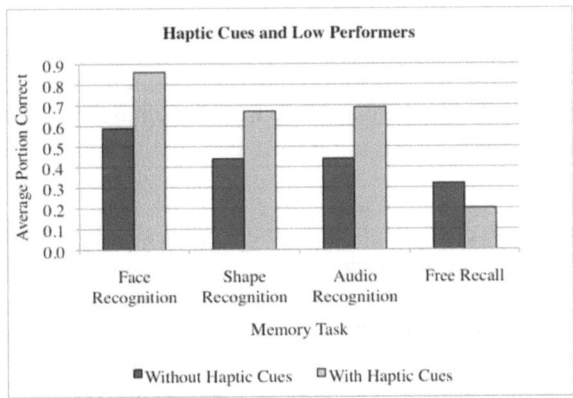

Fig. 3. Low performers showed improvement for all three recognition tasks with haptic cues

perceptual differences across participants, and consequently, some subjects were unable to accurately distinguish between pulses.

Free recall task performance was significantly lower with haptic cues, for most subjects. While recognition relies on perceptual memory where priming occurs subconsciously, free recall is influenced by declarative memory, which requires conscious, semantic processing [16, 17]. Hence, effective declarative memory cues must have semantic meaning that ties to the underlying concept. The haptic pulses employed in our study, however, were randomly assigned to nonsense syllables without semantic correlation. While non-semantic feedback successfully aided perceptual tasks such as visual and audio recognition, it failed for free recall, which hinges on explicit semantic memory. This explanation is consistent with participants' complaints about being unable to associate pulse frequencies and durations with specific letters of the syllables. Moreover, since participants were not screened by native language, semantic correlations between syllables may have been established based on subjects' linguistic backgrounds.

Our results suggest that high performers on non-haptic tasks perform differently from low performers on recognition tasks that are augmented with haptic cues. This interaction effect may be caused by a difference in cognitive processing and attention systems between high and low performers. Prior analysis of fMRI data implies that attention control systems vary between people with different working memory capacities [18]. Furthermore, people with good recognition of studied concepts show different event-related brain potentials (ERP's) than poor-recognizers [19]. Given this variance in attention and retrieval systems, the effectiveness of haptic cues may correlate with working memory capacity.

Since haptic cues benefitted participants with below-average performance on unassisted memory tasks, we propose haptic feedback as a memory aid for people with poor or impaired memory. It is not uncommon for memory-enhancing treatments to target poor performers while having neutral or negative effect on above-average subjects. For example, sabeluzole, a memory enhancing drug, has been shown to effectively improve consistent long-term retrieval in poor performers (below 50% long-term retrieval baseline), while having no effect on high-performers [20].

Furthermore, dextroamphetamine, a neurophysiologic inhibitor, has been shown to improve working-memory load only for participants with low baseline memory capacity, while worsening the performance of above-baseline participants [21]. Similarly, our data supports the use of haptic feedback as an assistive technology for low-performance participants.

Future Work – Improving Haptic Cues
We postulate that the adverse effects of haptic feedback can be attributed, in large, to a cognitive overload. Future work can focus on eliminating this effect through a refined implementation of the haptic feedback device. A more fine-grained motor or higher-resolution haptic display can offer more subtle cues. Moreover, an adaptive device can adjust intensity and frequency according to individual differences in perception. More importantly, however, the final wearable device must be context-aware and provide cues only in situations when the user is not suffering from cognitive overload. Since randomly assigned haptic cues proved detrimental for performance on free recall, an alternative approach could allow users to 'create' their own cues. That is, people may associate semantic meaning with different types of pulse signatures to aid personal recollection of declarative concepts.

6 Conclusion

We proposed haptic cues as an approach for improving human memory. While our user studies failed to validate haptic feedback as a universally effective aid for recall and recognition, we found significant interaction between performance on tasks with and without haptic cues. Poor performers improved by 20% or more on recognition tasks that were augmented with tactile pulses. It follows that vibro-tactile displays wield significant implications in the domain of assistive technology for memory-impaired users. As a ubiquitous, context-aware wearable device, haptic feedback has the potential to aid a multitude of people in overcoming memory challenges during everyday tasks.

References

1. Lin, W., Hauptmann, A.G.: A Wearable Digital Library of Personal Conversations. In: Proceedings of the Joint Conference on Digital Libraries, pp. 277–278 (2002)
2. Farringdon, J., Oni, V.: Visual Augmented Memory (VAM). In: The Fourth International Symposium on Wearable Computers, pp. 167–168 (2000)
3. DeVaul, R.W., Pentland, A., Corey, V.R.: The memory glasses: subliminal vs. overt memory support with imperfect information. In: Proceedings of the Seventh IEEE International Symposium on Wearable Computers, pp. 146–153 (2005)
4. Ballesteros, S., Reales, J.M., Mayas, J., Heller, M.A.: Selective Attention modulates visual and haptic repetition priming: effects in aging and Alzheimer's disease. Exp. Brain. Res., Epub., 473–483 (2008)
5. Ballesteros, S., Reales, J.M.: Intact haptic priming in normal aging and Alzheimer's disease: evidence for dissociable memory systems. Neuropsychologia 42(8), 1063–1070 (2004)

6. Piateski, J., Jones, L.: Vibrotactile Pattern Recognition on the Arm and Torso. In: First Joint Eurohaptics Conference and Symposium on Haptic Interfaces for Virtual Environment and Teleoperator Systems (2005)
7. Chêne, D., Zijp-Rouzier, S.: Haptic centred interface for geometry learning. In: Proceedings of the 1st International Conference on accessibility and assistive technology for people in disability situation (2007)
8. Wall, S., Brewster, S.: Providing External Memory Aids in Haptic Visualisations for Blind Computer Users. In: Proceedings of International Conference Series on Disability, Virtual Reality and Associated Technologies (ICDVRAT), pp. 157–164 (2004)
9. Young, J.J., Tan, H.Z., Gray, R.: Validity of haptic cues and its effect on priming visual spatial attention. In: Proceedings of the Haptic Interfaces for Virtual Environment and Teleoperator Systems, pp. 166–170 (2003)
10. Kildal, J., Brewster, S.A.: VibroTactile External Memory Aids in Non-Visual Browsing of Tabular Data. In: Proceedings of First International Workshop on Haptic and Audio Interaction Design, pp. 40–43 (2006)
11. Magnusson, C., Rassmus-Gröhn, K.: Force design for memory aids in haptic environments. In: ENACTIVE 2007, pp. 161–164 (2007)
12. Banzi, M.: Arduino (2005-2008), http://arduino.cc (last Accessed 20/09/2008)
13. Fauconnier, G., Turner, M.: Conceptual Blending and the Mind's Hidden Complexities, pp. 39–59. Basic Books (2002)
14. Database of Faces (2008), http://www.cl.cam.ac.uk/research/dtg/attarchive/facedatabase.html (last Accessed 20/09/2008)
15. Ebbinghaus, H.: Memory: A Contribution to Experimental Psychology. Teacher's College. Dover Publications (1913)
16. Tulving, E., Schacter, D.L.: Priming and human memory systems. Science 247(4940), 301–306 (1990)
17. Roediger, H.L.: Implicit Memory: Retention without Remembering. American Psychologist 45(19), 1043–1056 (1990)
18. Mariko, O., et al.: The neural basis of individual differences in working memory capacity: an fMRI study. NeuroImage 18(3), 789–797 (2003)
19. Curran, T., Cleary, A.M.: Using ERPs to dissociate recollection from familiarity in picture recognition. Cognitive Brain Research 15(2), 191–205 (2003)
20. Aldenkamp, A.P., et al.: Effect of Sabeluzole (R 58 735) on Memory Functions in Patients with Epilepsy. Neuropsychobilogy 32, 37–44 (1995)
21. Mattay, V.S., et al.: Effects of Dextroamphetamine on Cognitive Performance and Cortical Activation. NeuroImage 12(3), 268–275 (2000)

Exploring Privacy Concerns about Personal Sensing

Predrag Klasnja[1], Sunny Consolvo[2], Tanzeem Choudhury[3], Richard Beckwith[4],
and Jeffrey Hightower[2]

[1] The Information School, University of Washington, Seattle, WA, USA
[2] Intel Research Seattle, Seattle, WA, USA
[3] Department of Computer Science, Dartmouth College, Hanover, NH, USA
[4] Intel Research, Portland, OR, USA

Abstract. More and more personal devices such as mobile phones and multimedia players use embedded sensing. This means that people are wearing and carrying devices capable of sensing details about them such as their activity, location, and environment. In this paper, we explore privacy concerns about such *personal sensing* through interviews with 24 participants who took part in a three month study that used personal sensing to detect their physical activities. Our results show that concerns often depended on *what* was being recorded, the *context* in which participants worked and lived and thus would be sensed, and the *value* they perceived would be provided. We suggest ways in which personal sensing can be made more privacy-sensitive to address these concerns.

1 Introduction

Personal devices with embedded sensing are becoming pervasive. GPS units are present even in midrange mobile phones, and with the popularity of the iPhone and similar multimedia-oriented devices, accelerometers and proximity sensors are quickly moving into the consumer mainstream. Incorporating sensing into such personal, mobile devices enables a range of compelling applications, from location-based services—getting a restaurant recommendation near the user's current position, for example—to real-time detection of the user's physical activities.

However, having sensors embedded in devices that users wear or carry with them all day, every day can also be problematic. For example, having one's location constantly sensed can enable an unwanted person to learn where and when a user spends her time. Such information could potentially enable stalking or other types of criminal activity. In addition to such security considerations, people may simply be uncomfortable with others knowing their location, or even with their location being sensed in the first place. Mobile applications such as the Audio Loop [3], which continuously record raw audio, also raise concerns and introduce issues around how (or even if) to obtain consent to be recorded from others whose data might be captured by the user's device [5]. Such concerns could affect the adoption and use of devices that embed sensing and introduce problems into social relationships.

The usefulness of continuous sensing to enable a wide range of applications on the one hand and the potential privacy and security issues that accompany sensing on the other, raise a design challenge for pervasive computing researchers. That is, how can

H. Tokuda et al. (Eds.): Pervasive 2009, LNCS 5538, pp. 176–183, 2009.

we design such systems so that they use sensing when and where it is needed while respecting the privacy and comfort of users and others who may be monitored?

In this paper, we address this question by describing results from a study where 24 participants who used their mobile phone and a personal sensing device to track their physical activities for three months were interviewed about their reactions to and speculations about personal sensing in everyday contexts. As part of the study's exit interviews, participants were asked about any concerns they had with the sensing that was employed during the study and to speculate about other sensors that could be added to improve the system's activity inference capabilities. Our results reveal that privacy concerns varied greatly depending on *what* the sensor was recording, the *context* in which participants worked and lived and thus would be sensed, and the *value* they perceived in the capabilities that would be enabled. Participants often weighed the intrusiveness of what was being monitored about them and potentially others with whom they come in contact against perceived application benefits. If the latter were not seen as compelling enough, users would reject the use of sensing.

In what follows, we describe our method, results, and discuss implications of the results for the design of everyday personal sensing technologies. We then review related work and conclude.

2 Method

The data presented in this paper come from the exit interviews from a three-month field study of the UbiFit system which used mobile technology to encourage physical activity. Results that focused on how the system affected awareness and behavior related to physical activity and the effectiveness of the persuasive elements of the system have been presented elsewhere [2]. In this paper, we present results on participants' reactions to the sensors that were used during the study and other sensors that we were considering for a future version of the system to improve its activity inference capabilities. We describe the participants and our interview method below.

Twenty-eight people (15 female, aged 25-54), recruited from the Seattle metropolitan area's general public by a market research agency, participated in our three-month field study of the UbiFit system. The participants represented a range of professions, including real estate agent, personal care assistant, psychologist, teacher, comedian, public relations specialist and others. Twenty-four (14 female) participants took part in the sensing portion of the exit interview. The sensing questions were not asked of the remaining four because of time constraints.

During the study, 15 of the 24 participants wore a fitness device that used a 3-D accelerometer and a barometer to automatically infer walking, running, cycling, stair machine, and elliptical trainer activities. The remaining nine participants manually kept track of their physical activities using a journal on their mobile phones; participants who wore the fitness device also used the phone journal to record activities that the device was not trained to detect (e.g., swimming, yoga). Toward the end of the exit interview, we asked participants how they thought the fitness device inferred activities, and then explained to them how it did so. As part of our explanation, we provided a printout of the accelerometer and barometer data for three types of activities—sitting, walking, and running—to illustrate how the device could distinguish different activities. We then asked if they had any concerns about this data being recorded about them all day, every day and stored on their phones and, potentially, on a companion web site (which was not part of the study, but was something we were considering adding in future work).

We then suggested potential improvements, including providing flexibility on where the device could be worn (for the study, it had to be worn on the waistband), improving the accuracy of activity inference, inferring more information about activities, and inferring more types of activities. After getting participants' feedback about the usefulness of the suggested improvements, we explained that to implement the improvements would require the use of additional sensors, and we wanted to get their reactions to two sensors we were considering: (1) GPS and (2) a microphone. For each, we asked if the participants were familiar with the sensing technology and explained what it recorded and how it could be used to make the improvements. We also showed the participants a map of a run, along with the distance, elevation, and pace information for the run that was derived from GPS data. We then asked the participants to speculate on how they would feel about the sensors running all day, every day on their device, as the accelerometer and barometer had done in the study.

After they shared their initial thoughts, we probed about the implications of using the sensors. To ensure that the trade-offs were fully weighed, for people who were positive about the sensors, we brought their attention to possible concerns, and for people who were concerned, we pointed out the benefits that the sensors would enable. In addition, we suggested that the system would not have to keep the raw sensor data indefinitely, but could instead only maintain a small window of sensor data needed to calculate higher level measures such as distance and pace, after which the sensor data could be discarded. Finally, for audio, we noted that we might not need raw audio, but could record only certain frequencies within the audio stream that were needed for the inference. Such filtered audio would not contain enough information to ever reconstruct the content of a conversation, although an audio expert still might be able to determine that a conversation took place, how many people participated, what their genders were, and the general emotional tone of the conversation (e.g., whether it was an argument, a happy conversation, and so on). To clarify this idea, we played two audio recordings of the same nine second clip of a conversation between two males. One clip was raw audio, recorded by a microphone that was worn the same way the fitness device was worn for the study (i.e., clipped to the waistband). The other clip was the same recording, filtered to remove the unnecessary frequencies as described above [9]. We then asked participants how they felt about the filtered audio versus the raw audio and if and for how long they would be comfortable keeping that type of audio recording.

Interviews were audio-recorded and transcribed. The interview data was analyzed using open coding, a standard method of analyzing qualitative data.

3 Results

In what follows, we discuss how participants' reactions varied for different types of sensors, their speculations on data retention, the context in which the participants were likely to use the system, and the value that the participants saw in the functionality that the use of the sensing would enable.

3.1 Reactions to Different Sensor Types

Not surprisingly, participants reacted differently to the different types of sensors. None of the participants had any concerns about the accelerometer and barometer—the two sensors that were used during the field study. The participants did not consider this data to be particularly sensitive, and therefore had no problem with these sensors recording all day, every day, and for the data to be stored indefinitely on their mobile phone or on the fitness device. In addition, all participants who expressed wanting a companion fitness web site had no problem with the raw accelerometer and barometer data being stored there as well.

Reactions to GPS were more mixed. Unlike the accelerometer or barometer data, participants tended to think of the GPS data as being sensitive. 42% (14 of 24) of the participants had concerns about GPS being recorded all-day, every day. Concerns ranged from physical security—someone might get hold of the data and be able to figure out where the participants live or where their children go to school—to simply thinking that it was "creepy" or "big brother"-like. One participant commented that he does not like technologies that "*show where a human is exactly*" {Participant P3} and another one commented, "*I don't know about that...it can tell where you live and...that might be a little bit too much*" {P22}. When asked why GPS was different in terms of "being tracked" than the accelerometer, one participant explained that with the accelerometer "*it's not as specific, so the accelerometer isn't going to say that she is at* <the intersection of> *First and Pine*" {P22}. Nine of the 14 (64%) participants who were open to having the GPS data remain on their mobile phones also did not mind having the data stored on a companion web site.

Reactions to the raw audio were nearly unanimously negative. Only two of the 24 participants (8.3%) would consider a microphone that continuously recorded raw audio. Other participants adamantly replied that they were not willing to be recorded all the time, that they were uncomfortable being watched all the time (being recorded felt "*Big Brother-ish*" {P25}, "*I think I would feel too watched and too listened to*" {P22}), and even if they did not have a problem with being continuously recorded, they did not feel that it was okay to record those with whom they came into contact.

Recording audio in just the frequencies needed for activity inference was more acceptable. 25% (6 of 24) of the participants were willing to use the filtered audio and keep the data on their phones indefinitely (although only three were willing to upload this data to a companion web site). However, most participants remained uncomfortable. They simply found any audio recording to be too intrusive. The two most frequent classes of concern were 1) it made some participants feel as though they were being watched, and 2) even the filtered audio was seen as containing data that was too sensitive. One participant commented that "*I mean, even filtered, just I don't know. I would feel too exposed, but with this device* [with just the accelerometer and barometer], *I could care less*" {P22}. Another commented that filtered recording "*still just has that Big Brother effect to it*" {P27}. Regarding the sensitivity of the information, one participant commented that someone could still determine if you were with someone else, and another that the number of people in the conversation and its emotional tone were still *"a substantial amount of information"* {P2}.

3.2 Data Retention

In some cases, concerns about seemingly invasive sensors could be mitigated by changing the length of time that data were retained. While nearly half of the participants were unwilling to use GPS if the raw data (e.g., the latitude and longitude coordinates) were kept, all but one participant were willing to use it if the raw data were kept only for as long as was necessary to calculate the characteristics of detected physical activities (e.g., distance or pace of a run), and then promptly discarded. The exact length of the data window that the participants thought was acceptable varied, but most who wanted data purging thought that retaining one to 10 minutes of raw data at a time, unless a physical activity is being detected, was reasonable.

We found similar results for audio. A sliding data window of no more than one minute at a time of raw audio data was acceptable to 29% (7 of 24) of participants, although the majority (71%) found recording of any raw audio too invasive. Filtered audio fared better, however. If only a 10 minute sliding window of filtered audio was being saved, except for times when a physical activity is being detected, 62.5% (15 of 24) of participants were willing to use the microphone to get better activity detection.

3.3 Influence of Context of Use on Sensor Acceptability

Participants who worked in environments that required confidentiality unanimously objected to all forms of audio recording. When the use of a microphone was brought up, one participant explained that *"at work I'm privy to sensitive information that other people can't hear"* {P1}. Having a recording device of any sort was seen as completely unacceptable in that context, even if the audio was being continuously purged, often for fear that the device would be somehow compromised. Another participant, a psychologist, said that even the filtered audio contained too much information to be acceptable to use in patient consultations. She was afraid that such recordings could be subpoenaed in a potential legal case involving a patient, and that the risk was just not worth it. Even one of the two participants who would have been open to raw audio recording realized that his place of work would not have been okay with it, making the use of raw audio recording untenable for him.

The acceptability of using and storing GPS data was judged in a similar way. One participant commented that having data that shows where one is going makes her *"super uncomfortable"* since she has had friends with *"really controlling husbands,"* who would abuse such information {P3}. Another worked at *"a confidential site"* {P15}, and was concerned that someone could get access to the location data that he deemed confidential. GPS was generally more problematic for women who tended to feel more vulnerable than the men did. The characteristics of the context in which a sensing system would be used, such as the confidentiality requirements of a workplace, or the perceived vulnerability of the user—strongly influenced how the sensing technology was judged.

3.4   Value of Sensing-Enabled Applications

How much value participants perceive the data would provide was a factor in how they evaluated the acceptability of different types of sensors and data management. Runners who wanted to have maps of their running routes so they could plan future workouts were more willing to retain raw GPS data than were runners who did not think that those maps were particularly valuable. The latter group wanted to calculate the higher level information about their runs—the pace, distance, and elevation changes—but preferred not to keep the location information. Similarly, the psychologist reasoned that while she would really like the fitness device to automatically detect more types of activities, the risk and discomfort that any form of audio recording brought up for her far outweighed the benefit. She preferred to keep the device as it was and to continue to journal anything else manually. Another participant, who saw an additional benefit in the filtered audio (i.e., using the data to get feedback about her emotional reactions in different social situations such as being on a date), ultimately determined that she would not use it as she would feel obligated to explain to the people with whom she interacted that and how they were being recorded, and that would have been too complicated. Even though she found this emotional feedback idea really appealing, its value was not high enough to justify the potential harm it could do to her personal relationships.

The usefulness of a map of a run, the decrease in burden by having activities inferred automatically, and the value of other anticipated applications of the data such as emotional feedback, were weighed against other factors—legality, intrusiveness, and social etiquette—to determine whether the form of sensing needed to enable the desired functionality was deemed acceptable. If the benefits were not seen to clearly outweigh the perceived costs, the sensing method was rejected.

4   Discussion

The acceptability of personal sensing is a result of making trade-offs between the perceived value of an application and the costs—legal, social, and psychological—that the user perceives given the context in which she lives and works. While some researchers have argued that over time, changes in legal policy and social contract will decrease privacy concerns (e.g., [3]), it is unknown how far-reaching such changes are likely to be. For instance, many of the privacy issues raised in Warren and Brandeis's classic 1890 paper [8] remain relevant today. Moreover, there are clearly situations—such as attorney-client and doctor-patient interactions—where the need for confidentiality and privacy will not go away. Enabling users to make privacy trade-offs in an informed, educated way will be a key task for designers of sensor-enabled personal devices.

Our results suggest at least three ways in which the acceptability of sensing can be increased, while respecting privacy. First, sensor data should be saved only when relevant activities are taking place. Results for both GPS and audio revealed that continuously purging the raw data increased user acceptance of both sensors. Second, whenever possible, a system's core functionality should be based on minimally-invasive sensing. The users can then be given a choice to decide whether to enable additional functionality that might require more invasive sensors. Physical activity detection, much of which can be done with a simple 3-D accelerometer, is a good example of a domain where such graded sensing could be implemented. And third, researchers should explore ways to capture only those features of the sensor data that are truly necessary for a given application. This means, however, that sensor systems might need to have enough computational power to perform onboard processing so that each application that uses a sensor can capture only the information that it needs.

We also note that users can make informed privacy trade-offs only if they understand what the technology is doing, why, and what the potential privacy and security implications are. Building visibility into systems so that users can see and control what data is being recorded and for how long supports informed use. Determining how this can best be done is a difficult, but important, design challenge.

5 Related Work

In their work on the Audio Loop, a memory aid that continuously records a sliding window of 15 minutes of raw audio, Hayes et al [3] found that while over half of the participants in their lab study raised privacy concerns, the four participants in a field study were positive about the system. At least two reasons could explain why our data indicate lower acceptance of raw audio recording than Hayes et al found in their field study. First, the systems are different, and participants may have therefore valued their functionality differently. While the benefit of improving the physical activity inference in our system often did not warrant the use of invasive sensing, the Audio Loop's functionality might have been perceived as being valuable enough. Additionally, the core functionality of our system used sensors that did not raise privacy concerns; rather, the more invasive sensing would have been used to improve the functionality. Second, it is unclear if Hayes et al's participants encountered the types of confidential situations that were common for participants in our study. This potential difference in the context of use—what Hayes et al [4] call *social knowledge* affecting privacy perceptions—might explain why our results are different.

Iachello & Abowd's [6] *proportionality method* offers a design method aimed at ensuring that privacy is taken into account throughout the design process. Our data support their emphasis on *desirability*—making sure that the system's value makes any privacy compromises acceptable—and offer specific ways (e.g., graded sensing and sensor data filtering) to achieve design *appropriateness* and *adequacy*.

Nguyen et al [7] looked at people's responses to everyday tracking and recording technologies (TRTs), and found that people were highly concerned about privacy issues in the abstract but were simultaneously unconcerned with the TRTs of everyday life. They argue that familiar technologies provide known benefits and that risk is more abstract, rarely having been experienced. Their findings suggest that for users who already carry the sensors on a device they own (e.g., a GPS enabled phone), they may be more willing to adopt the services. Hayes et al [4] emphasize the role of users' experiences in shaping privacy perceptions of new pervasive systems.

Finally, Beckwith & Mainwaring [1] found that users' privacy concerns depend on their understanding of the technology they are using. Making good privacy decisions is difficult if the technologies are poorly understood. The high level of concern with

GPS and audio that we found in this study is likely due to the higher level of understanding that the participants had about these technologies.

6 Conclusion

This paper examined user reactions to four different types of sensors—accelerometer, barometer, GPS, and microphone—that can be used to infer physical activities. The reactions were obtained in interviews after 24 participants had used a mobile phone and/or personal sensing device to track their physical activity for three months, grounding their reactions and speculations in real world use. We found that *what* data is sensed and recorded, the *context* in which the sensing takes place, and the perceived *value* provided by the sensed data influenced the privacy trade-offs participants were willing to make. We suggest that conservative recording and data retention policies, graded functionality, and giving users visibility and control over which sensors are used could help with the adoption of systems that use continuous personal sensing.

References

1. Beckwith, R., Mainwaring, S.: Privacy: Personal information, threats, and technologies. In: Proc. ISTAS 2005, pp. 9–16. IEEE, Los Alamitos (2005)
2. Consolvo, S., Klasnja, P., McDonald, D.W., Avrahami, D., Froehlich, J., LeGrand, L., Libby, R., Mosher, K., Landay, J.: Flowers or robot armies? Encouraging awareness & activity with personal, mobile displays. In: Proc. UbiComp 2008, pp. 54–63. ACM Press, New York (2008)
3. Hayes, G.R., Patel, S.N., Truong, K.N., Iachello, G., Kientz, J.A., Farmer, R., Abowd, G.D.: The Personal Audio Loop: Designing a Ubiquitous Audio-Based Memory Aid. In: Dunlop, M.D. (ed.) Mobile HCI 2004. LNCS, vol. 3160, pp. 168–179. Springer, Heidelberg (2004)
4. Hayes, G.R., Poole, E.S., Iachello, G., Patel, S.N., Grimes, A., Abowd, G., Truong, K.N.: Physical, social, and experiential knowledge in pervasive computing environments. IEEE Pervasive Computing 6(4), 56–63 (2007)
5. Iachello, G., Truong, K.N., Abowd, G.D., Hayes, G.R., Stevens, M.: Prototyping and sampling experience to evaluate ubiquitous computing privacy in the real world. In: Proc. CHI 2006, pp. 1009–1018. ACM Press, New York (2006)
6. Iachello, G., Abowd, G.: From privacy methods to a privacy toolbox: Evaluation shows that heuristics are complementary. In: ACM Transactions of CHI, vol. 15(2), pp. 8:1–8:30 (2008)
7. Nguyen, D., Kobsa, A., Hayes, G.: An Empirical Investigation of Concerns of Everyday Tracking and Recording Technologies. In: Proc. UbiComp 2008, pp. 182–191. ACM Press, New York (2008)
8. Warren, S.D., Brandeis, L.D.: The right to privacy. Harvard Law Review 4, 5 (1890)
9. Wyatt, D., Choudhury, T., Kautz, H.: Capturing spontaneous conversation and social dynamics: A privacy sensitive data collection effort. In: Proc. ICASSP (2007)

Enabling Pervasive Collaboration
with Platform Composition

Trevor Pering, Roy Want, Barbara Rosario, Shivani Sud, and Kent Lyons

Intel Research Santa Clara
{trevor.pering,roy.want,barbara.rosario,shivani.a.sud,
kent.lyons}@intel.com

Abstract. Emerging pervasive computing technologies present many opportunities to aid ad-hoc collocated group collaboration. To better understand ad-hoc collaboration using pervasive technologies, or *Pervasive Collaboration*, a design space composed of three axes (composition granularity, sharing models, and resource references) is outlined, highlighting areas that are only partially covered by existing systems. Addressing some of these gaps, *Platform Composition* is a technique designed to overcome the usability limitations of small mobile devices and facilitate group activities in ad-hoc environments by enabling users to run legacy applications on a collection mobile devices. The associated *Composition Framework* prototype demonstrates a concrete implementation that explores the applicability of existing technologies, protocols, and applications to this model. Overall, Platform Composition promises to be an effective technique for supporting collaborative work on mobile devices, without requiring significant changes to the underlying computer platform or end-user applications.

Keywords: Pervasive technologies, ad-hoc collaboration, mobile devices, resource sharing, platform composition.

1 Introduction

Collocated collaboration is a highly dynamic and social activity where groups of people share information and create shared artifacts. *Pervasive Collaboration* enables users to share information using highly mobile and capable pervasive computing technologies such as smart phones, Mobile Internet Devices (MIDs), laptops, and large-screen displays. For example, the wealth of devices commonly found in corporate meeting environments could be used to enable more interactive group presentations. Studies of specific collaborative applications such as gaming [33], photo-sharing [4], and general file-sharing [35], detail how such platforms enable new collocated experiences for specific applications. However, in order to support a wide range of devices and generalized applications, collaborative middleware solutions have spanned a diverse design space consisting of collaboration granularity, sharing models, and resource referencing, leading to a number of different supporting middleware solutions.

H. Tokuda et al. (Eds.): Pervasive 2009, LNCS 5538, pp. 184–201, 2009.

Fig. 1. A Platform Composition showing the logically composed platform created out of the display and storage services from each user's individual device

Platform Composition is a technique that integrates standard computing components to support effective collaborative work by wirelessly combining the most suitable set of resources available on nearby devices. It is particularly well suited for ad-hoc tasks on wireless mobile devices such as laptops and advanced mobile phones: seamlessly incorporating fixed infrastructure such as projection displays. While these devices are becoming pervasive, their small form factors often impose usability challenges for supporting collocated group work. Composing devices enables them to act as a unified platform, enabling users to more easily overcome their individual limitations. The *Platform Composition* concept specifically refers to connecting devices together using standard distributed network protocols in such a way that existing familiar applications can be run unmodified. For example, it could be used to combine the file shares and displays of two laptops in order to run a standard application like Photoshop, enabling two users to more easily collaborate. This concept builds on the general notion of composition tailorability [34], but focuses on adapting the supporting computing platform, instead of changes to the end-user applications.

In order to understand how commonly available platforms and services support the general concept, the *Composition Framework* prototype provides a specific implementation of Platform Composition along with the user interface necessary to invoke standard platform services. By design, it is orchestrated around utilizing existing standards to support familiar applications on ad-hoc sets of devices. Although the underlying services and protocols used to share data among devices are not themselves new, the system instead focuses on the centralization and coordination of the sharing process. Experience with the Composition Framework highlights the efficacy of Platform Composition and informs a discussion of how existing systems can be modified to better support mobile ad-hoc collaboration.

The key contribution of this work is a focused understanding of how pervasive technologies can be used to support small-group collocated collaborations. The *Composition Framework* is the implementation of the *Platform Composition* concept, which is designed to support *Pervasive Collaboration*. Each of these three aspects provides an individual contribution to understanding pervasive collaboration applications:

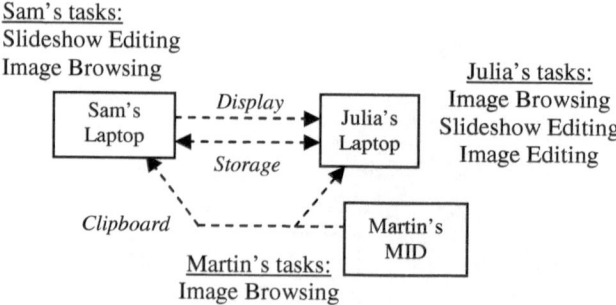

Fig. 2. Example sharing setup for the introductory scenario, showing how display, storage, and clipboard can be shared among mobile devices to form a composite collaborative working environment utilizing a variety of applications

1. *The Pervasive Collaboration Design Space* gives a concrete structure within which to understand how different pervasive technologies support collaboration.
2. *The Platform Composition concept* addresses several gaps in the pervasive collaboration design space by allowing the creation of dynamic ad-hoc collaborative device ensembles that support interaction using familiar applications.
3. *The Composition Framework prototype* highlights how existing services can be adapted to support collaboration as well as provides a system from which to collect observations on how users relate to the system.

1.1 Motivating Scenario

In the following detailed scenario, depicted in Figure 1, motivates Platform Composition by describing how a few friends meet in a café to create a birthday slideshow for their friend Kim:

> *Sam arrives first and starts to create a slideshow on his laptop adding some text and using some images he has on his system. By the time Julia arrives, he has a basic slideshow put together. To see his work, Julia mirrors Sam's display onto her laptop. She immediately thinks of a couple of funny captions she wants to include and adds the text while Sam discusses the images he selected.*

> *Julia mentions that on the bus ride over she had spent some time browsing through her photo collection and found a couple of additional images to include. She shares her file collection with Sam's computer and uses the shared display to show him the specific images she wants. Sam nimbly drags and drops the pictures into the document from the shared folder.*

> *At that point, Martin, a mutual friend, enters the café and stops by to say 'Hello.' He sees them working on the slideshow and realizes he has a great image in his email they could use. Using his Mobile Internet Device, he browses though his inbox, to find the image. They connect their clipboards together, allowing them to copy/paste the images between devices, both into the document Sam is working on and into Julia's personal collection.*

Sam and Julia continue working on the document as Martin goes off to order his coffee. Julia feels that some of the images need a little editing: Sam makes his file share available to her, and she uses an image editing application on her machine to touch-up the photos. All done, they put their systems away and head off to Kim's birthday party, implicitly disconnecting their shared services.

In this scenario, Sam, Julia and Martin compose their systems to create a logical aggregate platform using several different platform-level services, highlighted in Figure 2. They dynamically shift between sharing their display, file storage, and clipboard between their computers in order to accomplish their task as a group. Furthermore, they use a collection of standard applications to edit the main document, share personal content, and edit individual images using different resources from all three individuals' devices.

Currently, such interactions are problematic due to the numerous steps needed to share distributed content and difficulties resulting from interacting in mobile environments. For example, one commonly employed solution is using email to exchange images between users, even though they may be sitting right next to each other. Email requires people to *a priori* decide which specific images to share and incurs an interaction overhead through an external (to the task) email application, and routes all information through supporting infrastructure.

Alternately, collaborative sharing can be moved into the physical world, either through the motion of people or devices. Two people can share one display, but it can be difficult for multiple people to crowd around one screen. Similarly, sharing a USB memory stick between two computers would limit data sharing to a batch transfer of files, preventing any dynamic sharing of data between devices. As a result, these approaches are functional but do not necessarily offer the best collaborative experience.

2 The Pervasive Collaboration Design Space

This section highlights three system design issues (Figure 3) that characterize mobile collaboration systems. First, *composition granularity* impacts the nature of collaboration spaces formed by users and the steps they must take to integrate with legacy applications. Second, *sharing models* governs how users interact with the underlying data and how shared resources are managed. Third, *resource referencing* refers to the different ways resources can be named in a system, underscoring both the infrastructural needs of the system and mechanisms by which users relate to technological constructs. A clear understanding of these various design options is useful to understand how current systems exploit various aspects of pervasive technologies.

2.1 Composition Granularity

Within the context of modern operating systems, resources can be shared between systems at a number of different granularities. There are three primary levels of granularity which can be used to support the interactions between systems:

Events: Fine-grain sharing can be accomplished through events, which take on the form of small, individual occurrences that stream together to form higher-level

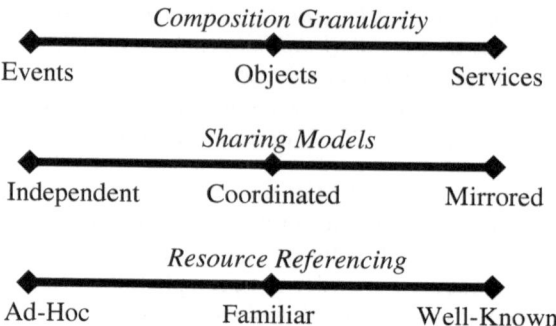

Fig. 3. Composition design spaces. Although representing distinct axes, each spectrum follows the general pattern ranging from small, individual elements to more encompassing coordinated constructs.

actions. User interface events, such as mouse movement or button presses are prototypical examples of event level sharing: each individual event is very small and short lived, but by combining multiple events together a cohesive stream of actions can be formed to interact with a remote system. The iRoom [13] system utilizes event level sharing in a collaborative room environment. It employs a centralized infrastructure for sharing events among mobile devices and fixed infrastructure to create a unified distributed work environment.

Objects: Medium-grain sharing relies upon the sharing of individual objects between systems. These objects represent persistent individually meaningful entities that can be acted upon in a variety of ways. Unlike event sharing, objects are persistent, and represent more data than is directly transferred. Casca [6] supports object-level sharing and enables users to create shared "converspaces" in which objects such as files and printers are placed to be made available to other users; i-Land [32] similarly allows objects to be managed within a collaborative space with many display surfaces. Multibrowsing [14] exposes web pages as objects layered on top of the iRoom event mechanism: persistent web objects are encoded as events, which incorporate transient properties such as the target display screen.

Services: Coarse-grain sharing is accomplished through sharing entire services from the platform, which are integrated with the underlying operating system to provide transparent application access to the associated resources. For example, access to a network file share represents a collection of file objects available to an application in a transparent manner. Like object sharing, service sharing has the quality of persistence; however, it is less specific in the semantics about what is being shared and more flexible in terms of legacy application support. Sharing services is currently possible piecemeal: users can use basic utilities to share services between their devices, but there is no coordinated application for this capability.

Collaboration systems such as the iRoom, Casca, and Platform Composition each provide the user with a different primary mechanism for sharing, encompassing a trade-off between fine-grain control using events and simpler construction using coarse-grain sharing. The iRoom, for example, makes it very easy to remote control systems with keyboard and mouse events, but will incur an extra step in order to share a file directory between systems. Casca makes sharing collections of files among users easy, but would require extra mechanisms to share keyboard and mouse events between systems. Platform Composition, on the other hand, provides transparent access to remote system level services, but requires user interaction in order to share events or objects between systems. Composition granularity, therefore, becomes a primary differentiator between these systems.

2.2 Application Sharing Models

Similar to the different levels in which sharing can be induced between systems, individual collaborative episodes can be managed on different levels [2]. These levels are similar to the composition granularity described in the previous section, but manifests themselves more directly in users' work practice in terms of mechanisms for conflict resolution and data synchronization. Application sharing models can be broken down into three levels:

Independent: An application can share data independently of other systems in such a way that changes are independently applied to each system and loosely propagated to other instances. For example, one implementation of a calendaring application might handle meeting requests between users but operate on different underlying databases (one per user). As a result, the instance of one appointment will not be inexorably linked to others. Independent sharing is supported by many variants of distributed single-user collaboration-transparent systems, such as email or personal calendars. Independent sharing offers no direct mechanism for conflict resolution, since there are two separate entities that only loosely correlate – relying on higher-level resolution mechanisms.

Coordinated: Sharing multiple views of the same underlying data enables coordinated sharing. Using the same calendaring example, the application could open up multiple views of a shared calendar, such that all views show the same underlying data but from different perspectives (e.g., maybe showing different days). Coordinated sharing allows multiple people flexible access to shared data, allowing shared context with independent interaction; conflict resolution is a key aspect of coordinated sharing, but is often handled on an application-specific basis. Coordinated sharing is supported by a wealth of *collaboration-aware* [17] systems, exemplified by projects such as the Coda filesystem [28] and TeamRooms [27].

Mirrored: Finally, applications can be shared by exactly duplicating all aspects of presentation, from the underlying data set to the visual representation and interaction. This level of sharing enables users to share interaction context while working on a common dataset. For example in a calendar application supporting mirroring, users could be using the mouse to point to a specific day and say "How about we schedule a meeting here?" Here, the system sharing is supplying the necessary related context. This interaction allows multiple people to coordinate directly, decreasing the interaction overhead but preventing independent operations. This level of application interaction is supported by services such as VNC [26] that support remote-desktop collaboration.

A challenge for collaborative systems is to provide support across multiple application sharing models [2], essentially enabling users to use the appropriate model at different stages of their overall task, exposing a trade-off between flexibility in manipulation through independent sharing with an emphasis on communication for mirrored sharing. For example, in a calendaring application, users may wish to initially work with *independent* systems to understand personal commitments, use a *mirrored* view to discuss potential scheduling options, and then a *coordinated* conclusion to record the group consensus. Furthermore, it is important for a system to reflect the mode of sharing to the user, so that they do not form incorrect mental models of the underlying system. Towards this end, the different services made available through Platform Composition provide a unified approach to encompassing multiple sharing models, and provide the user with the ability for independent, coordinated, and mirrored sharing.

2.3 System Resource Referencing

Individual resources, such as devices and services, can employ a variety of means to discover, address, and reference each other in distributed systems. Referencing is important because it directly effects how users interact with the resources made available by the system. There are three rough models for resource referencing:

Ad-Hoc: Systems which have never seen each other before can use *ad-hoc* mechanisms to identify resources. Conversationally, this translates to "use this device" where a user can select the device from a set of nearby devices, discovered by such techniques as UPnP, ZeroConf, or Bluetooth wireless scan. Ad-hoc references are advantageous in that they allow access without requiring prior setup, allowing for the *implicit* discovery of new resources, although it can suffer from confusion if there are many similar resources available.

Familiar: Referring to something as "the one I used before" invokes a *familiar* reference, which has been used in the past but is not necessarily well-known. Familiar references are advantageous in that they provide a mechanism to resolve ambiguity and complexity introduced by pure ad-hoc references, or ease *a-priori* set-up required by well-known references. Most systems will start from an ad-hoc or well-known foundation and construct familiar references using techniques such as bookmaking or machine learning as ways to mitigate the associated disadvantages.

Well-Known: Referring to something by a specific name, such as "fred.smith@mail.com," represents a *well-known* reference in that it works unambiguously and consistently in all contexts. The disadvantage of well-known references is that they require *a-priori* setup when they are first used and can be difficult to invoke in dynamic ad-hoc situations; in contrast to ad-hoc systems, well-known references require *explicit* discovery of new resources. Well-known references are commonly used with email and chat-client programs to send messages between people; also, referring to a specific remote network file-share by machine name or IP address would constitute a well-known reference.

Similar to Casca, Platform Composition itself does not inherently rely on any given reference mechanism, although the dynamics of ad-hoc mobile collaboration fundamentally starts with ad-hoc references. A challenge with both these systems is how to limit the scope of ad-hoc discovery to the most interesting or relevant devices: effectively relying on familiar references. Mobile systems can leverage the physical location or proximity of devices [19] in order to form associations, i.e., the "physical familiarity" of a device. Emerging mobile devices are starting to possess new input capabilities, such as accelerometers, cameras, and Near Field Communication (NFC), that may be useful in aiding mobile device composition. For example, DACS [7], bumping objects together [10], Relate [9], Gesture Connect [24], and Elope [25] are all sensor-based approaches to joining devices together, which could directly mitigate some of the problems associated with ad-hoc or well-known references. The trade-off for referencing is between the ease of accessing a new or unfamiliar object using ad-hoc referencing with a more reliable accessibility for common objects using well-known referencing.

3 Platform Composition

In contrast to the other composition systems mentioned above, Platform Composition enables users to interact by easily connecting their existing platforms' services together. For example, as highlighted in the motivating scenario and Figure 2, a file share from one device can be made available to another user's device; alternately, a user could remotely access the display of another user's device, allowing them to jointly view and edit content. This section covers the high-level concept of Platform Composition, covering the motivation and services that are applicable to pervasive collaboration applications.

3.1 Motivation

The concept of using Platform Composition to support mobile ad-hoc collaboration is motivated by four main factors: technology advances, end-user benefits, standard services, and available applications.

First, current improvements in mobile device processing, storage, and communication capabilities have created advanced mobile devices that can host significant applications. Mobile laptop computers are now the mainstay of many corporate environments, and the processing in some hand-held devices is capable of supporting high-quality touch-screen interaction; solid-state drives (SSD) and other high-capacity storage technologies provide ample capacity to store extensive caches of digital media; while advanced wireless standards such as ultra-wide band (UWB) and IEEE 802.11n provide the means for high-bandwidth inter-device communication. These trends present an opportunity in bringing these advances to cooperative work. Projects such as The Personal Server [36] and Dynamic Composable Computing [37] have explored how these same trends impact the mobile platforms themselves.

Next, composing systems from several mobile devices has the potential to improve several aspects of the collaborative user experience, addressing some of the fundamental limitations imposed by their small screens and limited I/O capabilities. Increasing the total available screen real-estate has been shown to improve group productivity by using multiple screens together [8], and also affords new opportunities for media consumption [30]. Not only do user's individual devices provide access to their personal resources, but they provide a readily available platform from which to access resources on other devices. The theme of orchestrating multiple devices to support collaboration is common in the research literature, having been addressed by systems such as Casca [6], iRoom [13], Pebbles [21], and The Display Mirror [22], among others.

From an application standpoint, there exists a wide variety of single-user applications that can be utilized for group collaboration, a notion known as *collaboration transparency* [17]. Platform Composition can directly leverage these existing applications, instead of relying upon special-purpose collaborative applications. Many previous systems have focused on collaboration transparency in the context of a single application, e.g. *intelligent collaboration transparency* [16] enables distributed cross-platform text-editing without application modification. In contrast, Platform Composition supports transparency at the *system* level, since it enables users to employ a variety of standard applications to accomplish their task.

Finally, there are several computer industry standards which provide functional abstractions for basic platform services, such as the file system, bit-mapped display, and socket-based networking. These standards provide a consistent programming model across diverse hardware platforms and utilize standard communication protocols, such as TCP/IP, enabling them to operate in distributed environments. Such distributed systems have been a cornerstone of fixed collaborative setups, which allow people to work collectively on distributed resources like desktop computers through shared file systems and remote desktop protocols. Platform Composition capitalizes on this capability by using these system abstractions as the primary building block supporting dynamic collaboration on wireless mobile devices.

3.2 Platform Services

Figure 4 shows how platform services are orchestrated by Platform Composition to enable existing applications to work with distributed resources. All these services are common platform components, with well-defined behaviors and control mechanisms: virtually every application can access these resources in a standard and consistent manner. There are three primary system resources supported by the concept:

Clipboard: By sharing the system's clipboard, users can easily transfer isolated pieces of information between systems. For example, a user can easily share the URL of a web page they are looking at by copying it to the clipboard – similarly, they can easily identify and share a specific image. As a standard system resource, the clipboard is closely integrated with virtually every application available on standard windowing systems (although it is not currently pervasive on all small mobile devices). The Remote Clip [20] project explored the impact of clipboard sharing between a personal mobile device and a desktop PC.

Storage: Groups of files, such as an image collection or current working documents, can be made available remotely using standard network file-share techniques. For example, if one user is working on a slideshow, as described in the introductory scenario, they can share their working directory to provide other users access to the files. This allows them to directly work with the content; similarly, a supporting

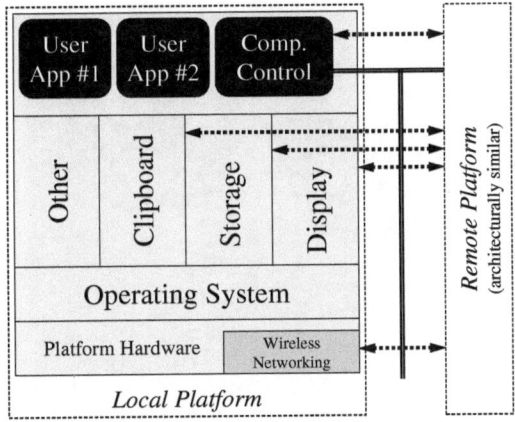

Fig. 4. Platform Composition architectural overview. Each symmetric platform is represented as a standard set of services implemented on top of the operating system and platform hardware. From the applications' perspective, the underlying system resources appear the same if they are local or remote.

user could make their local image collection available to the main document author enabling them to easily incorporate the images. Once a storage composition is formed, files are accessed through applications using their commonly available file open/save mechanisms. The Sharing Palette [35] and Coda [28] have explored the impact of storage sharing in distributed collaborative environments.

Display: Visual resources can be easily shared by replicating the user's visible display surface: either by remoting, pulling, or extending their display. Remoting their display to an external screen allows others to easily see (and potentially interact) with the primary user's data. Pulling a remote display to the local device enables a remote-control interaction with the other display. While this is fundamentally the same as pushing a display connection it changes who initiates the connection. Extending a system's display to utilize a supporting device enables applications running on the host system to transparently utilize additional screen real-estate, much in the same way as would be accomplished by physically attaching a second monitor. Several projects such as The Display Mirror [22], MobiUS, [30], and "The 22 Megapixel Laptop" [31] (among others) have explored the impact of display sharing in a number of different environments.

Additional services, such as the ability to share individual windows provided by Impromptu [3], arguably fit into this model as long as they provide a generic capability that would be available to any application. Sharing input devices, such as keyboards and mice, has also been incorporated into the system by sharing USB devices over the network based on commercially available USB-over-IP solutions. Likewise, other system services, such as dynamic networking, distributed sensing, and processor sharing also fall under the umbrella of Platform Composition, and represent areas for future work.

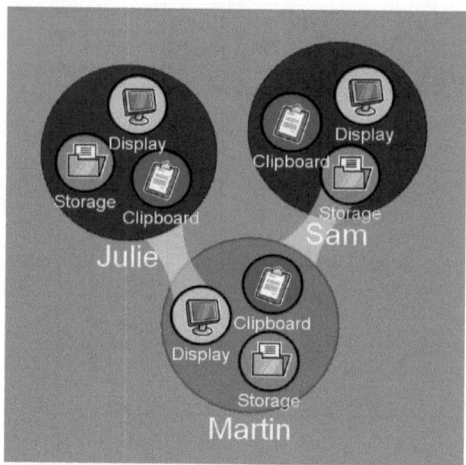

Fig. 5. Join-the-dots graphical user interface (GUI) used to manually create compositions by sharing resources with a simple line drawing metaphor. This interface has been designed to work well with touch-screen or pen interfaces, and does not rely heavily on detailed manipulation or textual input.

3.3 Supporting Platform Collaboration

In contrast to the examples cited under individual services above, Platform Composition provides an integrated framework from within which to access all these services. Additionally, since these services represent standard system capabilities, they are available to most applications without modification. These services present the user with very familiar mechanisms for sharing data, and they are compatible with virtually every available application, allowing Platform Composition to easily support pervasive collaborative activities. The key aspect of the underlying concept is to share coarse-grain resources in order to support familiar applications, instead of redesigning applications or trying to adapt applications into a highly constrained design space.

In essence, Platform Composition is creating a distributed logical platform that replaces the underlying individual pieces of pervasive computing technology. A modern laptop computer is made up of tightly-coupled storage, display, processing, and I/O capabilities, while a distributed composed platform provides these same basic resources, except sourced from a diverse set of devices. The potential downside, of course, is that the distribution process will either become overly confusing or complicated for users, or decrease system effectiveness due to increased communication latencies.

4 Composition Framework

The *Composition Framework* prototype presents a concrete implementation of the Platform Composition concept in order to better to understand its relationship with higher-level applications, underlying resource sharing services, and the overall user experience. At its core, the Composition Framework is a distributed message passing framework that has modules for user interfaces, device/service discovery, and service integration. The framework exists as a thin middleware layer that is used to orchestrate service connections among devices; however, once services are made available through the operating system, the composition interface does not play a role in using the applications themselves, which are layered directly on the exposed services.

The primary user interface employs a graphical join-the-dots metaphor, depicted in Figure 5. To effect a connection in the system, the user simply draws a line from the core service (small circles) to the target device (larger enclosing circles); likewise, in order to provide the user with feedback on the state of the system active, service connections are represented by directed links between the source service and device. Services can be disconnected by dragging a line across the service (metaphorically "cutting" the connection). Users are also able to invoke various configuration and diagnostic operations through specialized gestures. A system-tray icon is also available, primarily for use on laptop and desktop systems. The graphical interface is similar to others that have supported composition for both objects [5][23] and events [1][15]. Addiontally, the interface provides the ability save and restore predefined compositions, as well as automatically suggesting composition candidates.

The Composition Framework architecture consists of four major components: framework core, user interface, network discovery, and service modules. DBUS, a standard message passing infrastructure, is used to facilitate communication between the modules. The core components are implemented in Java, with various specific components utilizing a number of other languages and interfaces, as required. The system supports both Linux and Windows operating systems, and has been used on a variety of standard computing platforms such as Ultra-Mobile PCs (UMPCs), laptops, desktop systems, and projection displays.

Each individual service for sharing a resource is specified by an XML service descriptor file, which encodes basic properties of the service (name, icon, etc.), and provides details on how to probe, invoke, monitor, and disconnect the service. Some services, such as storage sharing, are implemented using asynchronous operating-system calls, while others, such as display sharing, are implemented by invoking a standalone client process. Services are handled using an explicit client/server model based on commonly available standard systems:

Clipboard sharing is realized using the *synergy* [29] system, which enables clipboard and mouse sharing among a group of systems. This system seamlessly integrates with the system clipboard, providing a virtually transparent mechanism for users to share information.

Storage sharing is implemented using standard SMB-based storage capability built-in to Windows and Linux platforms. Automation of standard command-line interfaces are used to access storage sharing, and the resulting client is presented to the system as an integrated drive or folder mount point.

Display sharing is built on the standard VNC protocol, using the standard VNC protocol [26], supporting multiple client and server implementations. This implementation allows easy access of the display between systems. Furthermore, on

Windows platforms, MaxiVista [11] is used to enable extending a display surface across multiple devices.

Peripheral sharing utilizes USB Redirector [12] to share USB devices between devices. For example, speakers attached to one device can be used to play music sourced from another, or a tablet input device wired into a tabletop can be used to augment traditional laptop input modalities.

These implementations each represent specific examples of how existing technologies can be used to implement the associated service, and could be easily replaced by alternate implementations if/when they become available.

4.1 Evaluation

The Composition Framework has been tested in both a laboratory based experiment and by the core research team over a period of several months. The laboratory experiment included eighteen participants recruited by convenience sampling from our corporation as well as personal friends. The participants were evenly distributed across gender, and 65% were between 20 and 29 years of age and 35% between 30 and 39. The experiment tested a well-structured sequence of tasks involving compositions between three systems. The extended core usage centered around five researchers using the system in a typical conference room environment. In both cases, the devices used consisted of a combination of laptops, ultra-mobile PCs (UMPCs), and desktop systems (attached to either a large-screen display or projector). These experiences have provided a wide range of users' reactions to the basic concept and allowed insights into the end-user benefits.

Overall, it is clear that users can relate to and understand the basic concept of Platform Composition, and find the Composition Framework user interface intuitive and easy to use. Participants from the user study were able to make use of the unified mechanisms provided by the Composition Framework to manipulate the state of the services involved in a composition. One user commented that the system was "*really easy to understand, learn and use,*" while another said that the GUI was "*a good visualization of what's going on: makes user more comfortable with sharing to know exactly what's shared and if it shared successfully.*"

Routine interactions with groups of up to five people have further revealed insights into the effectiveness of the system for collaborative tasks. The basic ad-hoc discovery mechanism and user interface have shown to be useful in lowering the barrier for initiating multi-device collaborative tasks. Based on this ease of sharing, users were able to export documents for collaboration in group settings, instead of requiring peers to gather around a single monitor or simply "talking to" a relevant document. The available alternative was generally using a standard VGA cable for projection, or emailing the document to all parties involved. In essence, users were able to share and engage their coworkers without as much interference from the underlying technology. One interesting use practice emerged as a result of being able to easily share *two* resources from the unified interface, discussed more fully in the next section.

Several conceptual difficulties with service sharing were observed relating to display and storage sharing. The most pronounced confusion was experienced with display sharing, where users would quickly become confused about which display was shown where – since many computer desktops looks very similar, and the physical separation of screens was no longer effective at differentiation. Similarly, users were somewhat confused at the difference between an underlying shared directory and the on-screen window that represented the share: the mistaken assumption was that closing the window prevented access to the underlying data. Accordingly, users commented *"All the connection lines can cause confusion,"* underscoring some of the difficulty with comprehending resource sharing.

Another deficiency apparent with shared services as they stand is the inability for a user to control a remote client, when they push a local service remotely: e.g., if they share their display, making it available on another computer. The basic problem is that client/server programs typically assume interaction from the client side, and generally don't support server-side control of the client. Specifically, the case of remote client control for VNC was an issue, since a user could push their screen to a remote display, but could not position, scroll, or maximize/minimize the client window. There are a number of immediate ways around this problem, specifically using another VNC connection to control the remote display or remoting a mouse or other input device over USB, but these solutions only offer limited relief to the underlying problem. This becomes more relevant in a collaborative environment because users may need to control the client to mediate limited client resources: e.g., switching between shared displays or tiling clients appropriately.

4.2 Observational Contributions

Based on experiences with the Composition Framework, described above, Platform Composition contributes a number of powerful properties to collaborative environments, even though it is constructed using a number of standard services and protocols. First, since users can easily create new connections to facilitate sharing, it becomes feasible for them to create multiple connections while previously they might only have created one due to the perceived overhead of interaction. Second, the interaction of standard system services with operating environments and applications allows users to fluidly adjust their composition space to meet immediate group goals. Essentially, these contributions stem from the lower bar for composition operations which otherwise are too onerous for users to consider in the midst of dynamic group interaction – a primary design goal of the system.

As mentioned above, empowering users with a unified and easy-to-use sharing interface crystallized new usage patterns that were not previously present. For example, when showing a presentation to a small group the most straightforward way to realize a composition system is to enter presentation mode on the local device and then create a composition with the projector (assuming the existence of a dedicated projection server). However, it was discovered that utilizing storage sharing to push a document to a remote display and then pull the remote display locally allowed for increased functionality. This shift allowed a user to effectively present a document to a group of people without unnecessarily exposing their system to the audience, e.g., their local screen was still private, and only the intended document was shared, although under their control. Furthermore, this sharing approach enabled other users to directly access the presentation locally on personal devices and could flip ahead or review pages, a usage not possible with a simple projection or display-service export. Another advantage was that the presentation program (PowerPoint) executed locally and therefore did not incur any display artifacts for the audience resulting from display sharing, typically present for animations and embedded videos. While this example utilizes infrastructure support, the shift in usage highlights the power of Platform Composition.

Since they are built on common existing standards, platform services can be fluidly configured in the system without significantly interfering with individual applications or the user's overall operating environment. Within the wealth of available platform services, sometimes it will be better to share individual files using the clipboard, other times they might want to access a complete file share, and sometimes they would want to interact using display sharing. Since Platform Composition utilizes resource sharing that is intimately integrated with standard system interfaces, users can quickly switch between modalities using familiar techniques like iconifying a window, pressing a keyboard shortcut, or drag-and-drop between windows. Once the initial connections are set-up, the composition mechanism itself does not play an active part in interactions. This fluidity of accessing sharing models transcends individual applications and allows users to share their resources in a manner that matches the sub-task of the overall group.

Satisfying another design goal, the Composition Framework was highly successful in supporting different applications. Photograph sharing was easily accomplished using built-in photo viewers and by sharing photographs through display sharing. Existing audio and video playback applications, such as Real Player or Windows Media Player, could easily operate on data exported through shared drives. Furthermore, Power Point easily ran in the environment and was used for shared presentations, while accompanying storage sharing was able to share the underlying presentation itself. However, the underlying services were not able to support video over a display-sharing channel, and interactions with more advanced applications such as Photoshop suffered from network latencies, both for the underlying storage and display sharing.

5 Future Work

Based on the concepts outlined in the previous sections, a number of avenues for future work become apparent. First, a more comprehensive evaluation of the various platform sharing techniques and how they interact would provide valuable additional insight on how sharing is realized using Platform Composition.

In addition to the individual underlying platform services to enable resource sharing, there is additional state present in the system that could be used to facilitate interactions among users. For example, data from the process table or files held open by programs, such as an image editor, could be used to provide a remote user with a more precise window into the first system's exported storage share (by opening up a window directly to a sub-folder in the larger hierarchy, instead of simply to the root of the storage share). Similarly, if one user is browsing a specific web page and would like to share it with their colleague, state sharing would allow another user to easily access the same page (currently, this can be accomplished by sharing the clipboard and using it to transfer the URL, but this introduces another step).

Currently, the security model employed by the Composition Framework relegates the security policy to individual services. Integrating authentication and access control with the core system has the advantage of presenting a unified front to the user as well as, similar to above, enabling more seamless switching between services. For example, successful authentication with one service might imply implicit authentication to another, reducing the number of unnecessary steps towards the final composition. Privacy falls into a similar situation where it is left up to social conventions or the pre-existing system configuration to manage users' privacy, an arrangement that could possibly benefit from a more managed approach.

As mentioned previously, an interesting aspect of resolving ad-hoc references is utilizing physical interfaces to invoke compositions – towards that end, realizing a multimodal interface, combining elements of an on-screen, physical, and speech interface, offers several compelling properties. One complaint with the GUI interface was *"When my hand is not free, I cannot use GUI. I have to rely on mouse a lot to use GUI, which is inconvenient in many environment,"* highlighting another problem with traditional on-screen interaction. Physical interfaces such as Near Field Communication (NFC) [24,23] can partially address the security problem between devices enabling physical access-based security policies. Furthermore, a speech interface would be applicable for small devices since it would have a reduced dependence on the screen. Overall, both physical and speech interfaces are attractive in a social environment because they provide transparency: one user can easily see or hear what another is doing.

Along these lines, recommendation systems can be used to help the user better decide which services to compose together. Such a module could look at the available devices and service, currently running applications, people involved or nearby, and other sources of context to suggest a specific sharing configuration. For example, if two colleagues often and typically share their storage and display when they are together, it would be natural for the system to recommend or automate the process.

6 Conclusions

The Composition Framework prototype has demonstrated the effectiveness of the Platform Composition concept, exploring how it fits into the framing of the Pervasive Collaboration application domain. Unlike many other collaborative systems, the necessary enabling concepts manifest themselves in a very lightweight manner, removing many potential barriers-to-adoption for platform-level composition. The basic principle of service composition, combined with an understanding of the spectrum of application sharing modes, makes Platform Composition well suited to support collaborative work on emerging mobile devices, and further integrating them with existing infrastructure to build effective systems that support work practice.

They key contributions outlined in this paper are a deeper understanding of collaborative systems by first providing a design space within which to place such systems, and then identifying several non-obvious benefits of the Platform Composition approach. These contributions highlight how emerging pervasive computing environments based on common mobile devices are capable of supporting highly dynamic group usage models. Furthermore, based on these experiences, a number of observed challenges highlight ways that existing pervasive systems can be evolved to better support collaborative usage models.

References

1. Ballagas, R., Szybalski, A., Fox, A.: Patch Panel: Enabling Control-Flow Interoperability in Ubicomp Environments. In: Proceedings of the 2nd int. Conf. on Pervasive Computing and Communications (2004)
2. Bardram, J.: Designing for the dynamics of cooperative work activities. In: Proceedings of the 1998 ACM Conf. on Computer Supported Cooperative Work (1998)
3. Biehl, J.T., Baker, W.T., Bailey, B.P., Tan, D.S., Inkpen, K.M., Czerwinski, M.: Impromptu: a new interaction framework for supporting collaboration in multiple display environments and its field evaluation for co-located software development. In: Proc. of the 26th Conference on Human Factors in Computing Systems (2008)
4. Clawson, J., Voida, A., Patel, N., Lyons, K.: Mobiphos: A collocated-synchronous mobile photo sharing application. In: Proc. of MobileHCI 2008 (2008)
5. Ducheneaut, N., Smith, T.F., Begole, J., Newman, M.W., Beckmann, C.: The orbital browser: composing ubicomp services using only rotation and selection. In: CHI 2006 (2006) (Extended Abstracts)
6. Edwards, W.K., Newman, M.W., Sedivy, J.Z., Smith, T.F., Balfanz, D., Smetters, D.K., Wong, H.C., Izadi, S.: Using speakeasy for ad hoc peer-to-peer collaboration. In: Proc. of the 2002 CSCW (2002)
7. Egi, H., Ohsuga, N., Nakada, A., Shigeno, H., Okada, K.: DACS: Distance Aware Collaboration System for Face-to-Face Meetings. In: Proc. of the 2004 Symposium on Applications and the internet-Workshops (2004)
8. Forlines, C., Shen, C., Wigdor, D., Balakrishnan, R.: Exploring the effects of group size and display configuration on visual search. In: Proceedings of the 2006 20th Conference on CSCW (2006)
9. Hazas, M., Kray, C., Gellersen, H., Agbota, H., Kortuem, G., Krohn, A.: A relative positioning system for co-located mobile devices. In: Proceedings of the 3rd international Conference on Mobile Systems, Applications, and Services (2005)
10. Hinckley, K.: Distributed and local sensing techniques for face-to-face collaboration. In: Proc. of the 5th international Conference on Multimodal interfaces (2003)
11. http://www.maxivista.com/
12. http://www.incentivespro.com/usb-redirector.html
13. Johanson, B., Fox, A.: Extending tuplespaces for coordination in interactive workspaces. Journal. System. Software (2004)
14. Johanson, B., Ponnekanti, S., Sengupta, C., Fox, A.: Multibrowsing: Moving Web Content across Multiple Displays. In: Abowd, G.D., Brumitt, B., Shafer, S. (eds.) UbiComp 2001. LNCS, vol. 2201, p. 346. Springer, Heidelberg (2001)
15. Johanson, B., Hutchins, G., Winograd, T., Stone, M.: PointRight: Experience with Flexible Input Redirection in Interactive Workspaces. In: Proc. of UIST (2000)
16. Ahn, J., Jeffrey, S., Pierce, J.S.: SEREFE: Serendipitous File Exchange Between Users and Devices. In: Proceedings of the 7th international conference on Human
17. Lauwers, J.C., Lantz, K.A.: Collaboration awareness in support of collaboration transparency: requirements for the next generation of shared window systems. In: Proc. of the SIGCHI Conference on Human Factors in Computing Systems: Empowering People (1990)
18. Li, D., Li, R.: Transparent sharing and interoperation of heterogeneous single-user applications. In: Proceedings of the 2002 ACM CSCW (2002)
19. Lyons, K., Want, R., Munday, D., He, J., Sud, S., Rosario, B., Pering, T.: Context–Aware Composition. In: HotMobile 2009 (2009)

20. Miller, R.C., Myers, B.A.: Synchronizing clipboards of multiple computers. In: Proceedings of the 12th Annual ACM Symposium on User interface Software and Technology (1999)
21. Myers, B.A., Stiel, H., Gargiulo, R.: Collaboration using multiple PDAs connected to a PC. In: Proceedings of the 1998 ACM Conference on Computer Supported Cooperative Work (1998)
22. Newman, M.W., Ducheneaut, N., Edwards, W.K., Sedivy, J.Z., Smith, T.F.: Supporting the unremarkable: experiences with the obje Display Mirror. Personal Ubiquitous Computing (2007)
23. Newman, M., Elliott, A., Smith, T.: Providing an Integrated User Experience of Networked Media, Devices, and Services Through End-User Composition. In: Indulska, J., Patterson, D.J., Rodden, T., Ott, M. (eds.) Pervasive 2008. LNCS, vol. 5013, pp. 213–227. Springer, Heidelberg (2008)
24. Pering, T., Anokwa, Y., Want, R.: Gesture connect: facilitating tangible interaction with a flick of the wrist. In: Proceedings of the 1st international Conference on Tangible and Embedded interaction (2007)
25. Pering, T., Ballagas, R., Want, R.: Spontaneous marriages of mobile devices and interactive spaces. Commun. ACM (2005)
26. Richardson, T., Stafford-Fraser, Q., Wood, K.R., Hopper, A.: Virtual Network Computing. IEEE Internet Computing (1998)
27. Roseman, M., Greenberg, S.: TeamRooms: network places for collaboration. In: Proceedings of the 1996 ACM Conference on CSCW (1996)
28. Satyanarayanan, M.: The evolution of Coda. ACM Trans. Comput. Syst. 20, 2 (2002)
29. Schoeneman, C.: Control everything from one place with Synergy. Linux J, 108 (April 2003)
30. Shen, G., Li, Y., Zhang, Y.: MobiUS: enable together-viewing video experience across two mobile devices. In: Proc. of the 5th international Conference on Mobile Systems, Applications and Services (2007)
31. Stødle, D., Bjørndalen, J.M., Anshus, O.J.: The 22 megapixel laptop. In: Proceedings of the 2007 Workshop on Emerging Displays Technologies: Images and Beyond: the Future of Displays and interacton (2007)
32. Streitz, N., Geißler, J., Holmer, T., Müller-Tomfelde, C., Reischl, W., Rexroth, P., Seitz, P., Steinmetz, R.: i-LAND: an interactive landscape for creativity and innovation. In: Proceedings of the SIGCHI conference on Human factors in computing systems, pp. 120–127. ACM, New York (1999)
33. Szentgyorgyi, C., Terry, M., Lank, E.: Renegade gaming: practices surrounding social use of the Nintendo DS handheld gaming system. In: Proceeding of CHI 2008 (2008)
34. Teege, G.: Users as Composers: Parts and Features as a Basis for Tailorability in CSCW Systems. Computer Supported Cooperative Work (October 1999)
35. Voida, S., Edwards, W., Newman, M.W., Grinter, R.E., Ducheneaut, N.: Share and share alike: exploring the user interface affordances of file sharing. In: Proceedings of the 2006 CHI (2006)
36. Want, R., Pering, T., Danneels, G., Kumar, M., Sundar, M., Light, J.: The Personal Server: Changing the Way We Think about Ubiquitous Computing. In: Borriello, G., Holmquist, L.E. (eds.) UbiComp 2002. LNCS, vol. 2498, p. 194. Springer, Heidelberg (2002)
37. Want, R., Pering, T., Sud, S., Rosario, B.: Dynamic Composable Computing. In: Proc. of HotMobile 2008 (2008)

Askus: Amplifying Mobile Actions

Shin'ichi Konomi[1,2], Niwat Thepvilojanapong[1], Ryohei Suzuki[3],
Susanna Pirttikangas[4], Kaoru Sezaki[2], and Yoshito Tobe[1]

[1] School of Science and Technology for Future Life, Tokyo Denki University,
and JST/CREST, Tokyo 101-8457, Japan
[2] Center for Spatial Information Science, University of Tokyo, Chiba 277-8568, Japan
[3] Institute of Industrial Science, University of Tokyo, Tokyo 153-8505, Japan
[4] Department of Electrical and Information Engineering, University of Oulu, Finland

Abstract. Information sharing has undeniably become ubiquitous in the Internet
age. The global village created on the Internet provides people with instant access to information and news on events occurring in a remote area, including access to video content on websites such as *YouTube*. Thus, the Internet has helped
us overcome barriers to information. However, we cannot conceive an event
happening in a remote area and respond to it with relevant actions in a real-time
fashion. To overcome this problem, we propose a system called *Askus*, a mobile
platform for supporting networked actions. *Askus* facilitates an extension of the
conceivable space and action by including humans in the loop. In *Askus*, a person's request is transmitted to a suitable person who will then act in accordance
with the request at a remote site. Based on a diary study that led to detailed understanding about mobile assistance needs in everyday life, we developed the
Askus platform and implemented the PC-based and mobile phone-based prototypes. We also present the results from our preliminary field trial.

1 Introduction

Today, a large number of people interact and share information through various media
such as blogs, wikis, social networking services, video sharing sites (e.g., *YouTube*),
and folksonomy-based services. The Web 2.0 phenomenon has shown that the World
Wide Web is no longer limited to being a platform for a passive consumption of information. Rather, it is now a networked medium that can amplify [18] a host of practices such as peer-to-peer interaction, participation, and community action.

Mobile and pervasive computing could, in a manner similar to that of Web 2.0,
provide a platform for active social practices. Existing trends of mobile phone usage
suggest the possibility of using mobile computing as a platform for networked actions. For example, in his discussion of *smart mobs*, Rheingold [21] describes the use
of mobile text messaging in collective activism in the Philippines while Ito and Okabe
[12] describe *keitai* communication practices in Japan; these examples suggest that
mobile phones can be used to quickly organize significant collective action, and to
connect strangers as well as friends.

Despite the ubiquity of wireless network access, we can easily imagine situations
in which physical constraints could be frustrating. Consider the following examples: a

H. Tokuda et al. (Eds.): Pervasive 2009, LNCS 5538, pp. 202–219, 2009.

participant of an academic conference cannot be physically present at all the interesting sessions that are being conducted simultaneously, it is not easy for travelers on a subway platform to locate the least-crowded car before the train arrives at the station, urbanites cannot operate the up/down arrow buttons outside an elevator until they walk up to the elevator door. These examples bring to the fore the challenge in integrating a user's physical and social contexts with the digital media's capability in order to connect people and spaces across physical boundaries.

In this paper, we propose an integrated mobile platform for supporting collective actions and information capture called *Askus*. This platform allows users to request friends and strangers in a relevant geographic area to capture information or perform other *lightweight* actions using mobile devices. In order to better understand how a technology like *Askus* can be integrated with our everyday life, we first discuss our diary study that had suggested the importance of awareness and privacy. The *Askus* platform considers these factors by the provision of a *task matching protocol* that incorporates location, time, and the users' current and historical characteristics.

We have implemented two prototypes of the *Askus* platform. The first prototype was an experimental system that operates on location-aware mobile computers. The second prototype was designed for scalability and consequently, operates on mobile phones. We tested this mobile-phone-based prototype with the aid of 20 users in the central area of Tokyo to examine user experiences, which led to our discussions on the issues related to the tool design for supporting lightweight mobile actions.

2 Amplification of Mobile Actions

2.1 Theoretical Framework

Distributed Cognition can provide a theoretical framework for understanding socially distributed, embodied, and contextually embedded human actions in a mobile environment. Distribution can take place among people, between human minds and artifacts, or as an integration of both these dimensions of distribution [8]. This framework emphasizes the importance of the observation of human activity "in the wild" and the analysis of distributions of cognitive processes [11].

According to McLuhan's theory [18], all the people from different levels of society would be connected through technology, that is, the extensions of a man. The advances of technology could enable us to form a distributed computing awareness without a centralized control center. Here, the people on the streets are acting as nodes in this awareness, much like the Borg [26] in the famous Star Trek series. To design a medium for networked mobile actions, we must understand how the medium *"shapes and controls the scale and form of human action"* [18,p.9], therefore, we carried out both a diary study and a field trial so as to understand not only current practices but also how the medium can change practices.

A graceful human-human communication is indispensable in socially coordinated distributed actions and information capture. *Social Translucence* [7] is an approach that can be used to support digitally mediated social activities by considering visibility, awareness, and accountability. In mobile and pervasive systems, social information can be made visible in both physical and digital spaces; this introduces the additional challenge of integrating interactions in the physical and digital spaces.

Finally, we are not only concerned with the manner in which people accomplish tasks efficiently but also with the meaningfulness of their experiences. To understand the impact of mobile tools on collective practices in a broader context, we need to consider the roles of place and space in collaborative environments [10], and also the manner in which mobile tools produce *alternative spatialities* [6].

2.2 Preliminary Diary Study

There are very few studies that have focused on the need for and the requirements of mobile assistance in everyday life. However, an in-depth analysis of such needs and requirements is indispensable for an informed design of mobile tools that support mutual assistance among users who may or may not be co-located. Therefore, we undertook a preliminary diary study to explore the patterns in which urban adults could meaningfully use networking tools to obtain mobile assistance. Our diary study focused on the social aspects of *mobile action needs*, which complement the existing studies on mobile information needs [25], and daily information needs and shares [9].

Method. In order to achieve a comprehensive capture of the *in-situ* needs and requirements, we combined an hourly experience sampling method and a diary study. We recruited 11 male undergraduate/graduate students whose ages ranged between 19 and 30 (mean: 22.5, SD: 2.94) and asked them to maintain an hourly diary for a day. This participant pool reflects the fact that young adults in their twenties use mobile internet the most in Japan [19]. The objective of this extensive hourly study that lasted for a day was to inform the design of *Askus*. We expect that future, in-depth investigations into various population segments will complement and extend the limits of our preliminary study that was based on this specific pool. We requested the participants to record a diary entry whenever an event occurred. An event can be something that happens in the world around them or in their minds. The diary entry was to include the event description, time, place, co-present people, busyness, and the participant's feelings along with anything they wanted to ask or state. A drawback of diary studies is that the participants may forget to record diary entries or be selective with their reporting. In order to overcome this drawback, we asked the participants to record an entry at the beginning of each hour. To capture daily lives of urban adults, which can potentially be dynamic and eventful, we leaned toward frequent reporting. Participants used the alarm clock functionality of their mobile phones so as to not forget the hourly reporting task. In addition, we asked the participants to record a diary entry even when they were not mobile since they would possibly want to interact with people who may be mobile at the time. We told participants that they did not have to record a diary entry while they were asleep, and that they were allowed to write "none" when there was nothing to report. Finally, we conducted a short survey and an interview[1]. In the interviews, we asked participants for clarification of any unclear entries, and how they might use a mobile phone-based tool for sending requests to relevant friends and strangers.

---

[1] Each interview lasted for an hour except for a half-hour interview with a participant who had to leave urgently (10.5 hours in total).

In each diary entry, participants recorded a relevant event along with additional information to answer the following eight questions[2]:

1.   Where are you?
2.   Who is around you?
3.   What do you want to say to some people around you? Who are they?
4.   What do you want to say to some people at a remote location? Who are they?
5.   What do you want to request of some people around you? Who are they?
6.   What do you want to request of some people at a remote location? Who are they?
7.   How busy are you?
8.   What is your mood?

We coded the data from questions 3-6 into 12 categories according to the following three dimensions:

(1)   *Physical distance*: close or remote
(2)   *Social relation*: friends, strangers, or anyone
(3)   *Content type*: requests for action[3], requests for information/data, or non-request messages such as greetings and comments

Results. Our study generated 321 diary entries with an average of 29.2 entries per person (min 23, max: 35, SD: 3.92). Participants articulated 240 messages in response to questions 3-6, with an average of 21.8 messages per person (min: 7, max: 63, SD: 17.6). These 240 messages included 119 (50%) *requests for actions*, 33 (14%) *requests for information/data*, and 88 (37%) *non-request messages*. This suggests that many of the participants' requests could not be addressed by merely improving information access. The frequencies of the types of messages, physical distances and social relations are shown in Table 1. Participants were able to record diary entries hourly; however, participants' comments suggested that they felt it rather demanding to record and entry every hour.

Table 1. Frequency of messages for content types, physical distances and social relations

Message types	Nearby			Remote		
	Friends	Strangers	Anyone	Friends	Strangers	Anyone
Requests for actions	43 (18%)	24 (10%)	0 (0%)	37 (15%)	7 (3.0%)	8 (3.3%)
Requests for information/data	6 (2.5%)	2 (0.83%)	2 (0.83%)	6 (2.5%)	0 (0%)	17 (7.1%)
Non-request messages	44 (18%)	11 (4.6%)	0 (0%)	24 (10%)	1 (0.42%)	8 (3.3%)

One of the largest message categories was requests for actions sent to nearby friends (18%). These messages may be requests to a specified friend, to any one of a group of nearby friends, or to a whole group, such as *"Please be quiet"* (to a friend), *"Can [any one of] you return the keys?"* (to fellow students), and *"Let's hurry up [and finish the meeting soon]"* (to a group of meeting participants). A related category is requests for information/data from nearby friends (2.5%). For example, one of the participants wanted to obtain information about how much progress his colleagues had made on their research project. Such requests were often directly prompted by ongoing conversations and interactions with friends.

---

[2] The original questions were posed in Japanese, and they, as well as any diary entries, have been translated into English for the paper.

[3] Requests that cannot be satisfied by merely providing information. They often ask for responses that involve physical efforts to go, make, find, buy, bring, wait, stop, call, etc.

The other largest category was non-request messages to nearby friends (18%). These are greetings, thanks, comments, complaints, and other messages, such as *"Thanks for the meal"* (to a friend), *"It is hot in this room, isn't it?"* (to a colleague), and *"Is it really my turn [to wash the dishes]?"* (to a younger brother). Although these messages are not requests, some of them could have the effect of influencing other people's actions. Non-request messages to remote friends (10%) were a similar assortment of greetings, thanks, comments, complaints, etc. These messages were often written when participants were not involved in interactions with nearby people.

The third largest category was requests for actions directed to remote friends (15%). These are requests to do something at a remote site, to join the requester and help with something, or to do something in the future, such as *"Please turn on the heater"* (to a mother), *"Please turn on a PC"* (to a colleague), *"Please keep the house unlocked"* (to parents), *"Come here, I'd like to play a game with you"* (to a friend), and *"Please take care of my part-time job tomorrow"* (to a fellow part-time worker). These requests were often prompted when participants needed to physically access remote people, things, and places; desired help from experts who had the knowledge and skills to accomplish difficult tasks; or were not interacting with nearby people. A related category is requests for information/data from remote friends (2.5%), including messages such as *"Do you want me to turn the lights off?"* (to colleagues who were out for a quick meal when a participant was leaving the office).

Participants also wanted to make requests of strangers. Actions requested of nearby strangers (10%) were often small things that could be done relatively easily and quickly. These requests included *"Please make room for me"* in a crowded train and *"Please have the elevator wait for me on the first floor."* Actions requested of remote strangers (3.0%) were often more complex and time-consuming. For example, one of the participants wanted strangers at a remote site to look for lost jewelry. There were only a couple of requests for information/data directed to strangers. The 12 non-request messages to strangers included compliments, warnings, and complaints, such as *"This book is expensive"* to a salesperson.

More remote requests and messages were directed to friends and familiar people than to strangers, and the requests made of remote strangers often dealt with things anyone could do or cases in which the participants did not know who would be able to perform an action (e.g., *"Buy tea and chips, and then bring them to me"*).

Three of the participants said it would be easier to communicate if they had more information about people, including their location, personal information, and status (e.g., how busy they were). At the same time, participants had privacy concerns about the obligatory disclosure of personal information. There were also concerns about receiving too many requests or irrelevant responses. Asking can be a difficult task if one must carefully and manually determine the right people to ask based on various types of information; it can be burdensome to explain what to do, find out a person's skills/motivations, and avoid any misunderstanding. Participants said that they sometimes think asking for help with something is more of a burden than doing it by themselves.

The participants had different expectations about a mobile phone-based tool for sending requests to friends and strangers. The most prominent centered around the possibility of easily and safely communicating with *strangers* and asking them to do various things. Five participants said they would or might respond to a request from a stranger to say whether a train is crowded. Major factors in providing such a voluntary contribution included a context in which people are not busy (i.e., a train ride), as well as the ease of responding to such a light request. Small contributions of this kind from strangers could collectively provide useful help for various people.

2.3 Scenario

Our preliminary diary study motivates design of a lightweight tool that allows people to easily ask things in the right way from the right people and respond to requests with ease. An interesting question is if such a tool can facilitate networked actions among friends and strangers in everyday life scenarios, by leveraging *situation experts and people with weak ties* [5]. We have developed the following scenario, which will be used to guide the design of the *Askus* platform:

Amy is a graduate student. She is paying a visit to the Jupiter Conference Center along with her colleague Meg to attend a large multi-track conference that pertains to her subject of study. The next sessions of the conference will start shortly, and Amy is interested in two sessions: Session A that is being held on the first floor and Session F that is being held on the fourth floor. Since Session A is very crowded, she wonders if Session F is less crowded. Meanwhile, Meg is attending the not-too-crowded Session E that is being held on the fourth floor.

Amy launches the Askus application on her mobile phone and inputs a query wanting to know if Session F is crowded. The system then initiates a search for relevant people based on location, busy/available status, social networks, past experiences, and reputation and recommends that she forward the request to Meg and BB. Meg is a friend of Amy, while BB is a friend of a friend of Amy. Amy does not know BB, but he is in Session F, his status is in the available mode, and he has a good reputation score. Amy chooses to ask BB, and he quickly responds with "it's pretty crowded, but looks like there are several seats still available." Amy asks BB if he can reserve a seat for her. He replies with "Sure," and places his conference bag on a vacant seat and informs her about the position of the seat.

Amy then starts walking toward the elevator thinking it would save her some time if someone was to push the elevator button even before she gets near it. She inputs a query for the same in Askus, which then forwards her request to people who are present near the elevator area. One of these people, Jim, following an interaction similar to the one described above, pushes the up-arrow button for her (in doing so, he earns a small number of points that are redeemable for purchase of books). When Amy reaches the elevator door, it opens just in time for her to enter. She walks into Session F, finds the "reserved" seat, thanks BB, and takes her seat.

The conference ends. Amy and Meg decide to go downtown for dinner. When they arrive at a subway station, Amy uses Askus to find the least-crowded car in the next subway train. Several strangers on the arriving train respond to her query. Amy and Meg board the 3rd car since a few people had replied that this car had vacant seats.

The scenario suggests that users must be able to easily search for relevant people considering various contextual factors. We therefore designed a client-server platform called *Askus*, which considers social matching techniques [27] so as to recommend people on the basis of distances, statuses, success rates, response time, and reputation scores. The scenario also motivates a design that combines automatic recommendation and visualization for supporting manual selection.

3 The *Askus* Platform

The *Askus* platform has a client-server architecture. An *Askus* server is responsible for many procedures including maintaining users' information, receiving and issuing requests, deciding or recommending agents who carry out the requested task, and so on. The server is capable of servicing multiple tasks simultaneously as shown in our implementation and experiments. If necessary, multiple servers can work in parallel in order to distribute heavy loads.

Each user who acts as both a requester and agent uses an Internet-enabled client device (e.g., mobile computer, PDA, cellular phone) to interact with the server. Although the user can use a mobile computer to access the *Askus* server, a small and light portable device such as a mobile phone is more appropriate for outdoor usages. It is desirable that the client device be equipped with GNSS-enabled devices and/or any outdoor localization systems [22]. Indoor localization systems [1,24] can be included, if indoor usages of *Askus* are preferred. Client devices are used to register personal information, request a task, receive a request, and report the task.

3.1 Task Matching Protocol

The *task matching protocol* finds appropriate agents that could potentially carry out each requested task. A *requester* uses a web API (Application Programming Interface) provided by *Askus* to input a *task* and a *place* (*L*) where the task should be carried out. The task needs to be input in the natural language thereby ensuring a high degree of flexibility while requesting. Numerous text processing tools (e.g., [17]) are available that extract the name of the place from the text input by the *requester*. In the scenario described in Section 2.3, the request could have been "Check whether session F which is held in room 405 is crowded," out of which "Room 405" is extracted as the name of the place.

In the event that multiple agents are available, which is possible for a given task, *Askus* chooses a particular agent on the basis of a *Score* that is calculated using the following equation:

$$Score = s \left(k_d d + k_p p + k_t/t + k_r r \right), \tag{1}$$

where k_d, k_p, k_t, and k_r are constants, d is the Euclidean distance between the center of the place L and the current position of a candidate agent, p is the success rate of a task performed by the candidate agent in the past, t is the average time taken by the candidate agent to finish the task, r is the reputation score of the candidate agent, and s is the status of the candidate agent. In our short field trial, we simply used two kinds of status, i.e., busy $(s = 0)$ and available $(s = 1)$ although a long-term user study would be needed to analyze users' practices with various status settings. Other status (e.g., "away," "out to lunch," "be right back") and the corresponding numeric values of s can be easily added to *Askus*. The initial values of p and t are set to *1*. The value of r, whose initial value could be any predetermined number, increases or decreases depending on whether an agent receives a positive or a negative reputation feedback.

There are two modes for choosing an agent, namely, a *manual* mode and an *automatic* mode. The former lists candidate agents in the descending order of their *Score*. The list may be categorized into friends and strangers if some of the candidate agents are registered as friends of the *requester*. In this mode, the *requester* may choose a candidate agent who is a friend as the agent, although his/her *Score* is not the highest. In the automatic mode, an agent with the highest *Score* is automatically chosen. *Askus* does not allow the *requester* from requesting from an agent whose *Score* is zero. If the *Score* value of all the listed candidate agents is zero, *Askus* informs the *requester* that no agent is available. If at least two candidate agents have the highest *Score*, *Askus* randomly chooses one of these candidate agents, or asks all the candidate agents with the highest *Score* to perform the task.

Askus forwards the request for a task to the selected agent through e-mail. After receiving the request, the selected agent replies whether he/she can carry out the task. Two modes can be used for task forwarding: *serial* forwarding and *parallel* forwarding. In the former, candidate agents are accessed one after the other, i.e., *Askus* will send the request to a new candidate agent, if a selected candidate agent either refuses to perform the task or fails to reply before the reply timer, before which a reply to the request has to be sent, expires. On the other hand, in parallel forwarding, *Askus* sends the request to multiple candidate agents simultaneously. The candidate agent who replies first will perform the requested task. In parallel forwarding, *Askus* also notifies all other candidate agents—except the candidate agent that replied first— that the agent has been determined; this is done so as to avoid redundant agents. In the case of serial forwarding, *Askus* notifies all requested candidate agents whose reply timer has expired after receiving a reply from a candidate agent.

The agent carries out the task and notifies the *requester* upon completing it. If, for some reason, the task is not completed, *Askus* can restart the *task matching* process. The details of the completed task are also included, if necessary, in the notification sent by the agent. In the scenario mentioned in Section 3, the notification could be "The second seat from the right in the last row has been reserved for you."

In addition, if a task can be divided into multiple sub-tasks, a requester may ask multiple agents to complete the entire task collaboratively. For example, imagine that Amy, who is attending the last session of the multi-track conference in Boston, realizes that she must submit the scholarship application at the department office in Los Angeles on the same day. In this scenario, *Askus* could help her by asking the first agent who could be working in the laboratory to print the document. Then, a second agent could deliver the printed document from the first agent's laboratory to the department office. In some cases, it may be better (faster or easier) to complete a task by relying on multiple agents.

4 Prototype Implementation

The *Askus* prototype has been developed as a client-server application. In the *Askus* server, we utilize MySQL to manage information about each client (location, status, user ID, nickname, and e-mail address) and task (task ID and place). We developed *Askus* clients that work on a mobile computer equipped with a GPS receiver, as well as a mobile phone. The *Askus* clients are responsible for: (1) registering/updating location and status, (2) requesting a task, and (3) responding to a requested task.

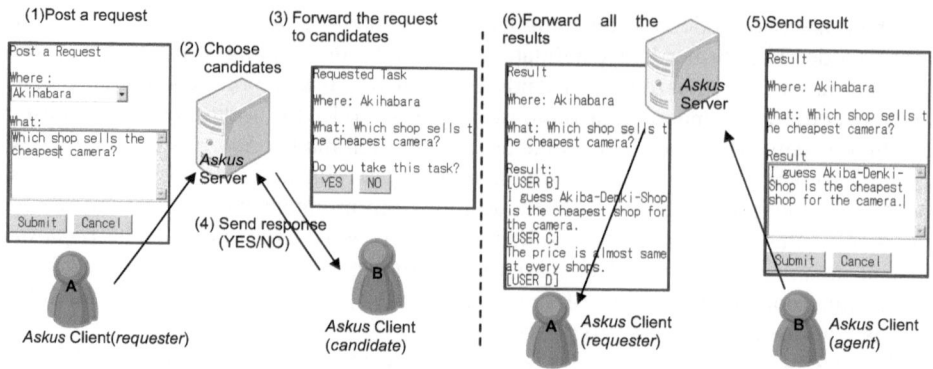

Fig. 1. Procedures to request/respond a task

Whenever a user changes his/her location, the mobile computer-based *Askus* client first determines the user's current area based on GPS coordinates and then sends the area description to the *Askus* server. To test the system with many users in the field, we also developed a mobile phone-based *Askus* client that allows users to update their location by choosing their current area/location from a drop-down list (see Fig. 1)[4].

Fig. 1 shows the procedures for requesting and responding to a task. As shown in Fig. 1, a requester uses a web API provided by the *Askus* server to submit a task and a place where the task should be carried out. After the client submits the request, the *Askus* server determines candidates to carry out the task using the Task Matching Protocol described in the previous section. In the prototype, the *Askus* server finds people who are not busy and are in the area of the requested task and lists the nicknames of those candidates. The requester looks through the list of candidates' nicknames and chooses at least one from the list using the Web API. This allows the requester to choose a friend or a user whose outcomes on previous tasks have satisfied him/her. Although we utilize location and status as the contextual information for Task Matching in our prototype, we can easily extend the prototype by storing historical information about task performance (e.g., task success rates and completion time) in a user profile, since such information can be extracted from the server log files. After obtaining the requester's list of chosen candidates, the *Askus* server sends the REQUEST message to the chosen candidates via e-mail[5]. A user who receives the REQUEST message sends his/her decision to either work on the task (YES) or not (NO). We call the user who submits a "YES" message and accepts the task an agent. When the agent finishes the task, he/she submits a result to the *Askus* server using the Web API. The *Askus* server then notifies the requester of the results with an e-mail message that links to a web page showing all the agents' results.

---

[4] Because of the restrictions imposed by the telecom industry, our software is currently unable to track users continuously by using mobile phones' GPS chips.

[5] Japanese mobile e-mail service is push-based, i.e., similarly to SMS, users receive immediate notification when a new message arrives.

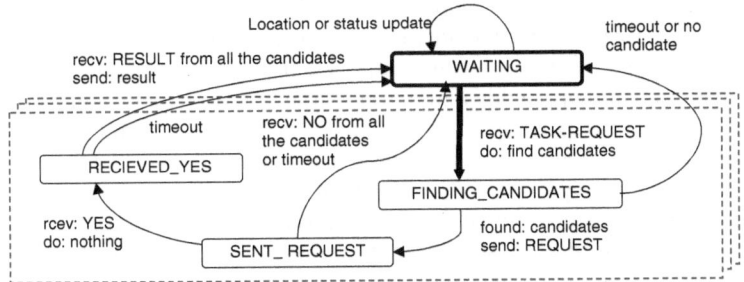

Fig. 2. State Transition Diagram for the *Askus* server

The *Askus* server operates according to the state-transition diagram illustrated in Fig. 2. The figure does not include the detailed descriptions of all possible fault situations. There are four states: WAITING, FINDING_CANDIDATES, SENT_REQUEST, and RECEIVED_YES, where WAITING is an initial state and the others in the dotted box are created for each task. When the *Askus* server receives a request for a task (TASK-REQUEST) from the *Askus* client, the server moves to the FINDING_CANDIDATES state and determines candidates according to the Task Matching Protocol. In the prototype, the *Askus* server chooses the candidates who are online and are in the acceptable range for the requested task. If there is no candidate in the acceptable range of the requested task, the state moves to WAITING; otherwise, the user selects nicknames from the list, the *Askus* server sends the request to the chosen candidate(s) via e-mail, and the state moves to SENT_REQUEST. Upon receiving a YES message from a candidate, the *Askus* server moves to the RECEIVED_YES state and waits for the result of the task. After receiving a result from the agent, the *Askus* server sends it to the client who requested the task, and the state moves back to WAITING. The prototype *Askus* server *immediately* sends a message to a client every time a result is generated for the client's request. In a future version of *Askus*, we can also incorporate a slower yet less obtrusive notification mechanism that waits for a certain amount of time before aggregating/summarizing multiple agents' results. Finally, the current prototype is designed to cope with problematic situations interactively (i.e., humans in the loop) rather than automatically. For example, if a requestor receives NO responses only, the system notifies it to the requestor although the *Askus* server could be extended with a smart mechanism that makes another round of requests automatically in such a case. We have successfully tested the mobile computer-based *Askus* using GPS receivers, with a small number of users distributed in two university campuses. Moreover, we experimentally deployed the mobile phone-based *Askus* to carry out a field trial.

5 Field Trial

5.1 Method

We recruited 20 participants through a course mailing list for computer science undergraduates and from among our personal contacts. All participants were required to own an internet-capable mobile phone, and their ages ranged between 19 and 25 (mean: 22.4, SD: 1.8). All participants but one were male. They constituted at least three separate groups of friends, which allowed us to simultaneously examine collaboration patterns among friends and among strangers in actual social networks. As in the diary study, this participant pool considers Japanese mobile internet demographics [19] and our expectation that the corresponding young adult population is where we may be able to pick "low-hanging fruits."

Participants gathered in front of a train station (JR Kanda Station) at 2 p.m. on a sunny Saturday afternoon. We used the first 20 minutes to introduce the system to the participants: The group practiced the operations to send, receive, and respond to requests. During this, there was little interaction across different groups of friends. We asked participants to go to one of five areas (Akihabara, Kanda, Ochanomizu, Jinboucho, and Awajicho) in an approximately *1km x 1km* region in central Tokyo, and then move freely within the specified region for about 100 minutes. All of the areas have many restaurants and shops, but each has a different image or specialty: electronics and *otaku* (Akihabara), businesspersons and bars (Kanda), universities and sports/music shops (Ochanomizu), secondhand bookstores (Jimboucho), and a place without a clear image (Awajicho). Participants interacted with one another using nicknames. Also, we told participants that they were allowed to ignore requests.

During the first half of the participants' time in the regions, one of the authors (Author 1) sent the following 10 requests to all participants so as to analyze participants' responses to different types of requests in different geographic areas[6]:

(R1) Check whether it is crowded around the entrance gate in Akihabara, Kanda, or Ochanomizu[7] station.

(R2) I get hungry. Recommend me a restaurant, which is not crowded now.

(R3) Find a trash can.

(R4) Find a restroom.

(R5) Push the up-arrow button of an elevator.

(R6) Get a free ad pocket tissues[8] from a distributor on the street.

(R7) Find a place to buy the morning edition of Asahi newspaper.

(R8) Check the price of iPhone at a nearby store.

(R9) Just walk around and enjoy yourself.

(R10) Put some money in a donation box at a convenience store.

Fig. 3. Using Askus on mobile phones

In this "structured session," participants merely responded to these requests, and did not send their own requests. Another author (Author 2) also participated in this session (21 users in total), accompanying and observing some participants.

When users respond to a request, they must push either a "YES" or a "NO" button. A "YES" response indicates that a user has accepted the requested task, and a "NO" response indicates otherwise. Then, the responding user can send text messages to report the result of the task or to explain why the request was rejected.

---

[6] The original requests were posed in Japanese, and they, as well as any participant comments, have been translated into English for the paper.

[7] From these three train stations, we selected the closest train station to a recipient.

[8] Pocket tissues bearing advertisements are often distributed free of charge in Japan.

The second half of the trial was a "free-form session" in which participants used the system as they liked. This session was used to have an initial look into participants' mobile practices including both requesting and responding. Three authors (Authors 1, 2 and 3) participated in this session (i.e., 23 users in total).

When participants returned to the train station at around 4 p.m., we asked them to fill out a survey about their experiences with *Askus*. Finally, we briefly interviewed each participant when we collected the survey sheet.

There are inherent limitations to this kind of short field trials because of the specific participant pool, the novelty effect, and the limited authenticity of the experimental settings. Yet, this preliminary trial is a first important step in the process of iteratively refining and improving the evaluation method and the design of the *Askus* platform. We expect that future long-term investigations into various population segments can complement and extend the limits of our preliminary look into this problem space.

5.2 Results

During the "structured session," which lasted about 46 minutes, 21 users responded to the 10 requests 180 times, with an average of 8.6 responses per person. Task acceptance rates varied according to the contents of the requests. As shown in Fig. 4, requests R1, R2, R3 and R4 had higher task acceptance rates (>70%) than the others. Requests R6 and R8, which required access to potentially hard-to-find people or things, had low acceptance rates (<10%). Request R10 had a higher acceptance rate than R8 even though donating money is more costly than checking a price. This may be because convenience stores are easier to find than mobile phone stores, and they usually have an easily located box for donations in front of the cashier. Fig. 5 shows the average response time by area and response type (YES/NO). Users often responded within several minutes of receiving a request (see Fig.6). The quickest response to each request was received in one or two minutes (see Fig. 7).

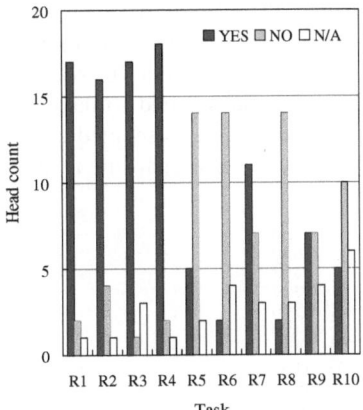

During the "free-form session," 23 participants generated 54 diverse requests with the average of 2.3 requests per person. The 23 participants received the 54 requests 165 times and responded to them 113 times, with the averages of 7.2 received requests and 4.9 responses per participant, and 2.1 responses per request. The 54 requests were targeted for the five different areas, with an average of 10.8 requests per area. Akihabara was the most frequent target and had 19 requests.

Fig. 4. Number of participants who responded YES/NO and did not respond (N/A) for each task

Ten of the free-form requests required physical actions: to search for things/people/shops, to go somewhere, or to talk to someone in person, and they generated 21 responses. Examples: *"Please get a flyer from a 'maid' in front of a train* station" (5 responses) *and "Please investigate the price of Blue Mountain coffee beans"* (no response). Thirteen free-form requests asked for information about people, and they generated 25 responses. Examples: *"Where are you?"*(1 response) *"Did you see anyone in a costume?"* (1 response) and *"Are you guys together?"* (1 response). Ten requests asked for information about a town, including

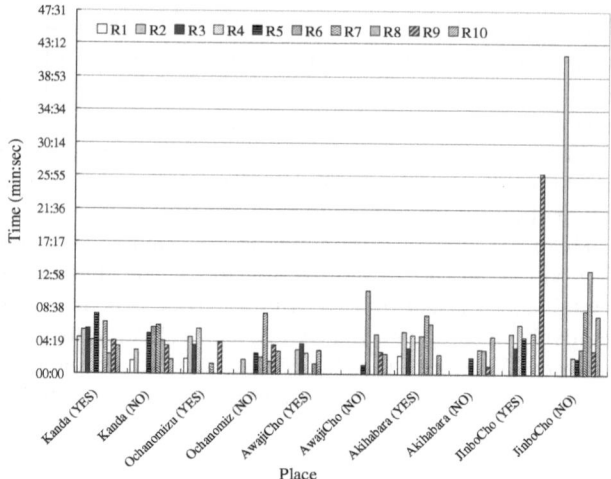

Fig. 5. Response time at each place

its people, shops, events, restrooms, smoking areas, special discount sales, places to have fun, and places to kill time, and they generated 18 responses. Seventeen requests asked about shops and other specific places in a town, including locations of vending machines, ATMs, karaoke, coffee shops, sushi restaurants, convenience stores, Japanese noodle restaurants, and mobile phone shops, and they generated 44 responses. A couple of these requests additionally sought information about how crowded a specific retail store or restaurant was.

Again, these are participants' initial reactions in a short field trial, and they must be interpreted with caution. Interestingly, participants used the system more creatively than we imagined. Playful social interactions and jokes were often observed. Also, one participant seemed to have appropriated the system as a location-enhanced chat tool. Some responses seemed contradictory, which raised the questions about trust and limited awareness regarding responders.

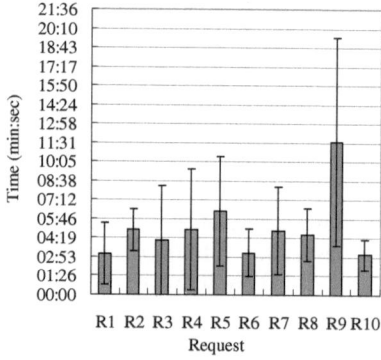

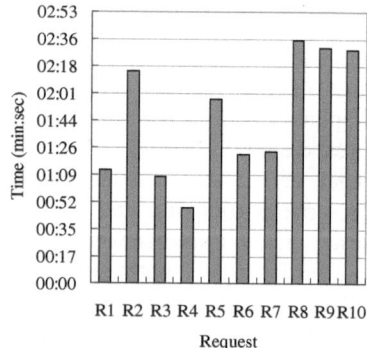

Fig. 6. Average response time of each task   **Fig. 7.** Minimum response time of each task

User Satisfaction. A five-point Likert scale (i.e., "very satisfied," "satisfied," "neutral," "dissatisfied," and "very dissatisfied") was used to rate users' satisfaction. We also asked the participants to rate future intent to use the system, perceived ease of use and usefulness. Eleven participants (55%) said they were satisfied with the system, five (25%) said they were dissatisfied, and four (20%) said they were neither satisfied nor dissatisfied. Participants' comments suggest that the system's usability, social experiences, and "request overload" influenced their perceived satisfaction. Their experiences seemed to be diverse: participants said *"I got responses for my request and thought [the system was] very good," "I was looking for a Japanese noodle restaurant (in Awajicho) and someone told me [where it was],"* and *"It was rather fun to respond to various people's requests,"* while others complained that *"I couldn't get the information I wanted"* and *"I received too many requests to respond."*

As for future intent to use the system, one participant (5%) said he strongly intended to use it and seven (35%) said they intended to use it. Ten (50%) said they were undecided about whether they would use the system, and two (10%) said they did not intend to use it. Participants commented that the system was *"interesting," "useful,"* or *"very useful when looking for something,"* that it *"can enhance the communications between people,"* and that one *"could do interesting things with it,"* but others said they received *"too many tasks,"* that their *"requests were all rejected,"* and that they *"can't really imagine the situations [in which I would] use it."*

More than half of the participants (55%) said the system was difficult to use, which seems likely to have affected their future intent to use the system. Six (30%) said it was easy to use, two (10%) said it was neither easy nor difficult to use, and one (5%) said it was very difficult to use. Text entry on mobile phones seemed to be burdensome, especially when participants received many requests at once. Because participants received notifications via mobile e-mail, they had to switch frequently between e-mail and web applications, which was perceived as a usability issue. Other concerns included battery life and the cognitive load of dealing with many requests.

Despite these usability concerns, more than half of the participants (55%) thought that the system was useful. Seven (35%) said that it was neither useful nor not useful, and only two (10%) said that it was not useful. The participants suggested that the system would be useful for obtaining weather and traffic information from remote sites, exchanging information with various people, and getting around in an unfamiliar city. In addition, one participant suggested that the system allows people to solve problems efficiently by relying on "nearby users." Participants also said that usefulness depended on the number of (kind) users and location resolution.

Response time. Six participants (30%) felt that people responded quickly to their requests and one (5%) felt that people responded very quickly. Seven participants (35%) said that people responded neither quickly nor slowly, and two (10%) said that the responses were slow. Thirteen participants (65%) said their requests were accomplished or accomplished very well. We also asked participants how many seconds they could wait for a response and still feel satisfied. Ten participants (50%) said 60 seconds and four (20%) said 120 seconds (mean: 156 seconds, min: 30 seconds, max: 1200 seconds, SD: 258 seconds).

Twelve participants (60%) said that they ignored other people's requests during the trial. Seven participants (35%) said that their requests were ignored. Participants ignored requests when they received too many requests or thought that requests were too ambiguous, not easy to do, or not likely to benefit the requester. Further, one participant thought it might be okay to ignore requests from strangers.

Communication Patterns. The system log files we captured during the "structured session" show that participants' communication patterns are complex. For example, some participants offered alternatives when they could not directly address a request. When one participant was walking away from the area of a requester's interest, he provided information about a slightly different, yet still relevant area. Another participant estimated how crowded a restaurant would be without actually visiting it. There was also a participant who responded with a promise to address the request in the future. Moreover, responses from many people can collectively allow a requestor to discover some generalized knowledge. For example, responses to request R3, "Find a trash can," mentioned trash cans in front of various convenience stores, which suggested that one could look for convenience stores when in need of a trash can.

Many participants felt comfortable about sending/receiving a request to/from strangers. Fourteen participants (70%) sent a request to strangers, and eleven of them (78% of the fourteen) felt very comfortable or comfortable about sending it to strangers. Several participants suggested that it was comfortable because of the use of nicknames and the feeling that, as they are using a server-based service, they are not forcing others to respond. Three (21% of the fourteen) felt neutral, and one felt uncomfortable. The participant said it was uncomfortable because the request was about his 'geeky' interests. Six participants (30%) sent no requests to strangers. Their comments indicate that they thought it was easier to ask friends than strangers or did not think of requests for strangers in the first place. Two of the six participants seem to have spent most of their time responding to request from others, and said they did not have time to send their own request. Eighteen participants (90%) received at least one request from a stranger, and ten of them (56% of the eighteen) felt very comfortable or comfortable receiving it. Several participants suggested that it was comfortable because they did not feel too obligated to respond. Seven (39% of the eighteen) felt neutral, and one felt uncomfortable. The participant said it was uncomfortable because there was a question that was difficult to answer. Two participants (10%) did not receive a request from strangers.

Finally, 9 participants (45%) said that they were satisfied or felt happy even when their requests were not fully addressed. Their comments suggest that this could be partly because they enjoyed social interactions through the system and were thankful for the efforts of friends and strangers. Comments also suggest that responding to requests can be rewarding if requested tasks are enjoyable and meaningful.

6   Related Works and Discussion

Increasing numbers of commercial services for mobile phones exploit GPS and cell-tower localization to support personal and group activities. In particular, location-based social networking services such as *loopt* [14], *brightkite* [2], and *loc8r* [13] allow for location-based information-sharing in one's own social network. Although some of these services have experimental features to meet and befriend strangers, they primarily focus on social networking, rather than collaboration among people who may not have strong social ties. Also, several researchers (e.g., [16]) have explored *mobile ad hoc collaboration* by focusing on spontaneous, opportunistic interactions with the aid of experimental devices, while some other have envisioned large-scale *participatory sensor networks* [3] that employ ubiquitous mobile phones. However, we still need a comprehensive analysis of a scalable platform that connects relevant people in relevant places and supports distributed mobile actions.

Though our field trial is limited, its results seem to suggest the roles of *awareness and accountability, changing costs,* and *privacy boundary control* in distributed mobile actions. First, awareness [4,7] makes it easier for people to imagine remote places, which facilitates the process of remotely asking friends and strangers to perform actions. Many participants of our field trial requested *concrete and lightweight tasks* of remote strangers using a mobile tool that supported awareness about who was in which area. This is a sharp contrast with what we observed in the diary study: Only a small number of tasks were requested of remote strangers and those requests were complex and time consuming to address. Also, the Web-based user interface of *Askus* seems to have affected participants' feelings of *accountability* [7] through which mechanisms for social control develop. Such mechanisms can influence people's expectations about collaboration with strangers.

Second, as a person's mobile context changes, so do the physical, social and cognitive costs of performing a task. For example, request R10, *"Put some money in a donation box at a convenience store,"* is relatively easy to do if one is near a convenience store; however, as a person walks away from such a store, the physical cost of addressing the request increases. Social cost is relevant to request R9, *"Just walk around and enjoy yourself,"* and this cost can change as a person moves from one place to another that may have a different social code. Systems that are unaware of these changing costs could suffer high request-rejection rates and, consequently, poor perceived usability.

Third, although our understanding of *Askus'* privacy implications is still limited, our field trial seems to suggest the need of supporting *privacy boundary regulation* [20]. In particular, it seems important that participants can ignore some requests and that the system considers *the human need for plausible deniability* [23].

Overall, the results of the field trial inspire the design of a lightweight mobile tool that considers meaningful places, incentives, extremely easy input and management of requests/responses, and good battery life. In our trial, the *Askus* prototype was perceived as useful despite the coarseness of its location resolution. We however think that finer identification of places is sometimes desirable. We also believe that the system should consider the image of a city [15], which inherently influences the ways place-relevant requests are articulated, shared, and understood.

Moreover, it is essential to provide participants with an incentive for using *Askus.* One possible approach is to use a points-based system that would award points to a person who has carried out a request. Points thus accumulated could be redeemed against some service. To achieve this, companies such as those that provide discounts at restaurants or hotels could support *Askus* and earn publicity in return. Other incentives that could be provided to encourage agents include enhancing a person's reputation in a social network or providing entertainment services. Also, participants' comments in our field trial reinforces that people do not act only for material rewards. We need to consider *intrinsic motivation for participation* [9] that could be influenced by providing a sense of mutual support and shared purpose.

The usability of *Askus* could be improved by integrating various interaction techniques besides text inputs, using cameras, microphones, 2D-barcodes, motion sensors, and touch screens. Informal input mechanisms such as scribbles reduce the burden of input; however, they make it difficult to aggregate information. Buttons and menus allow for quick input of simple information that can be easily aggregated.

It is a challenge for a user to deal with many requests and responses. Therefore, it would be desirable to reduce the user's cognitive burden through the provision of effective user interface tools. A related issue is that batteries can drain rather quickly if the user interacts with many messages on a mobile phone. In a large-scale deployment, it would be highly desirable that the system considers urgency and priorities of requests as well as aggregation and summarization of responses. In particular, the issues around urgency and timeliness should be studied further to design a useful system that scale.

7 Conclusion

In this paper, we explored amplification of human actions using a mobile platform that supports lightweight requests and responses. Based on a diary study that led to a detailed understanding of assistance needs and desires in everyday life, we designed the *Askus* platform and implemented PC-based and mobile phone-based prototypes. We also presented the results from a field trial in central Tokyo.

Although the mobile phones we used in our implementation of the *Askus* platform were not aware of as much context as we wished, they did provide a simple method to find relevant people based on manually-disclosed location information and user status. *Askus* can facilitate engaging reciprocal interactions when it is embedded in right places and social relations. A text-based simple interface supported small requests that could be articulated in a brief sentence. However, a simple request can lead to a need for more information and complex actions.

Our experiences with the *Askus* prototypes motivate an implementation of *Askus* on a technological substrate that allows for spontaneous interactions and context-rich sensing, such as networked wearable devices. In addition, we believe that long-term usage of *Askus* could generate rich historical data that could be used to enrich everyday life.

Acknowledgements. We thank the participants and the OSOITE members in our diary study and field experiment. We are grateful to our shepherds, Elaine Huang and Jin Nakazawa, and the anonymous reviewers for their valuable feedback. Yasuyuki Ishida and Hiroki Ishizuka integrated the GPS functionality with the *Askus* client.

References

1. Bahl, P., Padmanabhan, V.N.: RADAR: An In-Building RF-Based User Location and Tracking System. In: Proc. INFOCOM 2000, pp. 775–784. IEEE Press, New York (2000)

2. Brightkite, `http://brightkite.com/`
3. Burke, J., Estrin, D., Hansen, M., Parker, A., Ramanathan, N., Reddy, S., Srivastava, M.B.: Participatory Sensing. In: Proc. Workshop on World-Sensor-Web (2006)
4. Carroll, J.M., Neale, D.C., Isenhour, P.L., Rosson, M.B., McCrickard, D.S.: Notification and Awareness: Synchronizing Task-Oriented Collaborative Activity. J. Human-Computer Studies 58, 605–632 (2003)
5. Dearman, D., Kellar, M., Truong, K.N.: An Examination of Daily Information Needs and Sharing Opportunities. In: Proc. CSCW 2008, pp. 679–688. ACM Press, New York (2008)
6. Dourish, P.: Re-Space-ing Place: "Place" and "Space" Ten Years On. In: Proc. CSCW 2006, pp. 299–308. ACM Press, New York (2006)
7. Erickson, T., Kellogg, W.A.: Social Translucence: An Approach to Designing Systems that Support Social Processes. ACM TOCHI 7(1), 59–83 (2000)
8. Fischer, G.: Distributed Intelligence: Extending the Power of the Unaided, Individual Human Mind. In: Proc. Advanced Visual Interfaces (AVI), pp. 7–14. ACM Press, New York (2006)
9. Fischer, G.: End-User Development and Meta-Design: Foundations for Cultures of Participation. In: Proc. 2nd International Symposium on End User Development (2009)
10. Harrison, S., Dourish, P.: Re-Place-ing Space: The Roles of Place and Space in Collaborative Systems. In: Proc. CSCW 1996, pp. 67–75. ACM Press, New York (1996)
11. Hollan, J., Hutchins, E., Kirsh, D.: Distributed Cognition: Toward a New Foundation for Human-Computer Interaction Research. ACM TOCHI 7(2), 174–196 (2000)
12. Ito, M., Okabe, D., Matsuda, M.: Personal, Portable, Pedestrian: Mobile Phones in Japanese Life. MIT Press, Cambridge (2005)
13. Loc8r, `http://loc8r.jp/`
14. Loopt, `http://loopt.com/`
15. Lynch, K.: The Image of the City. MIT Press, Cambridge (1960)
16. Kortuem, G., Gellersen, H.-W., Billinghurst, M.: Mobile Ad Hoc Collaboration. In: Proc. CHI 2002, p. 931. ACM Press, New York (2002)
17. Kudo, T., Matsumoto, Y.: Fast Methods for Kernel-Based Text Analysis. In: Proc. Annual Meeting of the Association for Computational Linguistics (ACL), pp. 24–31 (2003)
18. Mcluhan, M.: Understanding Media: Extensions of Man. McGraw-Hill, New York (1964)
19. MIC: The Results of the 2008 Survey on Communications Technology Use (2008) (in Japanese), `http://www.soumu.go.jp/s-news/2008/pdf/080418_4_bt.pdf`
20. Palen, L., Dourish, P.: Unpacking "Privacy" for a Networked World. In: CHI 2003, pp. 129–136. ACM Press, New York (2003)
21. Rheingold, H.: Smart Mobs: The Next Social Revolution. Basic Books, New York (2003)
22. Sangratanachaikul, O., Konomi, S., Sezaki, K.: An Easy-to-Deploy RFID Location System. In: Adjunct Proc. Pervasive 2008, Austrian Computer Society, Vienna, pp. 36–40 (2008)
23. Smith, I., Consolvo, S., Lamarca, A., Hightower, J., Scott, J., Sohn, T., Hughes, J., Iachello, G., Abowd, G.D.: Social Disclosure of Place: From Location Technology to Communication Practices. In: Gellersen, H.-W., Want, R., Schmidt, A. (eds.) Pervasive 2005. LNCS, vol. 3468, pp. 134–151. Springer, Heidelberg (2005)
24. Smith, A., Balakrishnan, H., Goraczko, M., Yantha, N.P.: Tracking Moving Devices with the Cricket Location System,". In: Proc. MobiSys 2004, pp. 190–202. ACM Press, New York (2004)
25. Sohn, T., Li, K.A., Griswold, W.G., Hollan, J.D.: A Diary Study of Mobile Information Needs. In: Proc. CHI 2008, pp. 433–442. ACM Press, New York (2008)
26. Star Trek Wiki, `http://memory-alpha.org/`
27. Terveen, L., McDonald, D.: Social Matching: A Framework and Research Agenda. ACM TOCHI 12(3), 401–434 (2005)

Boxed Pervasive Games: An Experience with User-Created Pervasive Games

Richard Wetzel[1], Annika Waern[2], Staffan Jonsson[2], Irma Lindt[1],
Peter Ljungstrand[2], and Karl-Petter Åkesson[3]

[1] Fraunhofer Institute FIT
[2] Interactive Institute
[3] Swedish Inst. of Computer Science
richard.wetzel@fit.fraunhofer.de, annika@tii.se,
irma.lindt@web.de, staffanj@tii.se,
peterlju@tii.se, kalle@sics.se

Abstract. Pervasive games are rapidly maturing - from early research experiments with locative games we now start to see a range of commercial projects using locative and pervasive technology to create technology-supported pervasive games. In this paper we report on our experiences in transferring the successful involvement of players in computer games to 'modding' for pervasive games. We present the design process, the enabling tools and two sample games provided in boxes to end users. Finally we discuss how our findings inform the design of 'modding' tools for a pervasive game community of the future.

Keywords: User-centered design, Games and infotainment, Programming tools: Integrated environments, Pervasive computing, Pervasive games, End user programming.

1 Introduction

Pervasive games take the player out into the real world [13]. They are rapidly maturing – from early research experiments with locative games [2,3,8], we now start to see a range of commercial projects using locative and pervasive technology [1] to create technology-supported pervasive games. Designing and staging pervasive games requires however both high technical skills as well as in-depth design knowledge about what make pervasive games engaging and fun. Very few pervasive games can yet be created or staged by non-professionals.

Whereas computer games can be run just about anywhere (just insert the CD into a sufficiently powerful PC), pervasive games are typically created or configured for the particular location of the game. Many games require physical objects such as special markers or RFID tags to be placed in the gaming area; others can be staged by marking the interesting game locations on a map – places that the organizer has visited and decided it was suitable for the game. Pervasive games also tend to be technology-supported rather than fully implemented in technology. They may require game mastering, the creation of non-tech props or special-built technology, and trained actors participating in the game.

H. Tokuda et al. (Eds.): Pervasive 2009, LNCS 5538, pp. 220–237, 2009.

In this paper we report on our experiences in creating tools for 'modding' 'modding' pervasive games. In the context of computer games, 'modding' ranges from small modifications to existing games to the creation of entirely new games on top of the game engine of an existing one. To enable 'modding' we created an authoring and game-mastering tool and tested it in various ways with users that could be interested in – and able to – stage pervasive games and create their own pervasive games. The ultimate objective for the project is to create a 'modding' community for pervasive games.

2 Related Work

There is a long-standing tradition of end-user involvement in the computer science field. The vision of end-user programming [14], empowering end users to create their own applications without having to master abstract programming languages, has yet to come true. However, the increased availability of tools enabling user-generated content, popularized in the form of typical Web 2.0 applications such as blogs, wikis, and photo and video sharing sites provide a large step in this direction. The latest developments, in the form of easily manageable and configurable web services such as mashups, go beyond the pure management of non-computational media towards creating your own programs.

However, user-created or user-modified games are already legion in the domain of video and computer games. The first person shooter genre is famous for its active modding communities [11] and there also exist a large set of mods for the most popular games in other genres such as *World of Warcraft*[1]) and *Warcraft III*[2]. Modding communities tend to arise around highly attractive games which motivate the community members to dedicate time to create improved and innovative versions of the original game. The content that is produced by the players ranges from small modifications, such as the customized appearance of game artifacts and characters for *The Sims* game [21], via new game worlds and levels such as new maps for *Warcraft III*, to completely new games such as Counter-Strike[11], and Desert Combat [15]. Some games (see again *Warcraft III*) are released together with tools for scene development and scripting that together constitute a rich development environment. It is important to note that while most end users can benefit from the modding community, there are typically only a small number of players who are actively creating the mods.

End users are often able to invent ways of using the game assets that go beyond the intentions of the original designers; the modding communities are the 'lead users' [18] of the gaming industry. User-created modifications to a game can make the game richer or more interesting to play.

2.1 Pervasive Games

Pervasive games are games that take place in the real world rather than on-screen [13]. They leave the predefined playground, be it the screen or the tennis court, to be

---

[1] See http://www.warcraft-mods.com/ Although the World of Warcraft team discourages the development and use of mods, they are in frequent use in practice.

[2] Blizzard maintains an official Warcraft III mod site at http://www.battle.net/mod/

playable anywhere and at any time. There exist pervasive game designs that use no technology at all, but mobile and pervasive technology opens up new possibilities for pervasive game designs. Previous projects have investigated mobile Augmented Reality [9], GPS and RFID readers as well as custom-built hardware [17]. Advanced pervasive game designs often rely on a mixture of automatic and game-mastered game play, as this provides good support to deal with a changing environment and player improvisations. A good example of this is the success of the Alternate Reality Game genre [12], games that largely rely on semi-automatic gameplay in online sites and where the hidden 'puppet masters' constantly supply the players with new material.

The cost profile of a pervasive game is often that of quite low development costs combined with fairly high costs for staging and running the games. Despite the commercial success of Alternate Reality Games, this has hampered the development of commercial pervasive games. If players were involved in the authoring, staging and game mastering of pervasive games this would increase the number of games staged, dramatically reduce the costs of such games, and contribute to the attractiveness and adoption of this new type of game. Ultimately, the quality and innovation in the games as such would rise with an active modding community.

A first step towards this is to make the staging of the game and the localization of game content a game activity in itself. The game *Hitchers* [7] uses this approach by letting the players create game content and place it in the environment. Although this approach certainly is possible, it does not allow for the creation of full games that are tailored to a particular place or a particular player group; the game itself is not modified. *GeoCaching* is a similar activity that has grown out of the more and more commonly available GPS technology; a pervasive folk movement that comes very close to already being a pervasive game.

In this paper, we try to go one step further. We are looking for a model that allows players to stage games, contribute to the richness of their content, but also to create their own games. The most generic approach to date is probably the *Mediascapes* tool [10] which allows end users to create and place media content in a geographic landscape, either through interacting with a map online or on location through a mobile device. However, the *Mediascapes* system provides little or no support for creating the complex logics that typically are necessary to make a more complex game.

3   Boxed Pervasive Games

A future working model for 'modding' of pervasive games would rely on two things. Firstly, it would require access to technology and tools that can be used for modding. It also requires that there is a modding *community*; a community that knows about the game genre and is inspired to modify games and develop their own games. A central question for the project became what kind of tools and technology that potentially could inspire a modding community to form.

In order to select an overall approach, we considered two specific issues: that people typically don't own the technology needed to play an advanced pervasive game, and since they haven't played them, people don't know what pervasive games are.

Pervasive games do often (although not always) require unusual technology. As previously discussed, previous projects had used Augmented Reality, GPS and RFID readers, all of them generic but not commonly available technology requiring special hardware[3]. On numerous occasions, we had also developed custom-built hardware to fit the setting and theme of a special game. The solution to this problem was to create 'technology boxes', where the hardware and the development tools were packaged together and streamlined to work together. We also wanted to make it possible for a technology-savory user to build game-specific devices and integrate them with the same development tools.

The second issue cannot be completely addressed by just tools and technology. Understanding pervasive games typically requires that you have played them enough to understand something about their core game mechanics and what makes them fun. But this is not enough. A 'modding' community is driven by its own ideas and creativity, so the participants must feel that they both want to, and can, make truly great games. The latter is only likely to happen when pervasive games already are fairly common in society or there exists at least one blockbuster pervasive game.

In order to partially address the second issue, we settled for an approach where tools, technology and games were packaged together. The idea was to *package pervasive games in retail boxes together with the hardware needed to run the games*. The approach was inspired by the Lego boxes: we wanted each box to contain a game of its own much like Lego boxes contain instructions to build Lego model, but we also wanted it to contain the tools needed to build other games with the same hardware. Finally, we wanted the boxes to work together so that you could build more complex games if you combined several boxes. This approach lead to a number of design decisions concerning the hardware, software and games that we choose to put into our boxes. Towards the end of the article we will come back to these to discuss whether they were the right ones.

- The core system needed to rely on fairly standard hardware, and use only low complexity special-built hardware that could be packaged with the game. This was in order to keep the price of a box within reason. When special-built hardware was used, it needed to be semi-generic to support a range of games rather than just a single game.
- All software must be either pre-installed on the hardware or very easy to install. As much as possible should be available as on-line services.
- We assumed that the most common reason to buy a box would be to set up and play a game with your friends. This made it important to include games that were easy to grasp and easy to install.
- The content of a Box must inspire its buyer to start to create his or her own games. This issue mainly concerned the game; that should inspire novel game ideas.
- A Box must contain tools that allow people to build and stage their own games.

3.1 The Box Development Process

We have developed two example boxes. The *Magic Lens box* can enhance the physical world with invisible three-dimensional objects, which the player can see and interact

---

[3] Mobile Augmented Reality is a software solution, but it requires a particular type of powerful portable computer to run.

with through a portable computer. The *Location box* instead enhances the physical world with sound and hidden pictures, and builds on GPS and RFID technology.

The two boxes should be seen as examples of a generic approach. A core requirement was that each would contain technology to support a range of games, at the same time as it would offer an interaction metaphor that could be easily grasped. Finally, we wanted each box to be immediately useful through the inclusion of ready-made game, but at the same time be versatile enough to let its users develop their own games.

The boxes were developed in collaboration with three focus groups, ranging from high school students, university students in game and media studies, and a professional team producing a commercial production. The Gamecreator system discussed below was also used in this commercial production. The focus groups provided feedback on all aspects of the boxes, ranging from the selection of technology, the selection of games, and the functionality offered by Gamecreator.

The focus groups provided very good feedback during the development of the boxes. Towards the end of the development period, the focus groups as well as some other groups also got a chance to use the boxes for a limited period, to stage the games as well as develop their own games. This experiment was however less successful, as the participants had too little time to actually develop anything with the tools. Only one group managed to create a game of their own, and none of the groups staged the preinstalled games with friends. We will have reason to come back to the trials in the section on 'lessons learned'.

The boxes have so far been used in five different game projects. Firstly, the boxed games the *Alchemists* and *Crash* were developed using Gamecreator. Gamecreator has also been used to implement parts of a commercial production *"The Truth About Marika"* [6]. Although this was a truly enlightening experience, it was done in close collaboration with the Box development team and for commercial purposes. In this paper, we focus on the experience of using the boxes in external projects relying on volunteer efforts, as this usage situation more closely resembles that of a future modding community.

3.2 The Magic Lens Box

The Magic Lens box uses Augmented Reality technology [16] to create pervasive gaming experiences. The box contains a ultra-mobile PCs (UMPCs) with pre-installed software, printed markers for pose tracking and a user manual as seen in Figure 1. The software for the *Magic Lens box* is based on the MORGAN AR/VR Framework [16] and uses the ARToolkitPlus library [19] for marker based tracking. The system tracks specially designed black and white paper markers by calculating their exact position and orientation from a video stream. 3D models are overlaid on the video, creating the illusion of them really being where the marker is. The approach is particularly useful on a hand-held device with a camera mounted on the back, creating the illusion of a "magic lens" [5]. In contrast to other tracking methods this method is very easy to set up. All the user has to do is print out the markers or use the ones that come with the box and place them in the physical game world.

Fig. 1. The *Magic Lens box* (left) and its content (right)

Fig. 2. Augmented view from the *Magic Lens box* game *The Alchemists*

The sample game that comes with the box is called *The Alchemists*, a simple treasure hunt game (see [20]). The game master hides a variety of markers in the playing area. These markers are associated with 3D models of potentially powerful alchemistic ingredients, which need to be found by the players. The players can then pick them up with their magical bags (also represented by a marker) and brew them together with other ingredients in their alchemistic cauldron (again, a marker). If the ingredients fit well together, the players are able to brew different types of potions and are awarded points depending on the quality of the mixture (see Figure 2).

Modifying the existing game and creating new games is done in the *Gamecreator* system discussed below. For implementing game mechanics, the *Magic Lens box* software supports two different interaction models: looking with the webcam at a single marker and looking with the webcam at two markers at the same time.

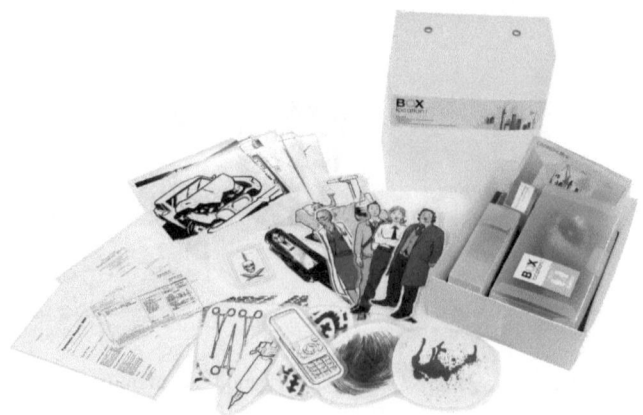

Fig. 3. The Location box and its content

3.3 The Location Box

The *Location Box* uses Internet-enabled mobile phones together with a GPS or a RFID reader connected via Bluetooth (Figure 3). The GPS units are ordinary off-the-shelf GPS receivers with Bluetooth, whereas the RFID readers are specifically built hardware that come in the shape of a wearable glove. The GPS allows to position game content to specific locations or areas while the RFID tags can either be placed at specific places or attached or hidden in artifacts. This provides flexible positioning of content with regards both to accuracy and scale.

The mobile phones primarily work as relay stations, gathering data from the GPS and RFID readers and sending it forth to a game server over Internet. They are also used as recipients of streamed content from the game server. The phone client is implemented in J2ME and enables images, texts, sound, and video files to be streamed to the phone and automatically displayed on the device. It is currently only available on one phone model, the Sony Ericsson K800.

The sample game that comes with the box is called *Crash* and is a simple detective clue hunt that can be set up and played as a party game. It uses only the RFID reader but can be extended by the organizer to also use the GPS. The players are invited into the game as junior police inspectors. Once they reach the place where the accident took place they start finding clues that lead them on into the story.

Game locations, characters and items are represented by printed images with RFID tags attached to them. The players use the RFID glove to interact with these images. This enables them to go to a place (and show where they are), examine evidence, and interact with game characters. The game responds by sending sound and image files to the player's phones. As the players progress through the game, the story unfolds further and gets more complex.

3.4 Gamecreator

A central decision in the project was to develop a joint tool for both boxes. This way, a person who had used one of the boxes would immediately recognize the development method for the next box. We also wanted the tool to support games that used hardware and software from both boxes in the same game.

The *Gamecreator* system was constructed to fill these requirements. It is directed towards the amateur developer of pervasive games, rather than towards the professional programmer or game designer. It still requires some level of programming or scripting proficiency. A target user could for example be a person who has some interest in pervasive games, and previous experience of Web 2.0 services (e.g. has set up and managed his or her own blog).

Gamecreator is developed in Ruby on Rails and runs as an online service. It supports both authoring of games, and the set up and game-mastering of existing games. Each box owner has an own account with Gamecreator, where he or she has access to a development area which contains a preinstalled game, and two empty game slots where the user can develop new games through an online interface. The pre-installed game is available with a full game script, so that the user can modify it at wish.

Authoring a new game
In order to make Gamecreator run with several different technologies and devices, we decided to make the core functionalities of Gamecreator device-independent. A central design decision for *Gamecreator* was to separate all information about what the technology could use as input (e.g. that an RFID tag had been read) or deliver as output (e.g. the 3D model that should be displayed) from its *meaning* in the game (e.g. that a player has entered into a game area, or found a game object).

The game object classes are not fully generic. This is done to support game developers to think about pervasive games, as the classes primarily represent aspects of the real-world. Gamecreator allows three types of objects: persons, places and items, primarily intended to represent real-world persons, places, and things. Each type of object can be identified by several different technologies. A location can e.g. be identified by a GPS area, or an RFID tag, or a Magic Lens marker, or even all of them. A person can be identified by a RFID tag and an item by an RFID tag of a Magic Lens marker. Game objects have game-specific properties that can be changed automatically (by scripted rules) or manually by a game master. Gamecreator contains an online system for scripting rules.

The Gamecreator objects will typically, but not necessarily, map to real-world locations, people and objects. The people category is often used to represent the different types of players (e.g. the team leader versus the team participants) or provide player roles depending on which technology they carry. It will sometimes be used to represent non-player characters (e.g. a paper doll, as in *Crash*).

The clear separation between technology-related input and output and in-game objects has several desirable effects. Gamecreator can fairly easily be integrated with new technology platforms. Gamecreator comes with a back-end API which can be used to query the current game state and trigger events from any external system. Also, the game can use hybrid methods to recognize in-game events. For example, the *Location box* allows indoor events to be triggered by reading an RFID tag and outdoor events by entering a GPS area. Finally, the in-game objects need not be recognizable by technology at all. An in-game location can be a virtual location (e.g. a website) or even a location that only exists as a game concept but never actually is visited by the players. Game objects that are not directly related to technology can still be manually managed by the game masters through changing their property values.

Fig. 4. Sample screenshot from Gamecreator. In the screenshot, a GPS location named 'A forest' has been added as the location of an in-game location named 'A place in the forest'.

Game Orchestration

More complex pervasive games often benefit from being game-mastered than fully automatic. As discussed above, Gamecreator also supports game mastering which was used extensively in the *Interference* game discussed below. There are two main functions in Gamecreator that support game orchestration. One is that Gamecreator maintains a log of all incoming events, and provides the game manager with an overview of the event history. Positioned events are plotted on a map using the Google map interface. Gamecreator also supports *manual* property changes; the game manager can select a subset of the game objects and change any of their properties. Finally, as the rules of a game are scripted in the Gamecreator system, even the rules can be changed while the game is running. This is seldom used in small games but can be very useful in more long-term games such as Alternate Reality Games.

4   Modding and Staging a Given Game: Interference

Our first example focusses on the restaging of a given game. The game *Interference* was developed within the project after *Crash* and the *Alchemists*, to show that the boxes can be combined to create more complex games. The game was restaged by a group of game design and media students at Mediadesign Hochschule Düsseldorf.

Fig. 5. Photo from the Interference stagings in Stockholm, January 2008

4.1 Gameplay

Interference [4] is a very different game compared to the two boxed games. Where *Crash* and *the Alchemists* are small and automatic games designed to be easy to set up and run, *Interference* was designed in collaboration with a professional artist as an artistic experience. It requires game mastering and a complicated staging consisting both of a technical setup and infrastructure, renting locales, and recruiting and training actors. The game had been staged four times in Kista, Stockholm before it was staged in Düsseldorf.

Interference is a game for six to eight players who play together as a single group throughout the whole game. The game is a pervasive mystery game, a kind of a 'treasure hunt' with a story line. The game starts out as in a fairly gamistic manner (the focus is on game mechanics rather than the narrative), as the players are given simple roles as electric engineers who are recruited by a company to fix their broken network. The background story unfolds through the players' meeting and communicating with the central characters, played by actors. Eventually, the players find the means to solve their original technical problem – but only at the expense of the death of one of the central characters. They are faced with a difficult decision and the game ends by letting the players collectively decide what to do.

Interference makes use of the Magic Lens device from the Magic Lens box as well as of GPS positioning from the location box. RFID tags and the RFID reader glove are not used in this game. The Magic Lens device is used to trace the network and find out spots where it is broken. The network is represented by 3D models visible only at certain locations (at the markers) and the players have to draw lines on a map to connect the nodes to each other. The GPS from the Location box is used to seek out magical memories in the landscape, located at emotionally interesting locations. The memories are sound and video files that are streamed to four different mobile phones. Finally, a special-built 'magic' doll was built for this game, which is able to recognize the tones played on a special bone flute. The magic doll was built from scratch, using audio circuits tuned to recognize the flute tones, and a Bluetooth circuit to connect the doll to the phone. Just as with the GPS and the RFID glove contained in the location box, the doll connects to a mobile phone that works as a relay station for communicating to a server. The flute is a real bone flute with no technology augmentation, made magic through the fact that the doll reacts to its tones. The players receive the flute only towards the end of the game, working its magic to mend the broken network.

Before the game can run in a new city, the organizers have to create a map of the fictional 'broken network' overlaying the new game area, and chose the 3D models to fit with the network intersections and lines. The network map should ideally fit the theme of the game, so that the memory traces picked up by GPS are placed in interesting and emotionally powerful locations. In particular, the last scene requires a dramatic location. Finally, there is the issue of language: if the game is replayed in another country the memories may need translation. In Düsseldorf, the sounds were re-recorded in German by the actors.

The staging team must also prepare several non-technological aspects of the game. As *Interference* is a technology-supported game, the players' experience relies largely on non-technological activities such as plotting the network lines on a paper map, visiting interesting locations, meeting with actors, and playing the flute (see Figure 5). The staging team has to rent locales, draw and print maps[4] and visual markers, and training the actors. The last issue is perhaps the biggest challenge, as a successful staging depends on gradually change the attitude of the gamers from "happy questing" in the beginning, to deep involvement in a difficult choice towards the end. The three actors are central in this process; their involvement as game characters must engage the players sufficiently to make them dig deep into the story line. An additional complication is that the actors often will relay information between the game masters and the players. The task of, at the same time, acting out a role in an improvised theatre with the players and relaying messages from the game-masters is far from simple.

While the game is running, *Gamecreator* is used as an orchestration tool, and offers several ways to control the game progress. In particular, the game master can trace the players' location and help them back on track in case they get lost, but they can also directly manipulate the game state to speed up or slow down the game progress. To enable the students to restage the game, two of the students participated in one of the Stockholm stagings, where they had a chance to play the game as well as participate in the briefing of actors and the game-mastering. We also ensured that there was at least one person from the development team onsite during stagings; although these people tried to be as little involved as possible this was necessary both to train the full student group and to help out with technical problems should any occur.

---

[4] This is a larger issue than it might seem. The physical map was one of the devices that actually broke when staging *Interference* in Stockholm - it survived much, but not a snowstorm.

4.2 Observations

The Düsseldorf student team was very active and enthusiastic about staging Interference. Two of them participated in the Stockholm stagings, to get an idea of the particular game as well as a general understanding of technology-supported pervasive games. Based on the experiences from the Stockholm stagings, they started out by doing some changes to the game. The first one was that they created a much more interconnected virtual network overlay for their selected game area. This change was done in order to make the first game task (scouting the network) more interesting to the players and minimize the risk of getting lost. They also modified the use of the Magic Lens markers, placing them on the flat on the ground instead for on poster areas and walls. This made it easier to trace the direction of lines for the players, again contributing to making the initial part of the game more fun.

The group also spent a lot of effort on scouting the area and selecting interesting places for the players to visit. They were particularly happy about the location for the final scene. In Stockholm, this scene had been placed in an office in a high tower overlooking Kista. In Düsseldorf, the student group managed to get access to the *rooftop* of the highest house in the vicinity and staged the final dramatic scene there.

Learning to use Gamecreator as an orchestration tool took some time. However, once the student team learned to use it, they appreciated its versatility:

"I slowly start to understand Gamecreator. You can really do a lot with it – if you understand which is which." (Student comment)

Interference was staged on three occasions in Düsseldorf with a total of 22 players. The stagings in Düsseldorf received positive feedback from the majority of the players. These had no previous experience in pervasive games, and were very excited about the real world involvement in the game.

"The best moment was when we met Matilda for the very first time. She was standing on a dark bridge in the dimly lit park and seemed to be watching us. I was not sure if she belonged to the game and it took us a while to build up the courage to talk to her."
(Player comment, from postgame interviews)

There were some technical malfunctions during the stagings with GPS breaking down and loss of Internet connection for the cell phones. Ironically this was probably caused by some real life interference in downtown Düsseldorf. This however showed one of the strong points of Gamecreator. As it allows for both automatic and manual control of a game, the students were able to trigger rules manually when the some of the automatic functions did not work. During the restaging process another strong point of Gamecreator emerged: The separation between the actual game content and the game logic proved to be extremely helpful. Exchanging the 3d models, adapting the game to the new location and setting was possible without too much of an effort. Additionally, the ability to trace where the players were moving in real-time as part of the orchestration interface was very valuable to the game masters. This way they were always aware of the players position which made it much easier to react and assure a smooth flow of the game (e.g. when the players were spending too much time in one space discussion, the students would have one of the actors call the players to put more pressure on them).

Afterwards we were told by the organizing student group that both preparing and staging the game was a truly great experience for them. In the end, their teachers also complained that the students were spending too much time preparing *Interference -* skipping class way too often. While this on the one hand shows that the process of staging a pervasive game is still very demanding, the fact that the students were highly motivated throughout the process is very promising. Staging pervasive games can be a lot of fun, perhaps in particular when the game is demanding and the staging itself requires creative engagement.

5 Creating a New Game: *The Treasure*

The game *The Treasure* was created by three B-IT Bonn students over the course of one semester. This game was built using the Magic Lens box. This was a supervised student project, where the students were first introduced to the concept of pervasive games by trying out and experimenting with the sample game *The Alchemists*. The students were then asked to come up with a game design by themselves.

The students created a variety of ideas before their choice became *The Treasure* which they then developed over a couple of weeks and finally staged in March 2008. The students were given free hands with the development, but could always contact the Box development team in case they ran into technical problems.

5.1 Gameplay

The Treasure was inspired by classical adventure games where the main character has to travel to different locations, solve riddles and collect and combine objects with each other. The students settled for developing a game that was tied to a particular place, the castle Birlinghoven and its surrounding gardens where the FIT research group is located. The choice of location matched the team of the game, as the castle Birlinghoven is a twentieth century 'remake' castle that has somewhat of a 'fairytale' ambience.

According to the game storyline, a long dead king has hidden away his treasure to protect it from greedy, undeserving people. The king has set up several challenges to test the mind and heart of the brave who venture to retrieve this treasure. The challenges are associated to statues and images found in the castle and the garden: for example in one case the players have to find meat for an eagle and his starving offspring, while in another they have to solve a riddle given to them by an old monk.

The students combined the real world and the virtual *Magic* Lens augmentation in a highly engaging manner throughout the game. Instead of having the players interact only with virtual characters, they made use of the real statues and paintings in and outside the castle. These would come "alive" when being inspected with the UMPC and start to talk to the players. Else, the interaction model was the same as in *The Alchemists*: the players could interact with a marker through viewing it through the UMPC, and pick up and drop items through placing two markers close to each other and inspect them together through the UMPC.

Fig. 6. Three players of *The Treasure* bringing a requested game artefact to the stone hunter

To solve the riddles, the players need to take a visual marker and bring it to the non-player character from who the player had received the task (see Figure 6). If the marker and its content did not solve the task, players would get some feedback and some further hints. The game ended when the player found a physical treasure map and could locate the treasure – which was a box of sweets.

5.2 Observations

Both boxes have an initially steep learning curve. Part of the problem is that the Gamecreator interface is far from self-explanatory (after all, it is a first prototype), but users also have a problem in understanding how to set up the boxed games. However, in this project the students rapidly overcame these issues and the process was smoother than in the *Interference* stagings, showing that Gamecreator is somewhat better at supporting authoring than orchestration. After only a short while the students had grasped the concept behind the *Magic Lens* box and *Gamecreator* and were already able to test small parts of their game. The game could then be developed and tested iteratively, contributing to the successful final design.

> "I personally liked this way to create the game, mostly because it was really feasible to create an Augmented Reality game at this short time period we had. If we would have to develop our own functionality it would not be possible at all, but with Gamecreator and the Magic Lens box we could just focus on the game design." (comment by another student developer)

The students ran into some problems with the built-in limitations of the box. The problems were a combination of the limited interaction model that the box implements and some restrictions in the scripting language implemented in Gamecreator.

"For the more of less complex games or games which have some new interaction concepts using Game Creator can be problematic because some functions can be just unavailable in it. This happened also during our game design and we were forced to use some "hacks" or change a game design in order to use Game Creator. I think that even after including more and more functionality into Game Creator such a situation ca accrue. One solution for such a problem is to allow user by some means (programming Game Creator core or something else) to extend Game Creator functionality according to their needs."

An interesting observation is that the game *The Treasure* manages to deliver a game experience that is quite different from *The Alchemists*, even though both games use the same way to model and specify a game, the same gaming hardware, and the same basic interaction techniques. The treasure game uses sound files to tell the background story and to guide the player through the game. In *The Alchemist* the background story is rather brief and told in a text document as part of the setup. Secondly, the treasure game suggests different walking paths. In *The Alchemist* players pick up ingredients and have to go back to their cauldron every time, but in *The Treasure* players need to bring game artifacts to different non-player characters, which makes the movement through the landscape more interesting. Finally, in *The Treasure* the game was closely adapted to the physical gaming area. This makes *The Treasure* much more of a true "mixed reality" experience. Apparently, the box and Gamecreator manage to create a good balance, providing guidance through its restrictions in interaction and scripting model but at the same time providing some degree of freedom that spurs the designers' creativity.

The play-tests with *The Treasure* show that the game was engaging also to players. The student team was happy with the outcome.

"During the playing session we were surprised how the players interacted with a new technique as if they had known it for a long time. What we were afraid of didn't appear, because the players were very enthusiastic and tried to hurry and run, even though we didn't mention something like having a deadline or time limit." (comment by one of the students developing the game)

6   Analysis

We now turn back to an analysis of our box idea. Could our boxes actually work as tools for a modding community? To answer this question, we reflect on the original design decisions to give a direction for future work in the area.

6.1 Designing Pervasive Game Boxes

Our first assumption was that the hardware should be cheap and off the shelf, or if not, at least semi-generic. This assumption was supported by the focus group feedback, which was a bit skeptic to the UltraMobile PC due to its price (to the extent that they were afraid to borrow and use it).

We also assumed that software should be pre-installed or easy to install, and that as much as possible should be available as on-line services, including the tools to create your own games. This proved to be very important and only partially fulfilled by the boxes. Gamecreator comes close to realizing this through its online scripting system, but its ease of use is compromised by its generality and a slightly obscure interface. The boxes have a steep initial learning curve: it is not easy to understand even how to set up and run the preinstalled games. To some extent this is due to quirks in the (prototype) interface and the setup procedures, but more it has to do with that the general concept of pervasive games is novel to the users. However, the experiences with *Interference* and *The Treasure* show that once a user has grasped the functionality and purpose of the system, they start to appreciate the versatile support offered by the system. The overall system seems to have created a good balance, providing guidance through its restrictions while at the same time providing enough freedom to spur the designers' creativity. Gamecreator is currently under re-development with an increased focus on configurability; something that also was desired by the Treasure developers. The new setup will allow games to be configured in a more generic way, and then scripted in a way that uses in-game concepts rather than generic concepts.

One of our assumptions turned out to be false: we assumed that the consumer would buy a box to play games with their friends, and that the games for this reason should be easy to grasp and easy to install. However, the boxed games were perceived as too simple to be interesting to stage by our focus groups, whereas the rather complex *Interference* game inspired intense engagement by the organizers and was successfully staged. We can expect the same to be true for a true 'modding' community for pervasive games: to inspire modding, the original game needs to be a rich, complex and inspiring game.

7 Conclusions

The most important lesson learned from this project is that the pervasive game boxes are not really targeted for a consumer market. Our focus on supplying small, easily staged, and easily understood games with the boxes was not successful, whereas providing a rich and complex game that required large investments to stage was. This is in line with the experiences from computer games: modding communities do not consist of ordinary consumers; they are formed out of a small percentage of the full player community for a game.

To create a modding community, the boxes need to be reshaped to support a truly attractive and rich game. One possible approach is to include a pervasive variant of a table top role playing game with a box. Such games are based on a very rich story space and rely extensively on game mastering. Game mastering such a game is a creative and skilful task which might attract 'modders'. The boxes could also be turned into a game themselves. The ideal game would be a continuously running massively multiplayer game service that requires special hardware. Players would buy a box to allow them to enter into the game, and they could be responsible for managing their own setup within the game. The Gamecreator system could then support players in creating modifications to their game clients, but also to build their own games.

With the ongoing advancement in technology, devices suitable for being used in pervasive games are already becoming more powerful and standard. The new iPhone3G by Apple for example already offers a decent sized screen and built-in GPS. This will make it easier in the future to package a box that makes full use of the hardware the potential players already have at their disposal, thus making the boxes more accessible and cheaper.

Acknowledgments

We thank our colleagues at the Interactive Institute, the Collaborative Virtual and Augmented Environments Department at Fraunhofer FIT and the Swedish Institute of Computer Science for their comments and contributions. We further wish to thank our project partners of the IPerG project for their ideas, cooperation, and support. IPerG (FP6-2003-IST-3-004457) was partially funded by the European Commission as part of the 6th Framework.IPerG. Finally we like to thank our focus groups for their insight, the students from the Mediadesign Hochschule Düsseldorf for doing a great job at restaging Interference and the student group from B-IT Bonn for developing The Treasure.

References

1. Ballagas, R.A., Kratz, S.G., Borchers, J., Yu, E., Walz, S.P., Fuhr, C., Tann, M., Hovestadt, L.: RExplorer: A mobile, pervasive spell-casting game for tourists. In: CHI 2007 extended abstracts on Human factors in computing systems, San Jose, CA, USA (2007)
2. Barkhuus, I., Chalmers, M., Tennent, P., Hall, M., Bell, M., Sherwood, S., Brown, B.: Picking Pockets on the Lawn: The Development of Tactics and Strategies in a Mobile Game. In: Beigl, M., Intille, S.S., Rekimoto, J., Tokuda, H. (eds.) UbiComp 2005. LNCS, vol. 3660, pp. 358–374. Springer, Heidelberg (2005)
3. Benford, S., Crabtree, A., Flintham, M., Drozd, A., Anastasi, A., Paxton, M., Tandavanitj, N., Adams, M., Row-Farr, J.: Can you see me now? ACM TOCHI 13(1), 100–133 (2006)
4. Bichard, J.P., Waern, A.: Pervasive Play, Immersion and Story: Designing Interference. In: Proceedings of DIMEA2008, Athens, Greece (2008)
5. Bier, E., et al.: Toolglass and Magic Lenses: The See-Through Interface. In: Proc. ACM Conf. Computer Graphics and Interactive Techniques (Proc. Siggraph), pp. 73–80. ACM Press, New York (1993)
6. Denward, M.: Broadcast Culture Meets Role-Playing Culture: Consequences for audience participation in a cross-media production. In: Proceedings of International Association for Media & Communication Research, IAMCR, Stockholm, Sweden (July 2008)
7. Drozd, A., et al.: Hitchers: Designing for Cellular Positioning. In: Dourish, P., Friday, A. (eds.) UbiComp 2006. LNCS, vol. 4206, pp. 279–296. Springer, Heidelberg (2006)
8. Falk, J., Ljungstrand, P., Björk, S., Hansson, R.: Pirates: proximity-triggered interaction in a multi-player game. In: Proc. ACM CHI, pp. 119–120 (2001)
9. Fischer, J.E., Lindt, I., Stenros, J.: Evaluation of Crossmedia Gaming Experinces in Epidemic Menace. In: Magerkurth, C., et al. (eds.) Proceedings of the 4th International Symposium on Pervasive Gaming Applications PerGames, Salzburg, Austria (2007)
10. Hull, R., Clayton, B., Melamed, T.: Rapid Authoring of Mediascapes. In: Davies, N., Mynatt, E.D., Siio, I. (eds.) UbiComp 2004. LNCS, vol. 3205, pp. 125–142. Springer, Heidelberg (2004)

11. Kücklich, J.: Precarious Playbour: Modders and the Digital Games Industry. Fibreculture Journal, ISSN: 1449 – 1443 Issue 5 Australia (2005)
12. McGonigal, J.: This Is Not a Game: Immersive Aesthetics and Collective Play. In: Proceedings of Digital Arts & Culture (2003)
13. Montola, M.: Exploring the Edge of the Magic Circle. Defining Pervasive Games. In: Proc. Of Digital Experience: Design, Aesthetics, Practice conference, Copenhagen (2005)
14. Nardi, B.: A small matter of programming. MIT Press, Cambridge (1993)
15. Nieborg, D.B.: Am I Mod or Not? - an Analysis of First Person Shooter Modification Culture. In: Creative Gamers Seminar - Exploring Participatory Culture in Gaming. Hypermedia Laboratory (University of Tampere) (2005)
16. Ohlenburg, J., Broll, W., Braun, A.: MORGAN: A Framework for Realizing Interactive Real-Time AR and VR applications. In: SEARIS, Workshop on Software Engineering and Architecture for Realtime Interactive Systems at IEEE VR (2008)
17. Stenros, J., Montola, M., Waern, A., Jonsson, S.: Play it for Real: Sustained Seamless Life/Game Merger in Momentum. In: Proceedings of the DIGRA conference (2007)
18. Von Hippel, E.: The Sources of Innovation. Oxford University Press, Oxford (1998)
19. Wagner, D., Schmalstieg, D.: ARToolKitPlus for Pose Tracking on Mobile Devices. In: Proceedings of 12th Computer Vision Winter Workshop CVWW 2007 (2007)
20. Wetzel, R., Lindt, I., Waern, A., Jonsson, S.: The Magic Lens Box: Simplifying the Development of Mixed Reality. In: Proceedings of DIMEA, Athens, Greece (2008)
21. Wright, W.: Triangulation: A schizophrenic approach to game design. In: Proceedings of the Game Developers Conference, San Jose, CA (2004)

RF-Based Initialisation for Inertial Pedestrian Tracking

Oliver Woodman and Robert Harle

University of Cambridge Computer Laboratory
{ojw28,rkh32}@cam.ac.uk
http://www.cl.cam.ac.uk

Abstract. Location information is an important source of context for ubiquitous computing systems. We have previously developed a wearable location system that combines a foot-mounted inertial unit, a detailed building model and a particle filter to locate and track humans in indoor environments. In this paper we present an algorithm in which a map of radio beacon signal strengths is used to solve two of the major problems with the original system: scalability to large environments and uncertainty due to environmental symmetry.

We show that the algorithm allows the deployment of the system in arbitrarily large buildings, and that uncertainty due to environmental symmetry is reduced. This reduction allows a user to be located after taking an average of 38 steps in a $8725\,\mathrm{m}^2$ three-storey building, compared with 76 steps in the original system. Finally, we show that radio maps such as those required by the algorithm can be generated quickly and automatically using the wearable location system itself. We demonstrate this by building a radio map for the $8725\,\mathrm{m}^2$ building in under two and a half hours.

Keywords: Radio, localisation, inertial tracking, particle filters.

1 Introduction

Some of the first context-aware computing systems made use of location information as the primary source of context. Today, GPS provides localisation outdoors, but precise indoor tracking of people remains an open research problem. We have seen indoor location systems based on infra-red [1], ultrasound [2], WiFi signal strength [3], ultra-wideband[1] (UWB), vision [4], and many others [5]. Nearly all location systems based on these technologies require the physical installation of fixed infrastructure in the environment. Furthermore, there is often a correlation between the amount of infrastructure and the positioning accuracy achieved, as shown in Figure 1. The amount of infrastructure is often prohibitively expensive to deploy and maintain.

---

[1] http://www.ubisense.net

H. Tokuda et al. (Eds.): Pervasive 2009, LNCS 5538, pp. 238–255, 2009.
© Springer-Verlag Berlin Heidelberg 2009

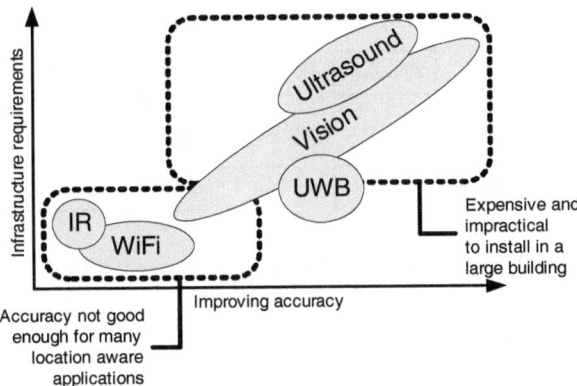

Fig. 1. The correlation between accuracy and infrastructure requirements in existing indoor location systems

In a related paper [6] we developed a wearable location system, in which a foot-mounted inertial measurement unit (IMU), a detailed building model and a particle filter were combined to locate and track a user with sub-metre accuracy. This system suffered from two major problems. Firstly, the computational power required by the system was relatively high and scaled as $\mathcal{O}(A\log_2(A))$, where A was the floor area of the building in which it was deployed. This made it impractical to deploy the system in large buildings. Secondly, ambiguity caused by symmetry in the environment could delay or even prevent the system from determining the user's true location.

It has been shown that radio beacons can be used to seed our location system with an approximate initial position for the user [6], helping to alleviate the problems described above. In this paper we present an improved algorithm in which a map of radio beacon signal strengths (a radio map) is used throughout the localisation process. We show that when using this algorithm, the user is localised after taking an average of 38 steps in our building, compared with 55 when radio beacons are only used to seed the initial position, and 76 when they are not used at all. We also show that the algorithm makes it feasible to deploy the system in arbitrarily large buildings.

In addition to their use in this paper, radio maps are also used by many existing RF-based location systems. Until now the construction of such maps has been done manually, which is very time consuming. The second contribution of this paper is to show how our wearable location system can be used to construct such maps quickly and automatically.

The structure of this paper is as follows. Section 2 introduces our wearable location system. Existing work in the field of RF-based location systems is outlined in Section 3. Section 4 describes the automatic construction of radio maps. Section 5 presents an algorithm for using such maps during the localisation process. Finally, we evaluate the effectiveness of the algorithm in Section 6.

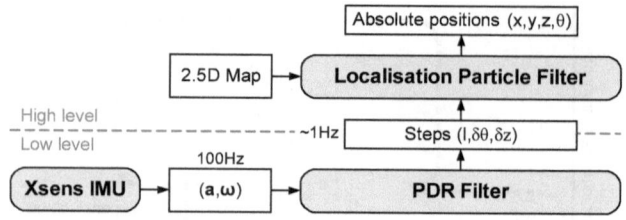

Fig. 2. Wearable location system overview

2   Wearable Location System

An overview of our existing wearable location system is shown in Figure 2. The user wears a small and light inertial measurement unit (IMU), which is mounted on the foot and connected to a hip-mounted ultra-mobile PC (UMPC). The IMU measures acceleration and angular velocity at a frequency of 100 Hz. These measurements are processed by a pedestrian-dead-reckoning (PDR) filter, which generates a noisy step event of the form $(l, \delta z, \delta\theta)$ for each step taken by the pedestrian, where l is the computed step length, δz is the change in height of the foot and $\delta\theta$ is the change in heading relative to the previous step. The sequence of step events generated by the PDR filter describes the approximate path followed by the pedestrian, relative to an unknown starting position and orientation.

The localisation particle filter combines the step events with knowledge of constraints that exist naturally in indoor environments (specifically walls and floors) to determine the user's absolute position. This process is known as localisation, and is described in Section 2.1. The user's position is then tracked as described in Section 2.2. The building constraints are defined by a 2.5D map, which consists of a collection of rooms. Each room is defined by one or more planar floor polygons, which correspond to surfaces within the room on which a user's foot may be grounded. Each edge of a floor polygon is either an impassable wall or a connection to the edge of another polygon. It is possible to represent even complex rooms using this format, such as the lecture theatre shown in Figure 3.

2.1   Localisation

Initially it is assumed that the user can be located anywhere in the building. Many particles (each of which represents a possible location and orientation of the user) are generated uniformly over all floor polygons in the map. Every particle is assigned an identical weight, which is the probability that the particle corresponds to the user's actual position. Each time a new step event is received, a new set of particles is generated to represent the user's updated position. The new set is generated by randomly sampling old particles in proportion to their weights (this is known as re-sampling). The position and orientation of each particle in the new set is then updated according to the received step event,

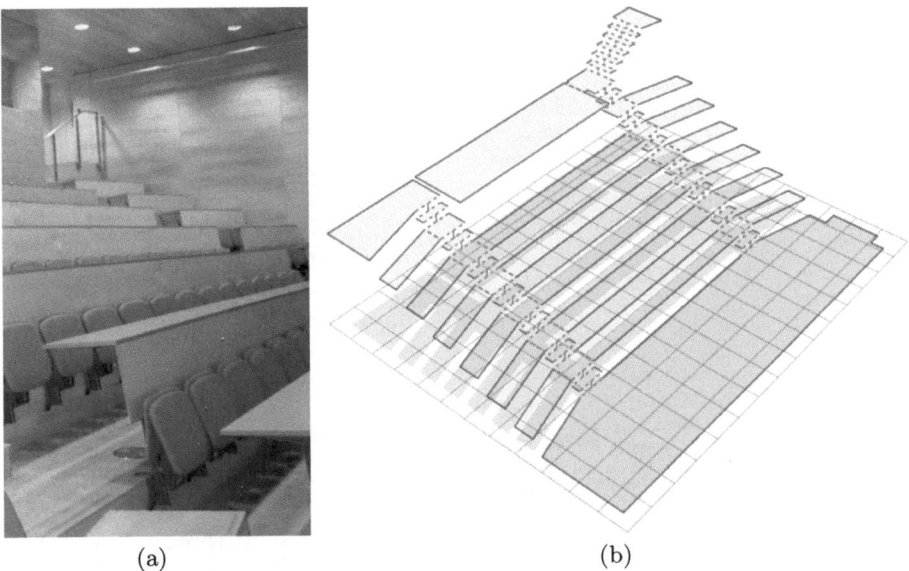

(a)   (b)

Fig. 3. (a) A lecture theatre. (b) Its 2.5D representation. The edges separating adjacent rows of seating are treated as walls (solid blue lines), whereas the edges between stairs in the aisle are connections (dashed green lines). Grid size $= 1\,\mathrm{m}^2$.

perturbed by some noise to model uncertainty. If the perturbed step causes a particle to pass through a wall then that particle is assigned a weight of zero, which prevents it from being re-sampled during the next update. This is equivalent to removing the particle. Hence with each step, the user's possible location is narrowed down as shown in Figure 4. This continues until only a single cluster of particles around the user's true position remains. There are two major problems faced during the process of localisation, which are outlined below.

Scalability. The localisation particle filter requires $\mathcal{O}(n\log_2(n))$ time to update the set of particles when a new step event is generated, where n is the number of particles [6]. The number of particles is varied at each step and is proportional to the floor area within which the user may be located. Since the user may initially be located anywhere in the building, it is clear that for a large enough building it will not be possible to perform localisation in real time. Our lab is an example of such a building, with a floor area of $8725\,\mathrm{m}^2$. We found that around 4,530,000 particles were required to achieve reliable localisation. Our current implementation (written in Java and run on a 2.6 GHz Linux machine) is only able to update 800,000 particles during the average step duration.

Environmental symmetry. Symmetry of the environment can delay or prevent convergence to a single cluster of particles. If the layout of a building exhibits translational or rotational symmetry then the relative path described by the step events is often consistent with multiple different locations. In such cases

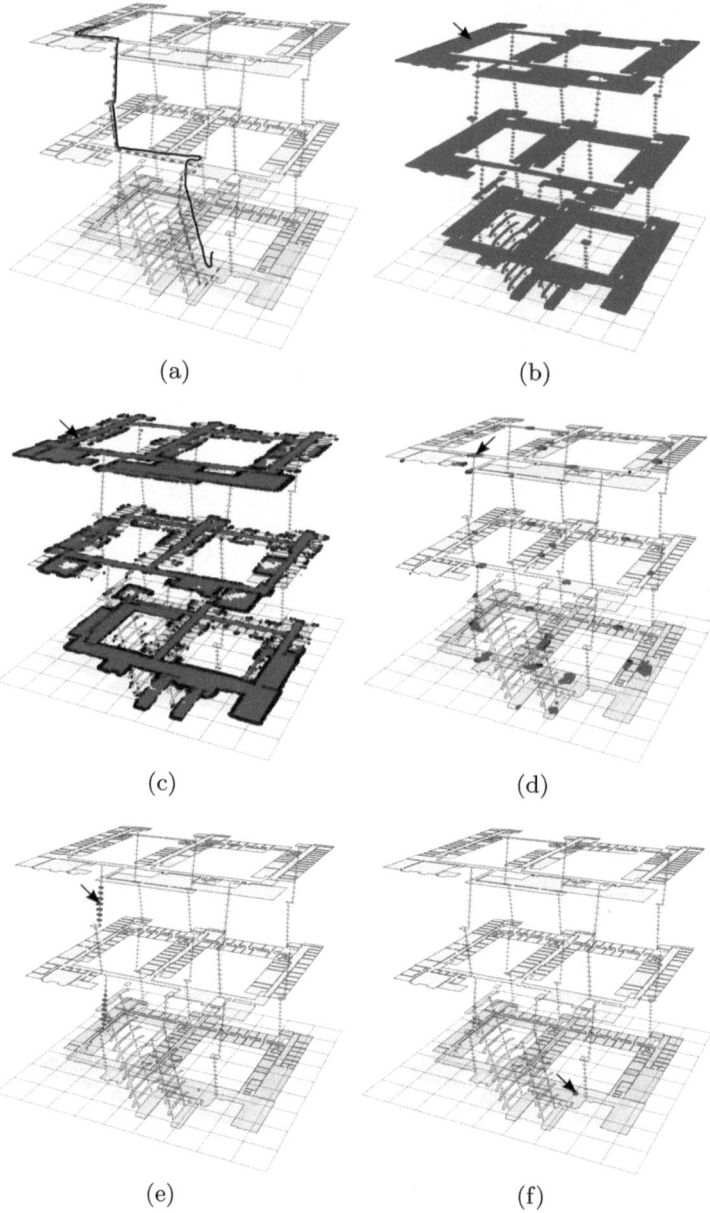

Fig. 4. An example of localisation in a three-storey building. (a) The route taken by the pedestrian (dashed red line) and a manually-aligned overlay of the steps generated by the inertial navigation component (solid black line); (b) The initial distribution of particles; (c-f) The particle distribution at four points during localisation. Particles which have passed through walls in the previous step are coloured black. An arrow indicates the actual position of the pedestrian in each figure. Grid size $= 10\,\mathrm{m}^2$. Diagrams are exploded 10x in the z-axis.

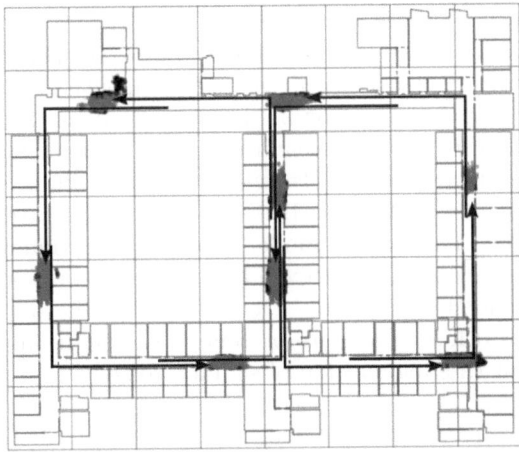

Fig. 5. Multiple clusters of particles caused by symmetry of the environment. The user has walked in a straight line before making a 90° turn to the left. This movement is consistent with eight different locations which arise due to both translational and rotational symmetry. Each arrow indicates the path taken by a cluster to reach its current position. The cluster at the bottom right corresponds to the user's true position. Grid size = $10 \, \text{m}^2$.

a cluster of particles forms at each location, only one of which corresponds to the user's true position. An example of this problem is shown in Figure 5.

2.2 Tracking

When the number of particles falls below a threshold (10,000), a clustering algorithm originally used to cluster GPS positions [7] is used to identify distinct clusters of particles. First a particle is drawn at random and its location is taken as the centre point of a new cluster. All particles within a 7 m radius of the centre (and less than 1 m vertical displacement) are marked and added to the cluster. The mean is then taken as the new centre point. This process is repeated until the mean no longer moves. The procedure is repeated on the remaining particles until all particles have been assigned to a cluster.

The localiastion process is said to be complete when all particles are assigned to a single cluster. The position of the user is then tracked as the centre of this cluster. It has been shown that once the user has been localised, they can be tracked to within 0.73 m 95% of the time [6].

3   RF-Based Location Systems

The development of RF-based location systems is an active research area [3,8,9,10]. Such systems can be divided into model-based and map-based techniques. In the former, the user's location is calculated from a radio propagation model and the known locations of radio beacons. In the latter, a radio map of the environment is constructed during an offline phase, in which beacon strengths are recorded at positions throughout the environment. The radio map is then used to locate the user during an online phase via pattern matching.

Radio propagation in an indoor environment is dependent on many factors, such as the wall and floor materials, which are difficult to accurately model. Complex propagation models have been devised which attempt to take such factors into account [11]; however model-based RF-location systems have been unable to match the accuracies achieved by map-based systems [3]. Radio propagation is also dependent on dynamic factors such as whether doors are open or closed, and whether other people are present in the environment. Hence it cannot be assumed that the distribution of signal strengths at a particular location is constant.

For human location systems (in which beacon strengths are measured by a device attached to the user), the user's body has a significant impact on the measured signal strengths. Obstruction of the line-of-sight to a beacon by the body causes signal attenuation, which can result in a reduction of up to 9 dBm in the received signal strength for WLAN access points [12]. The overall effect of the user's presence is that mean signal strengths are reduced and variances are increased. Hence if a radio map is to be used for human location, then the map itself should be constructed when a user is present.

A number of different algorithms exist for computing the user's position from a beacon scan made by the user at his current position and a pre-constructed radio map. In RADAR [3] the Euclidian distances between the measured signal strengths and those in each entry of the radio map are calculated. Taking the location for which the smallest distance was calculated to be the position of the user results in an error of 2.94 m 50% of the time, and 9 m 95% of the time. Wang et al. [13] used a similar algorithm to obtain a mean error of 6.44 m, which was improved to 4.30 m by using relative movement information obtained with an accelerometer.

The systems described above are deterministic, in that they compute a best guess of the user's position. An alternative is to compute a probability distribution $p(s|z)$, which describes the probability that the user is located at position s given the scan z. The best guess location is then the position s which maximises $p(s|z)$. The Horus system [8] used a probabilistic approach, with a reported accuracy of 1.4 m 95% of the time. It is unclear how much control was exerted on the environment within which these results were obtained (e.g. whether other humans were present in the environment), and how vulnerable the system was to changes in the environment such as the opening and closing of doors.

Until now most radio maps used in RF-based location systems have been constructed manually. To do this a human stands at different locations in the building, manually marking his position on a map and then performing a beacon scan at each location. This method is very time consuming and is also prone to error, since the user must estimate their true position each time a scan is performed. Furthermore, the map would have to be manually updated if new radio beacons were added, or if existing beacons were moved to new locations. As a result many deployments have been small [8], or restricted to the corridors of larger buildings [3]. Castro et al. [10] describe map-building as "tedious". In one of the few large-scale deployments of a RF-based location system, 28 man-hours were required to construct a radio map covering a 12,000 m^2 building [14]. Robots have been used to acquire radio maps for use in robot location systems [15]; however such a map would not be well suited for use in a human location system since it would not take the presence of the user's body into account.

4   Constructing Radio Maps

Radio maps can be constructed quickly and easily by a user who is being tracked using our wearable location system. In this paper we focus on mapping WiFi access points, however the same approach could be used to map other radio beacons. The WiFi hardware embedded within the hip-mounted UMPC is used to scan repeatedly for visible access points and their received signal strength indications (RSSIs). When the PDR-filter signals that a new step has been taken, the localisation filter updates the set of particles and clusters them to obtain a new position and orientation. The most recent WiFi scan to have been completed since the last position update (if one exists) is then incorpoated into the radio map at this location. Note that radio map construction does not start until after the user has been localised. If environmental symmetry prevents localisation, then the user may manually select the cluster of particles which corresponds to his true position in order to complete the localisation process.

The radio map data structure (and the amount of data stored within it) can be changed according to the desired application. In this paper we construct a radio map for use during the localisation step of our location system. To do this we divide the 2.5D map into cells. A cell is a region in space within which radio measurements are grouped, with the cell size defining the granularity of the map. Each cell contains a histogram of measured RSSI values for each access point that has, at some point, been visible from the cell. A new measurement is added to a cell by simply updating the histogram for each access point that was sighted in the scan. If an access point is sighted for the first time in a cell then a new histogram is created. We considered two factors when deciding how to divide a 2.5D map into cells:

1. The radio map is most useful when the variability of radio measurements within cells is minimised, and variability between cells is maximised. Since walls attenuate radio signals more than air, we should not allow cells to span multiple rooms.
2. To get a complete radio map it is necessary to take measurements in every cell. Some rooms such as the lecture theatre shown in Figure 3 are made up of many small floor polygons. Therefore it should be possible for cells to span multiple floor polygons within a room.

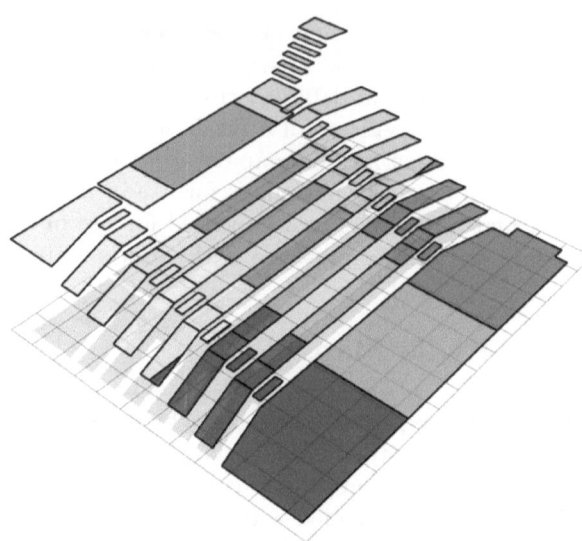

Fig. 6. The cells generated for the lecture theatre shown in Figure 3 with a cell granularity of 6 m. Each cell is shaded with a different colour.

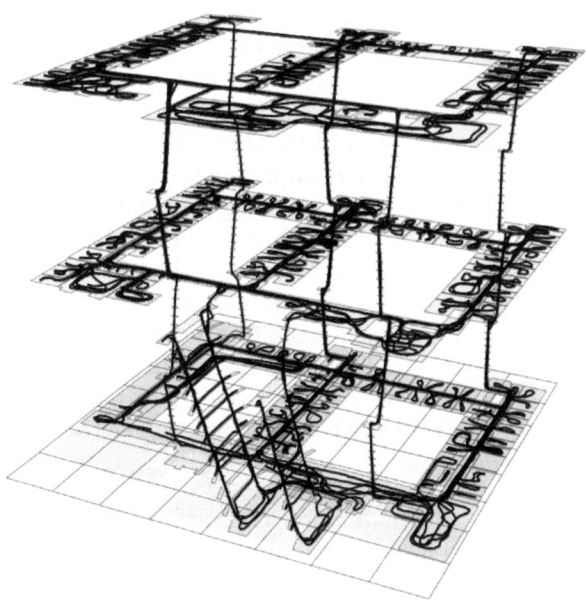

Fig. 7. Tracks generated by a user over 2 hours and 28 minutes, during which a radiomap was constructed.

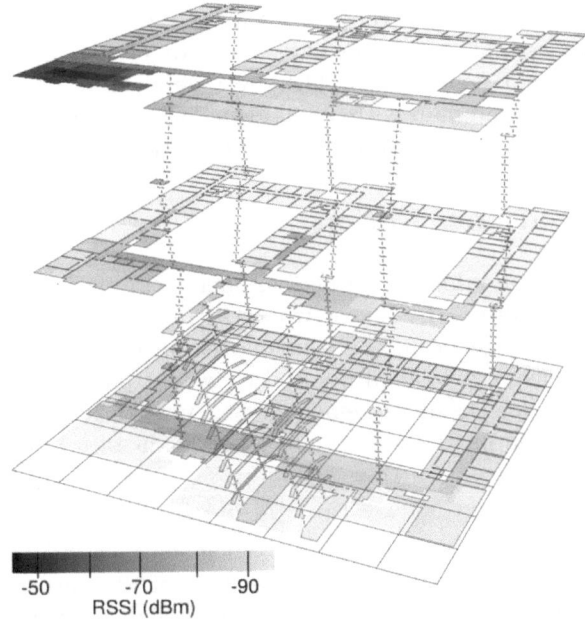

Fig. 8. The radio map generated for a single access point. Cell colour indicates the mean RSSI value. Cell granularity = 5 m.

Table 1. The map coverage and mean scans per cell at different cell granularities

Cell granularity (m)	Coverage (%)	Mean scans per cell
1	59	1.3
2	85	4.3
3	96	9.14
4	99	12.9
5	99	18.9

Hence we divide each room into cells as follows:

1. Compute the 3-dimensional bounding box of each room.
2. Divide the bounding box into the minimum number of identical cuboids such that no edge is longer than the specified cell granularity.
3. A cell is defined as the fragments of floor polygon which are contained within a single cuboid.

The Weiler-Atherton clipping algorithm [16] is used to compute the fragments of each floor polygon that are contained in a cell, as shown in Figure 6.

We used our location system to construct a map of WiFi access points in a large (8725 m²) three storey building. A user was tracked for 2 hours and 28 minutes (split into three sessions), travelling a total distance of 8.7 km as shown in Figure 7. In total 33 access points were observed, one of which is shown in

Figure 8. We define coverage as the percentage of the total floor area covered by cells in which at least one measurement was recorded (excluding restricted areas to which we could not gain access). The coverage at different cell granularities is shown in Table 1. The map with a cell granularity of 5 m is used in the next section to improve the localistion performance of the system.

5   RF-Assisted Localisation

There are two stages at which WiFi measurements can be incorporated into the localisation particle filter. The first is when the initial distribution of particles, known as the prior, is generated. The second is during the localisation process itself, in which measurements can be used to weight particles according to the probability of observing the measurement given a particle's state (i.e. the position and orientation of the particle). Formally, the weight of a particle at time t is calculated as

$$w_t = w_{t-1} \cdot p(\boldsymbol{z}_t|\boldsymbol{s}_t) \tag{1}$$

where w_{t-1} is the previous weight of the particle and $p(\boldsymbol{z}_t|\boldsymbol{s}_t)$ is the probability of observing the measurement $\boldsymbol{z}_t$ given the state of the particle $\boldsymbol{s}_t$. A particle which is consistent with the measurement will be assigned a high weight, making it more likely to be re-sampled during the next update relative to a particle that is not consistent with the measurement. The result is that over multiple updates more and more of the particles that closely match the observed measurements are generated, and particles which do not match the measurements die out.

5.1   Prior Generation

In previous work, a simple uniform prior was generated based on visible access points in order to solve the scalability problem [6]. Here we present an improved algorithm which takes into account the observed RSSI values of each visible access point. During initialisation the WiFi hardware is used to scan for visible access points and obtain corresponding RSSI measurements. Let the vector of observed RSSI values be $\boldsymbol{z}_0 = (\mathrm{ap}_1, \mathrm{ap}_2, ..., \mathrm{ap}_n)$, where ap_n is the RSSI observed for access point n.

To constrain the initial location of the user we first compute the Euclidian distance between $\boldsymbol{z}_0$ and the vector $\boldsymbol{c}_i$ of mean RSSI values for each cell (indexed by i) in the radio map. A prior region is then generated, consisting of all cells for which the distance is under a threshold $\epsilon = 9.5\,\mathrm{dBm}$, which was determined experimentally to ensure a low probability of failure as described in Section 5.3. Formally the prior region is given by

$$R_{\mathrm{prior}} = \bigcup_i \mathrm{Cell}_i \mid \mathrm{ndist}(\boldsymbol{z}_0, \boldsymbol{c}_i) < \epsilon \tag{2}$$

where Cell_i is the set of polygon fragments in the ith cell and $\mathrm{ndist}(\boldsymbol{z}_0, \boldsymbol{c}_i)$ is the normalised Euclidian distance between $\boldsymbol{z}_0$ and $\boldsymbol{c}_i$ as described below. The initial distribution of particles is generated uniformly across the prior region.

Since we are deploying our tracking system in a large building, careful consideration must be paid to computing the distance when different access points are visible at different locations in the building. This is a problem rarely addressed in existing literature, since most experimental deployments of RF-based location systems are small and implicitly assume that all access points are visible at all locations in the building. Consider the following scan and cell vectors, where a question mark indicates that an access point was not observed:

$$z_0 = (-82, -73, ?) \tag{3}$$
$$c_1 = (-84, -75, ?) \tag{4}$$
$$c_2 = (-84, -75, -75) \tag{5}$$

As a first attempt we might choose to compute the distance between the observation and a cell using only the access points for which measurements are present in both vectors. This approach computes equal distances to both cells, yet intuitively the distance to cell 1 should be smaller, since an access point that is strongly visible in cell 2 was not observed by the user. Our solution is to compute the Euclidian distance over all access points, filling in missing readings in both the cell and the measurement vectors with the weakest RSSI value that can be measured by the WiFi hardware, in this case -96. Each distance measurement is then normalised by dividing by the number of access points that are visible from the cell according to the radio map. This makes it possible to compare distance measurements between areas where different numbers of access points are visible. Note that this normalisation penalises a missing value in the cell vector more than a missing value in the measurement vector:

$$z_{0a} = (-82, -73, ?) \tag{6}$$
$$c_{1a} = (-84, -75, -75) \tag{7}$$
$$\text{ndist}(z_{0a}, c_{1a}) = 7.1\,\text{dBm} \tag{8}$$
$$z_{0b} = (-84, -75, -75) \tag{9}$$
$$c_{1b} = (-82, -73, ?) \tag{10}$$
$$\text{ndist}(z_{0b}, c_{1b}) = 10.6\,\text{dBm} \tag{11}$$

This is a desirable property, since each cell in the radio map typically contains many measurements. If an access point were visible from a cell, then you would expect that it would have been sighted in at least one of them. In contrast the measurement vector z_0 is from a single scan, and an access point that it is possible to sight from the cell may have been obscured by the user's body or some other object in the environment.

Figure 9 shows the result of using R_{prior} in the same example as shown in Figure 5. Note that the prior region covers only a small proportion of the total floor area. Since the number of particles is proportional to the floor area that they cover, the computational cost of localisation has been reduced. By using the prior the number of distinct clusters is reduced from eight to three, since only three of the clusters shown in Figure 5 originated from inside the prior region. Hence

Fig. 9. The remaining clusters for the same example as in Figure 5 when R_{prior} (indicated by the shaded region) is used. The cluster at the bottom right corresponds to the user's true position.

the uncertainty due to symmetry in the environment has also been reduced, but the user's location has not been uniquely identified. One problem which a prior region does not solve is distinct clusters that arise due to rotational symmetry and originate from the same location. For example the top left and bottom right clusters shown in Figure 9.

5.2 Localisation

To remove clusters which arise due to only rotational symmetry, we generate containment regions for each step during the localiasation process. For a step at time t we generate the region

$$R_t = \bigcup_i \text{Cell}_i \mid \text{ndist}(\boldsymbol{z}_t, \boldsymbol{c}_i) < \epsilon \tag{12}$$

where $\boldsymbol{z}_t$ is the last WiFi scan to be completed prior to the completion of the step. We use this as a containment measurement with distribution

$$p(\boldsymbol{z}_t|\boldsymbol{s}_t) = \begin{cases} 1 & \text{if } \boldsymbol{s}_t \text{ is in } R_t \\ 0 & \text{otherwise} \end{cases} \tag{13}$$

which sets the weight of all particles outside the region to 0. These measurements remove clusters of particles which arise due to rotational symmetry in the environment. Figure 10 shows the result of using both the prior and containment measurements. Only a single cluster remains, which corresponds to the true location of the user.

Fig. 10. The remaining clusters for the same example as in Figure 5 when both R_{prior} and containment measurements are used. The shaded area indicates the most recently applied containment measurement.

5.3   Robustness

In the solution presented above we have opted to use containment regions, where the probability $p(z_t|s_t)$ is uniform across the entire region. An alternative (and perhaps more conventional) approach would be to compute a probability $p(z_t|c_i)$ for each cell and then apply a measurement with the distribution:

$$p(z_t|s_t) = p(z_t|c_i) \quad \text{where } s_t \text{ is in Cell}_i \qquad (14)$$

The distributions $p(z_t|c_i)$ could be computed directly from the histograms in the radio map and the measurement vector z_t. We chose to avoid the use of probabilistic measurements because they assume that the distribution of RSSI measurements recorded for a given access point at a given location is constant. It has been noted elsewhere that this assumption is not valid [12], since the distribution can be affected by a change in the environment such as the opening or closing of a door. Such a change can introduce a systematic error into the resulting probability distributions, which can in turn draw the particle cloud away from the user's true position.

In our approach we avoid introducing systematic errors into the resulting probability distributions by making the distributions uniform across containment regions. The only problem that can arise is for the cluster corresponding to the user's true position to fall outside the containment region. To ensure that this is unlikely to occur, we processed just under one hour of trace data (distinct from that used to construct the radio map) gathered by the system. After each step during tracking, we calculated the normalised distance ndist(z_t, c_i) for each cell overlapped by the single cluster corresponding to the user's position. The cumulative distribution of the calculated distances is shown in Figure 11. This

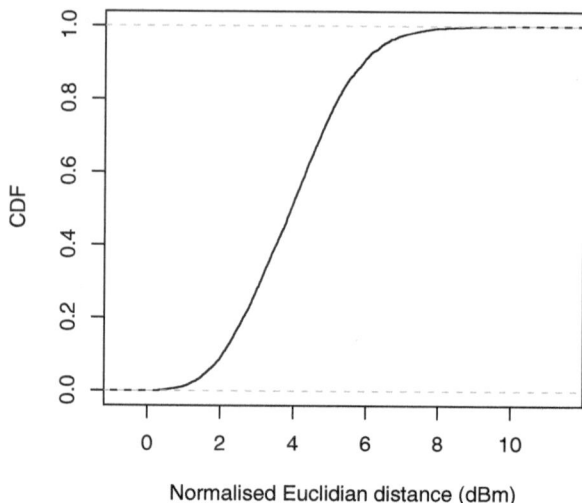

Fig. 11. The cumulative distribution of ndist(z_t, c_i), for cells overlapped by the correct cluster of particles

distribution shows that for a threshold $\epsilon = 9.5$ dBm, the probability of the cluster falling outside the containment area is 0.00028 for a single step. We have not found any localisation to require more than 60 steps at this threshold. If 60 steps are required, the overall probability of failure is 1.7%. When a failure does occur, the localisation process is simply restarted.

6   Localisation Performance

In this section we quantify the localisation improvements provided by containment measurements generated using a radio map. To do this we randomly select 50 starting points from the first 30 minutes of a 35 minute continuous trace of the user moving throughout the building. For each starting position, we run the localisation process without using the radio map, using only R_{prior}, and finally using both R_{prior} and containment measurements.

The results obtained for a single starting position are shown in Figure 12. In this example 4,478,049 particles were used in the prior and 82 steps were required to localise the user without the radio map. When R_{prior} was used to constrain the initial location of the user, the prior consisted of 190,850 particles and 68 steps were taken before localisation. When both R_{prior} and containment measurements were used, the prior contained 228,001 particles and localisation required only 31 steps. The number of particles in the prior for the two cases where R_{prior} was used differ only due to the random process which is used to generate the priors [6]. Note that the number of particles is well under 800,000 when the radio map is used, which is the number of particles that can be updated in real time by our current implementation.

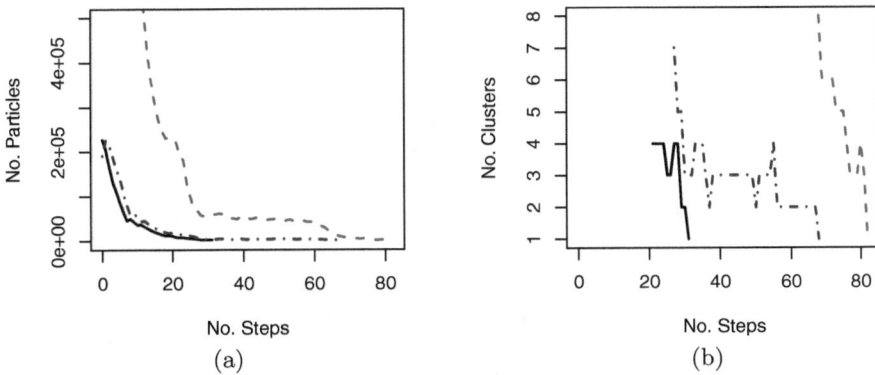

Fig. 12. The data gathered for a single localisation without using the radio map (red dashed), using R_{prior} only (blue dot-dashed), and using both R_{prior} and containment measurements (black solid). (a) The number of particles generated at each step. (b) The number of clusters at each step (clustering is only performed when the number of particles is below 10,000).

Table 2. Summary of localisation performance

	Radio map use	Mean	Std-dev	Max	Min
	None	4,449,156	44,195	4,482,151	4,371,901
Particles in prior	Prior only	535,493	446,751	1,435,550	211,850
	Throughout	539,644	462,869	1,493,701	228,001
	None	76	14.1	96	63
Steps until localisation	Prior only	55	24.7	96	24
	Throughout	38	12.2	53	23

Table 2 summarises the results. Using R_{prior} reduces the average number of particles to 535,493. There is however a large variation between the number of particles generated in different runs, with one run generating a prior of 1,435,550 particles. This is explained by the presence of large open plan areas in our building within which radio signals propagate much more freely than in areas where there are many walls. Hence larger priors are generated in such areas. In all cases the number of particles fell under the real-time limit of the current implementation within 5 steps, allowing the algorithm to catch up before the user was located. Hence using R_{prior} allows localisation in real time. The number of steps taken by the user before localisation is also reduced by using the constrained prior; however the worst case of 96 steps was not reduced, indicating that R_{prior} alone did not always reduce the problem of environmental symmetry.

Using containment measurements in addition to R_{prior} reduced the average number of steps before localisation to 38. The worst case was reduced to 53 steps from 96. Hence the use of containment measurements throughout the localisation process helps to solve the problems which arise due to environmental symmetry.

7   Conclusions and Future Work

In this paper we have developed an algorithm for using radio beacons during the localisation phase of our wearable location system. A radio map is used to generate containment regions within which the user is almost certainly located. These regions constrain the location of the user, allowing localisation to be performed in real time in arbitrarily large buildings. We have also shown that the algorithm reduces uncertainty caused by symmetry in the environment. In tests performed in a large building, a user was located after taking an average of 38 steps, compared with 55 steps when the beacons were only used to generate the prior, and 76 steps when the beacons were not used at all. Hence we have shown the benefit of using containment regions throughout the localisation process.

Our second contribution has been to demonstrate how the existing system can be used to construct radio maps quickly and automatically. We demonstrated our approach by constructing a radio map of a $8725\,m^2$ three-storey building in 2 hours and 28 minutes, with map coverage of 96% at a cell granularity of 3 m. This is of particular interest to developers of existing RF-based location systems, for which radio maps are usually constructed manually in a far more time consuming process.

In the future we plan to investigate how a radio map could be built, shared and updated by multiple users. This would allow the system to respond to changes in the environment including the installation, relocation and removal of radio beacons. We also plan to use our ability to quickly construct radio maps to investivate the temporal and spatial properties of radio beacon signal strengths throughout a large building.

Acknowledgements

The authors would like to thank Andy Hopper and David Cottingham for their insightful comments. This work has been funded by EPSRC.

References

1. Want, R., Hopper, A., Falcao, V., Gibbons, J.: The active badge location system. ACM Trans. Inf. Syst. 10, 91–102 (1992)
2. Addlesee, M., Curwen, R., Hodges, S., Newman, J., Steggles, P., Ward, A., Hopper, A.: Implementing a sentient computing system. Computer 34, 50–56 (2001)
3. Bahl, P., Padmanabhan, V.: RADAR: an in-building RF-based user location and tracking system. In: INFOCOM 2000. Nineteenth Annual Joint Conference of the IEEE Computer and Communications Societies. Proceedings, vol. 2, pp. 775–784. IEEE, Los Alamitos (2000)
4. López de Ipina, D., Mendonça, P.R.S., Hopper, A.: Trip: A low-cost vision-based location system for ubiquitous computing. Personal Ubiquitous Comput 6, 206–219 (2002)
5. Hightower, J., Borriello, G.: Location systems for ubiquitous computing. Computer 34, 57–66 (2001)

6. Woodman, O., Harle, R.: Pedestrian localisation for indoor environments. In: Ubi-Comp 2008: Proceedings of the 10th international conference on Ubiquitous computing, pp. 114–123. ACM, New York (2008)
7. Ashbrook, D., Starner, T.: Learning significant locations and predicting user movement with GPS. In: Proceedings of Sixth International Symposium on Wearable Computers (ISWC 2002), pp. 101–108 (2002)
8. Youssef, M., Agrawala, A.: The horus location determination system. Wirel. Netw. 14, 357–374 (2008)
9. Smailagic, A., Siewiorek, D.P., Anhalt, J., Kogan, D., Wang, Y.: Location sensing and privacy in a context-aware computing environment. IEEE Wireless Communications 9, 10–17 (2001)
10. Castro, P., Chiu, P., Kremenek, T., Muntz, R.R.: A probabilistic room location service for wireless networked environments. In: Abowd, G.D., Brumitt, B., Shafer, S. (eds.) UbiComp 2001. LNCS, vol. 2201, pp. 18–34. Springer, Heidelberg (2001)
11. Lott, M., Forkel, I.: A multi-wall-and-floor model for indoor radio propagation. In: Vehicular Technology Conference, 2001. VTC 2001 Spring. IEEE VTS 53rd, vol. 1, pp. 464–468 (2001)
12. Kaemarungsi, K.: Distribution of WLAN received signal strength indication for indoor location determination. In: 1st International Symposium on Wireless Pervasive Computing, pp. 1–6 (2006)
13. Wang, H., Lenz, H., Szabo, A., Bamberger, J., Hanebeck, U.: WLAN-based pedestrian tracking using particle filters and low-cost MEMS sensors. In: 4th Workshop on Positioning, Navigation and Communication, 2007. WPNC 2007, pp. 1–7 (2007)
14. Haeberlen, A., Flannery, E., Ladd, A.M., Rudys, A., Wallach, D.S., Kavraki, L.E.: Practical robust localization over large-scale 802.11 wireless networks. In: Mobi-Com 2004: Proceedings of the 10th annual international conference on Mobile computing and networking, pp. 70–84. ACM, New York (2004)
15. Ocana, M., Bergasa, L., Sotelo, M., Nuevo, J., Flores, R.: Indoor robot localization system using WiFi signal measure and minimizing calibration effort. In: ISIE 2005. Proceedings of the IEEE International Symposium on Industrial Electronics, vol. 4, pp. 1545–1550 (2005)
16. Weiler, K., Atherton, P.: Hidden surface removal using polygon area sorting. SIG-GRAPH Comput. Graph. 11, 214–222 (1977)

PL-Tags: Detecting Batteryless Tags through the Power Lines in a Building

Shwetak N. Patel[1], Erich P. Stuntebeck[2], and Thomas Robertson[2]

[1] Computer Science & Engineering, Electrical Engineering, & Dub Group
University of Washington
Seattle, WA 98195 USA
shwetak@cs.washington.edu
[2] College of Computing, School of Interactive Computing
Georgia Institute of Technology
Atlanta GA 30332 USA
{eps,troomb1}@cc.gatech.edu

Abstract. We present a system, called PL-Tags, for detecting the presence of batteryless tags in a building or home through the power lines. The excitation (or interrogation) and detection of these tags occurs wirelessly entirely using the powerline infrastructure in a building. The PL-Tags proof-of-concept consists of a single plug-in module that monitors the power line for the presence of these tags when they are excited. A principal advantage of this approach is that it requires very little additional infrastructure to be added to a space, whereas current solutions like RFID require the deployment of readers and antennas for triggering tags. An additional benefit of PL-Tags is that the tags are wirelessly excited using an existing phenomenon over the power line, namely electrical transient pulses that result from the switching of electrical loads over the power line. We show how these energy rich transients, which occur by simply turning on a light switch, fan, television, *etc.*, excite these tags and how they are detected wirelessly over the power line. We contend that the PL-Tag system is another class of potential batteryfree approaches researchers can use for building pervasive computing applications that require minimal additional infrastructure.

Keywords: Ubiquitous computing, Pervasive computing, Tagging, Sensing, Sensors, Power lines, Hardware.

1 Introduction and Motivation

Today's computers have seen an astonishing increase in computational and storage capabilities. Battery technology, however, has not followed the same desirable trend. The success of many pervasive and ubiquitous computing systems will be contingent on their self-sustainability. It is neither practical nor desirable to replace hundreds of batteries in large distributed sensor deployments. Scavenging energy available in the environment for use by low-power sensors and electronics offers an ideal solution for many mobile and ubicomp applications.

Motivated by the success of RFID technology, researchers have developed ways to wirelessly transmit energy for the purpose of powering identification tags and sensors

H. Tokuda et al. (Eds.): Pervasive 2009, LNCS 5538, pp. 256–273, 2009.

[15, 20, 25]. The advantages of these batteryless solutions are their potential to be long-lived and their reduction in size requirements. Although these solutions are attractive for future ubicomp applications, they still require the deployment of custom hardware throughout a home or building (antennas and readers in the case of RFID-based systems). In addition, although these solutions may eventually become more cost-effective on an individual unit basis, they are not without some drawbacks. For example, having to install and maintain a collection of readers is time-consuming, and the large number of readers required for coverage of an entire space increases the number of potential failure points.

Inspired by this desire to provide a practical batteryless solution and building on the current trend of leveraging already existing infrastructure [6, 17, 18, 19, 24], we present a proof-of-concept system, called PL-Tags, for detecting the presence of batteryless tags in a building or home through the power lines. The PL-Tag system uses a building's electrical infrastructure to wirelessly trigger and interrogate those tags. The excitation and detection of these tags occurs entirely using the powerline infrastructure. Electrical transient pulses produced by the switching of electrical loads over the power line trigger the tags. These transients occur naturally through the switching of certain electrical loads in a building or home. The power lines naturally radiate these broadband transients as RF energy, which excite the tags, causing them to inductively couple their response back over the power line. A plug-in module that continually monitors the power lines senses the tags' responses.

The PL-Tag system requires only the installation of a single plug-in module that connects to an embedded or personal computer. Unlike RFID, PL-Tags does not require the addition of readers or antennas in the space since the power lines act as the antenna and the plug-in module acts as the reader. In addition, the interrogating signal that excites the tags is a naturally occurring phenomenon found in any electrical system. This opens up the possibility for new interaction where human-initiated electrical events, such as turning on a light or television, cause the interrogation of the tags. Although our current proof-of-concept is limited in its read range (~30-50 cm from the power line), we highlight a new strategy in batteryless sensing and spur interest in this research area.

In this paper, we report the results of a series of experiments to determine the viability of this approach in a variety of settings. We also report the performance results in terms of distance and detection accuracy in a home and office building. Results show that the PL-Tag system can trigger and detect tags up to 30-50 cm from the nearest electrical wiring. Finally, we discuss potential applications, the limitations and potential future improvements for this approach.

2   Related Work

Common examples of battery-less systems are those that harvest energy from the environment. Solar power is the most popular and well-known example. Many calculators, landscape lights, and backup power systems use solar energy. Researchers have created a solar-powered node composed of sensors, a processor, an RF transmitter, and solar cells, to collect, process, and forward sensor data to a central wearable device in [3]. They showed the possibilities of energy harvesting through ordinary exposure to sunlight and indoor light for an entire day. Although this solution shows promise in a variety of applications, there may be concerns in dimly lit areas and the sensor may not be enclosed within other objects. Recent research in this area has also targeted temperature differentials [9, 23], low-level mechanical vibrations [1, 2, 14, 26], and other sources of kinetic energy, such as the natural motion of the human body [11, 21, 22].

Tagging systems, such as passive RFID systems, use inductive coupling to remove the need for batteries in the tags. RFID systems consist of an antenna and reader combination that interrogates nearby tags. The MediaCup [7] and commercial projects like electric toothbrushes use inductive coupling to charge a small battery or capacitor for extended use. Other researchers have extended the concept of inductive coupling to remotely charge sensor packs [4]. Researchers have demonstrated wirelessly powering RFID-based sensors using long-range RFID readers [15, 20, 25]. For example, the Wireless Identification and Sensing Platform (WISP) is a passive UHF RFID tag that uses a low-power microcontroller for sensing and communication. More general-purpose wireless power transfer techniques have been demonstrated to power consumer electronic devices [10]. Although these solutions are encouraging for practical ubicomp sensor installations, our focus is to present a different strategy for potentially powering sensors - namely, leveraging an existing infrastructure like the power line as the interrogator. Both approaches have advantages and disadvantages. Thus, specific applications will ultimately dictate the necessary complexity of the additional infrastructure. One can imagine RFID sensors being more appropriate for applications that require the detection of a large number of tags and very high accuracies, where as PL-Tags would be more appropriate for home applications, where there might not be as many tags and costs plays an important role.

Alternatives, such as vibration and motion-powered devices, do not require external powering sources, but the harvesters usually result in extra bulk. Physical motion and actuation, which convert kinetic energy to electrical has proven to be successful. MIT's self-powered switch is capable of wirelessly transmitting a static identification number using the power generated from pressing a spring-loaded igniter switch [16]. The flexing of a piezo plate mounted on shoes can produce enough energy to power a simple wireless transmitter [11]. Commercial products like the Faraday Flashlight use continuous physical motion and magnetic induction to power an LED [5]. Hand cranks have also been integrated into small flashlights and radios to produce temporary power. An off-body example of physical actuation is the Electro-Kinetic Road Ramp in the UK, which generates power each time a car drives over its metal plates [2]. This power is stored to run streetlights and traffic signals.

Detecting tagged objects has also been a long studied problem in ubicomp. Traditional examples include static labels (*e.g.*, a barcode or 2-dimensional glyph) that are placed on an object and read or scanned by some form of reader device. Barcode solutions are typically limited to line-of-sight locations of a camera or optical reader. MIT's FindIT Flashlight uses active tags that respond to an incident laser [12]. A modulated laser signal is used to wake-up and interrogate the tags. An indicator light on the tag notifies the user that the desired object has been found.

Previous work has focused on leveraging the existing infrastructure in a home or building to collect signals at a single location, called infrastructure mediated sensing [17]. Researchers have recently begun exploring the use of existing home infrastructure to detect human originated events, such as using the power lines [18, 19, 24]. A single plug-in sensor can classify events, such as the actuation of a light switch, through the analysis of noise, transduced along the power line, from the switching and operation of electrical devices [18]. Other prior work includes the PowerLine Positioning system [19, 24], which uses existing power line infrastructure to do practical localization within a home.

3   PL-Tags Approach and System Details

The PL-Tags system consists of wireless tuned passive tags and a power line interface module (seeFigure 4) that is plugged into any electrical outlet in a building or home. The power line interface is attached to a data collection apparatus and a PC that performs the analysis on the electrical signal. Tags are triggered wirelessly by electrical transients naturally produced over the power line from the switching of electrical loads. These transients are radiated as RF energy that is received by the tags. Tags resonate at their tuned frequency and inductively couple over the power line in response to the electrical transients. The powerline interface isolates and detects both the transient pulse and the resulting "ringing" waveform from the coupled tags over the power line. For this initial proof-of-concept, we tuned each tag to a different frequency in order to identify the triggered tag. The broadband nature of the electrical transient facilitates this effect.

3.1   Theory of Operation

Electrical transient pulses are short-lived, high-frequency increases in voltage on the power lines (see Figure 1). Some sources of these transients are the result of the arcing contacts in switches and relays, which are caused by the abrupt switching of high current loads and the sudden actuation of motors and other inductive loads [8,13]. These types of transients are usually normal mode (occurring between hot and neutral). In addition, they are both broadband and high in energy across the entire band, with voltage surges exceeding hundreds of volts. Typically, the rise times are on the order of a few microseconds, with the decay times approximately tens to hundreds of microseconds (see Figure 2). The frequency components of the waveform range from a few kHz to over 100 MHz. Peak voltages for typical electrical devices found in a home are hundreds of volts to a few thousand volts. For example, a motor-type load, such as a fan, will create a transient noise pulse when it is first turned on and when it is turned off. In addition, a light switch connected to a 100 W incandescent light or the relay switch on a television will produce an electrical transient pulse.

Patel *et al.* showed an approach that can be used to train on the duration and frequency characteristics of these transient electrical noise pulses to later determine which device has been actuated when a transient is seen on the power line [18]. Depending on the switching mechanism, the load characteristics, and length of the transmission line, these impulses can be very different. Fig. 2 shows the energy distribution of an electrical transient pulse caused by a light switch connected to a 1000 watt load being actuated. Note the rich number of high amplitude frequency components for each pulse and their relative strengths. If we take a different electrical load, such as a television, the resulting electrical transient pulse is different, but there is still considerable energy across a broad portion of the frequency spectrum.

The operating principle of PL-Tags is that the transient surge can be used to wirelessly induce an RC circuit to resonate (*i.e.* near-field effect), and in turn cause the resonator to inductively couple back over the power line. By tuning the resonator to a known frequency that is both within the high energy band of the transient and is not a harmonic of another signal source, such as a radio station, the presence the frequency can be seen over the power line as an exponentially decaying "ringing" waveform (see Figure 3). In this case, the decay rate is related to the resistance between the resonator and the electrical line. Since there may still be potential ambient noise present at the tuned frequency of the resonator, we use the bounds of the electrical transient to help detect the presence of the ringing waveform. Fig. 2 shows an example of a transient with and without the excited resonator. Note the flat tail at the end of the second transient at 180 kHz, which is the ringing.

The onset of the transient is used to determine the noise floor of the power line right before the onset of the pulse and the end of the transient is used to mark where the ringing waveform should occur. Tuning each PL-Tag resonator to a different frequency that is not a harmonic of another is a simple way to differentiate between tags. We can imagine extending this concept to directly harvesting the power and implementing a low-power load modulation scheme or a DTMF-based scheme for identifying the tags, but our intent with this work was simply to demonstrate the concept. However, the frequency-differentiation approach could be sufficient for applications that only require a small number of tags. Also, note that because PL-Tags relies on the near-field effect, coupling occurs only over short distances (tens of centimeters), with actual distance being dependent on the strength of the transient.

For devices that produce continuous noise, they are still bounded by some transient phenomena, but also exhibit electrical noise during their powered operation. For this class of noise, we are interested in the bounding transients although certain inductive loads might provide continuous energy sources during their operation. For this paper, we only look at these electrical transients.

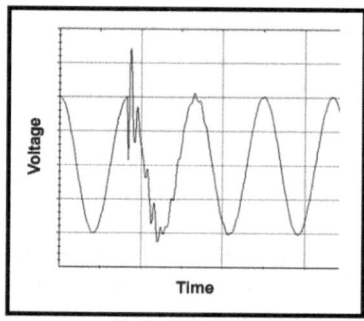

Fig. 1. Example of an electrical transient voltage spike

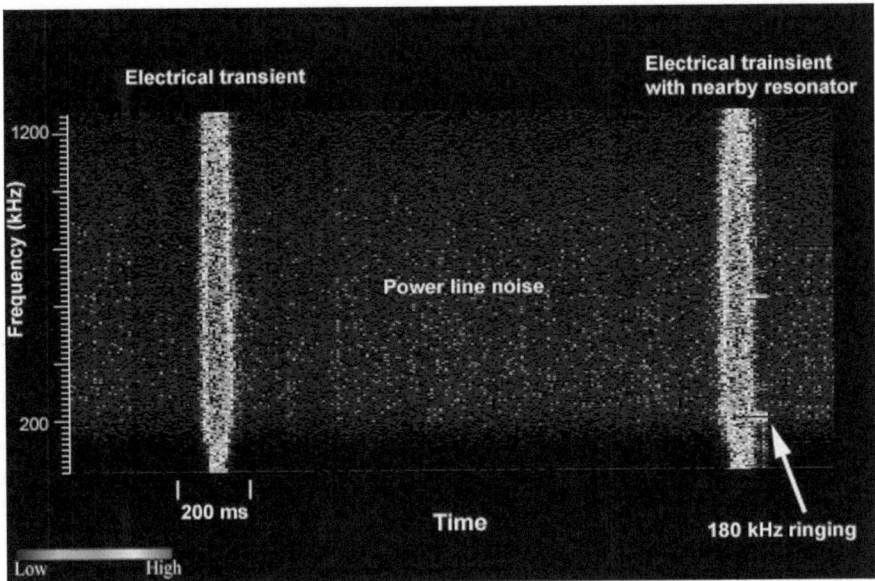

Fig. 2. Frequency spectrum of transients with and without a 180 kHz resonator placed near the power line. The second transient shows the ringing at the 180 kHz at the end of the transient.

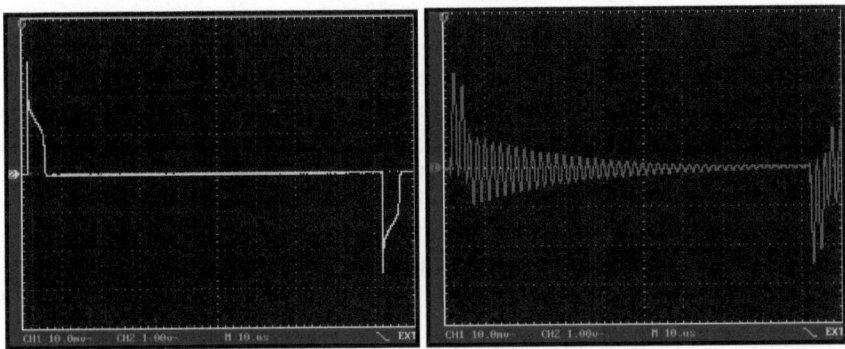

Fig. 3. Zoomed in view of a transient pulse (left) inducing the ringing waveform (right) of a tuned resonator

Finally, it is interesting to look at the practicality of using the transients as the "triggering" source from a regulatory point of view. In the United States, the Federal Communications Commission (FCC) sets guidelines as to how much electrical noise AC-powered electronic devices can conduct back onto the power line (Part 15 section of the FCC regulations). Device-generated noise at frequencies between 150 kHz-30 MHz cannot exceed certain limits. Regulatory agencies in other countries set similar guidelines on electronic devices. However, this mainly applies to continuous noise output of electronic devices, such as those that have solid state switching power supplies. Controlling high power electrical transients is a more challenging task and is

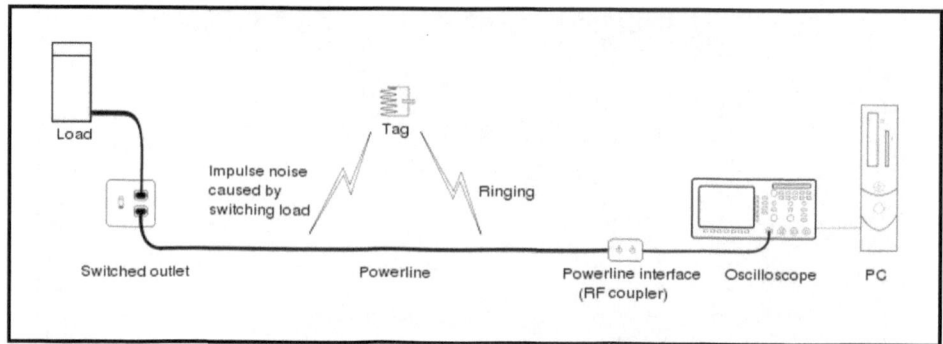

Fig. 4. Overview and high-level diagram of the PL-Tag system

often not directly regulated. The generation of electrical transients extends beyond the control of an individual device such as a light switch or television. It is the result of the electrical load of other devices on the power line and the transmission lines themselves. Regulatory agencies usually require sensitive electronic devices to implement proper transient filters within their own devices (IEC 61000-4-4). Thus, there is value in using this naturally occurring phenomenon for interrogating at power levels that may not be allowed by certain regulatory agencies.

3.2 Experimental Apparatus Details

The experimental setup consisted of a plug-in power line interface module, an electrical transient pulse generator, and prototypes of PL-Tags tuned at different resonant frequencies (see Figure 4). The plug-in power line interface was custom hardware consisting of two filtered outputs (see Figure 5). The first was a bandpass-filtered output with a passband of 100 Hz to 250 kHz. The second output was bandpass-filtered with a 100 kHz to 50 MHz passband. Both filtered outputs incorporated a 60 Hz notch filter in front of their bandpass filters to remove the AC power frequency and increase the effective dynamic range of the sampled data.

The outputs of the powerline interface were connected to a multi-input 2.5 GHz oscilloscope with a PC interface, which allowed us to capture and transfer signal traces for post processing. Although most of our analysis can be performed in real-time, this method allowed for flexibility during the experimentation. The device produced 14-bit resolution samples with a sampling rate of 40 GS/sec.

The tags consisted of very minimal hardware in which each tag had a tuned LC circuit (see Figure 6 and Figure 7). The simplicity is encouraging from a size and cost point of view. Each tag was equipped with an adjustable capacitor for tuning and a custom wound antenna. We used both a signal generator connected to a transmission antenna and our transient pulse generator to tune the antenna to the desired frequency. For our experiments, we built three different tags. Two of the tags incorporated an air core antenna design, tuned at 135 kHz and 180 kHz. The third was a ferrite rod-based antenna tuned at 135 kHz (see Figure 7). For additional experimentation, we also built a full wave rectifier circuit and loaded it down with a 10k resistor to measure the actual voltage transfer from the transient pulses (see Figure 6).

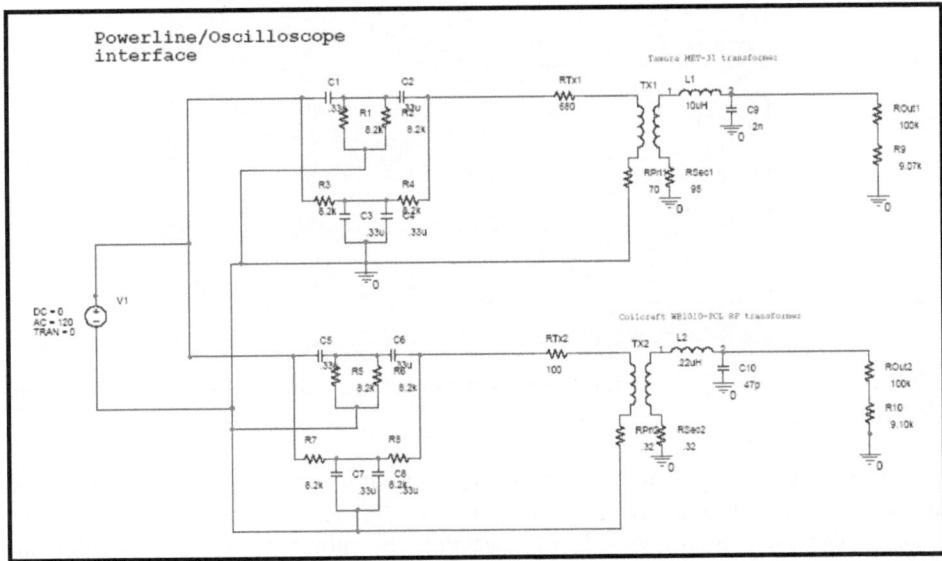

Fig. 5. The schematic of our powerline interface device used in detecting the transients and the presence of a PL-Tag

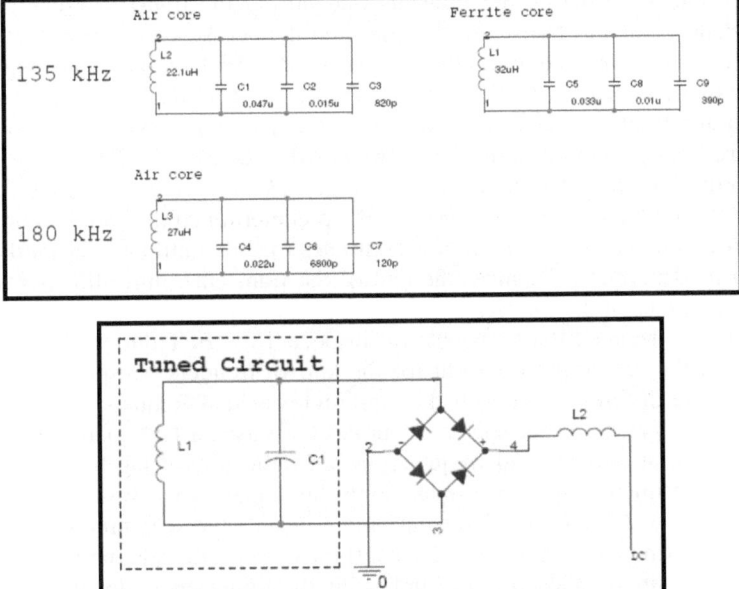

Fig. 6. Top: Schematic for the LC resonators used for the tags. Bottom: Setup for measuring the power being coupled to the tag in response to the electrical transients

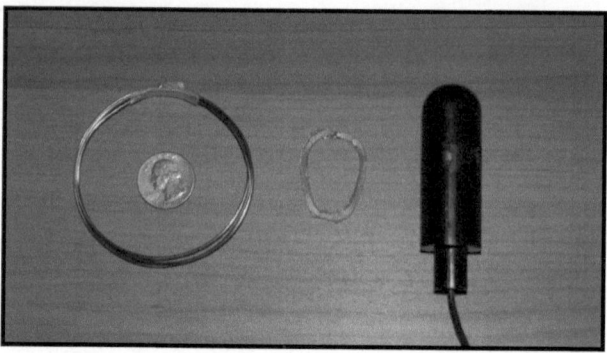

Fig. 7. Example PL-Tags. We experimented with both open air coil and ferrite rod antenna designs.

For the software components of our system, we built a C++ application to sample the digital oscilloscope and perform a Fast Fourier Transform (FFT) on the recorded signal. The application allowed us to visualize the entire spectrum and produce the plots presented in this paper. In addition, the application also detected the transient pulses for the recorded signal. We used a simple sliding window algorithm to look for abrupt changes in the input line noise using a threshold value determined through experimentation. The threshold was proportional to the difference between the FFT of the power line noise before the transient and during the transient noise pulse. The sliding window for the transient detector used 1-microsecond samples, which was averaged from the data acquired after performing the FFT on data collected from the powerline interface hardware. Each sample consisted of frequency components and its associated amplitude values in a vector form. Each vector consists of amplitude values for frequency intervals ranging between 50 kHz and 5 MHz. Then a simple Euclidean distance between the previous vector and the current window's vector was computed. When the distance first exceeded a predetermined threshold value, the start of the transient is marked. The window continues to slide until there is another drastic change in the Euclidean distance (the end of the transient), thus allowing us to find the bounding times of the transient.

After having isolated the transient, we inspected the FFT during the onset of the transient and the end of the transient for the known frequency response of the tuned tags. Using the difference between the amplitudes at that frequency, we determined whether a tag is present in the space. In our case, we used a 10% increase in the peak-to-peak voltage at the centered frequency as an indicator for tag being present. Because our experiments were in a quasi-controlled space, this was sufficient for our detection scheme. A future implementation would involve searching over the end of the transient to find the ringing waveforms. If the decay rate and length of the decay is know, those features can also be used in finding the signal (more details in Section 5).

3.3 Creating the Transients

We built a portable test apparatus that would reliably and consistently produce an electrical transient noise pulse generated by standard electrical devices found in a building (see Figure 8). Although we could have use almost any device connected to the power line, our apparatus allowed us the flexibility to move the test equipment around. We could have used a signal generator connected to a linear amp to produce similar high-powered noise pulse. However, we wanted to take care to try to match the transient characteristics of standard devices.

The transient generator consisted of a self-contained galvanized gang box with 2 grounded electrical outlets and a 120 V 20A light switch. There was a power cable, which plugged into any electrical outlet power the gang box. The two electrical outlets on the box allowed us to apply various loads in generating the transients. We experimented with a variety of different resistive loads, such as a standard 200 watt light, 1000 watt flood lights, and a space heater to confirm its operation.

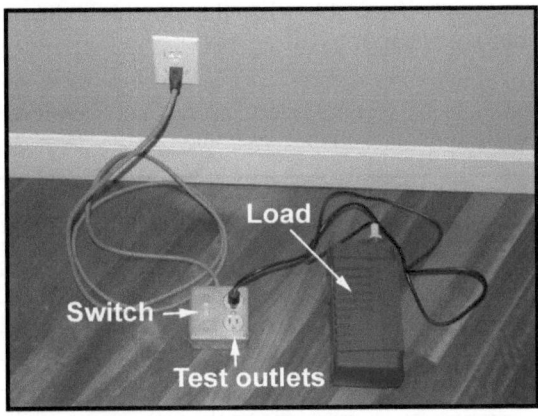

Fig. 8. Experimental electrical transient noise generator box. The gang box includes a switch connected to test outlets where the loads are plugged in.

4   Feasibility and Performance Results

In this section, we discuss a collection of experiments conducted to determine the feasibility of our transient triggered resonator approach. We attempt to address the following important questions of the approach, including (1) what types of devices generate the necessary electrical transient pulses for this approach, (2) what is the response of the tuned circuit to the electrical transient pulses, and (3) how well can the ringing waveform be detected over the power line?

4.1   Transient Generators

Table 1 shows the various household devices used during our experiments to explore the devices that would produce a transient impulse capable of triggering a PL-Tag. The table shows the devices that produced a discernable ringing waveform at the powerline interface module. For these experiments, a PL-Tag was placed approximately a few centimeters from where the device was actuated and the receiver module was plugged in the outlet in the same or nearby room. We found that most loads that drew under 1 A of current did produce detectable transients, but the ringing waveform was not present with our test equipment. A different weak signal detection approach may help discern the signal over the noise. This is what accounts for devices like the microwave door and refrigerator door not being able excite the PL-Tags. We expect other low-power loads would exhibit the same problem. In addition, some microcontroller-based appliances may not directly produce a transient pulse until an internal relay is switched. However, most high current loads like an oven or dryer and inductive loads like an exhaust fan produced enough power to allow us to observe the ringing waveform of the resonator at the powerline interface. We also observed that transients from an inductive load such as a motor create a very strong kickback spike when the load is turned off.

The 1 A load was borderline - in most of the cases it created a detectable signal, however, the signal was only detectable when the plug-in module was located close to the tag location (about 1-2 meters).

Table 1. Examples of devices that produced electrical transients capable of causing a PL-Tag to resonate and being able to be detected off the power line

Devices Observed	Transient Produced Detectable Resonating Signal
100 W Incandescent light and wall switch	Y
Microwave door light	N
Oven light/door	Y
Electric stove	Y
Refrigerator door	N
Electric Oven	Y
Bathroom exhaust fan	Y
Ceiling fan	Y
Garage door opener	Y
Dryer	Y
Dishwasher	Y
Refrigerator compressor	Y
Lights via a dimmer wall switch	Y
Garbage disposal	Y
Drill	Y
Microwave Oven	Y
Television (CRT, plasma, or LCD)	Y

4.2 Antenna Response of Tuned Tags

Figure 9 shows the frequency response of the tuned resonator in response to an electrical transient 30 cm from the power line for the 180 kHz tag. The peak indicates the resonate frequency of the tag. Clearly, there is noise across the entire band from the transient, but the response is strong enough to resonate and be detected at the receiver module depending on the distance of the tags to the power line and the distance to the module. Between the air core and ferrite rod antenna design, no one design stood out to be significantly better than the other. Thus, we conducted most of our experiments with the air core, because of its more compact size.

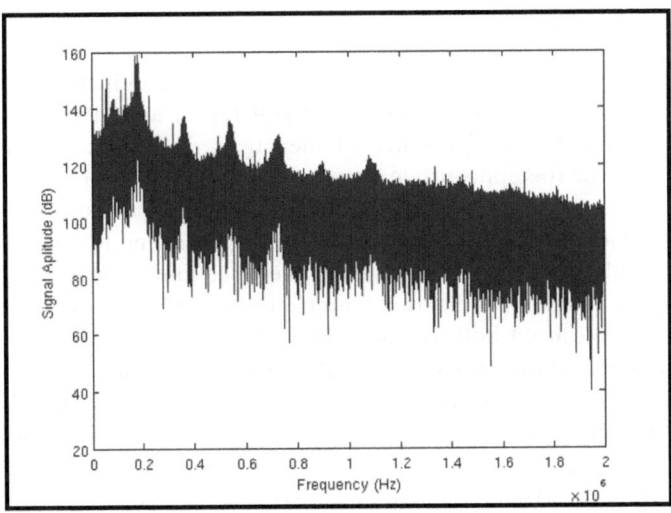

Fig. 9. Frequency response of a 180 kHz tuned resonator excited by an electrical transient. Note the peak at the resonate frequency

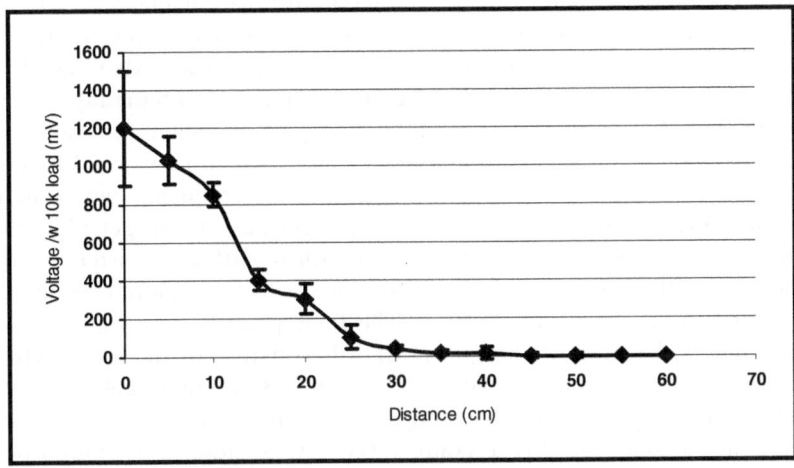

Fig. 10. Voltage across tuned tag loaded down with a 10k ohm load at various distances from the power line

In Fig. 10, we show the voltage induced off the PL-Tag when run through a Schottky-based full wave rectifier and a 10k ohm load. Readings were taken at increasing distances and there is a quick drop off as the tag moves beyond 30 cm from the closest power line. Although these are very short-lived responses from the transient pulse, there might be some hope of actually harvesting power for potential onboard processing with a new antenna design and using very low noise components.

4.3 Tag Identification over the Power Lines

In order to stay with a batteryless tag, we use a simple inductively coupling approach to detect the presence of the tag. Each tag is assumed to be tuned at a different frequency and the presence of the tag is determined by searching the frequency space for the appropriate ringing tags at the end of an electrical transient pulse. We used this simple approach to demonstrate the proof-of-concept, but could imagine extending it incorporate a low-power modulation scheme.

For the tag identification experiments, we used our transient generator box connected to an inductive load (a 15-amp power drill) and a resistive load (a 500-watt light). We used both of these loads to generate transients in different locations around a home and office building to determine how accurately we could detect the ringing response of the tags over the power line. We used our transient detection application (see Section 3) to determine the absence or presence of the tags. Figure 11 shows the results of the experiment at various locations using the two different transient generators. We report the read accuracy (the percentage of time the tag was detected) as various read distances. We controlled both the distance between the tag and the electrical system as well as the distance between the tag and the power line interface module. With the power-line interface a fixed 2 meters away, the maximum read distance for the tag was about 40-50 cm, but at very low detection accuracies. The usable range appears to be about 30-40 cm away from the closest electrical wiring. Also, the closer the plug-in module is installed the higher the detection rate. With the tag 20 cm from the power line, the farthest the module could be installed for our proof-of-concept was about 4 meters. Reliable distances at which tags could be detected were in the 2-3m range. Moving to a higher resonate frequency would help with longer read ranges. However, there is a limit, because as we get higher on the frequency band the shorter the ringing waveform will last (faster decay rate) after the end of the transient. This tradeoff needs to be explored in more detail for extending the range of this system. The false positive rate was quite low in our experiments, because of the quasi-controlled nature of the space. However, of the approximately 120 total read cycles, we had 8 unintentional triggers. Most of these occurred in the office building, where other loads were being actuated that were not under our direct control (HVAC system and motion-powered lights). Although we did not experience these, another potential false triggers could result from lightning strikes.

The results also indicate a slight difference in the read accuracies based on the kind of load generating the electrical transient pulse. We would have expected the greater load (the 500-watt light) to produce better results, but it turns out the lower powered inductive load performed better in this case. Possible reasons for this might be that the continuous noise generated by the device contributed a better chance for the tag to be detected or the noise pulse generated by the drill is considerably wider than that of the resistive load.

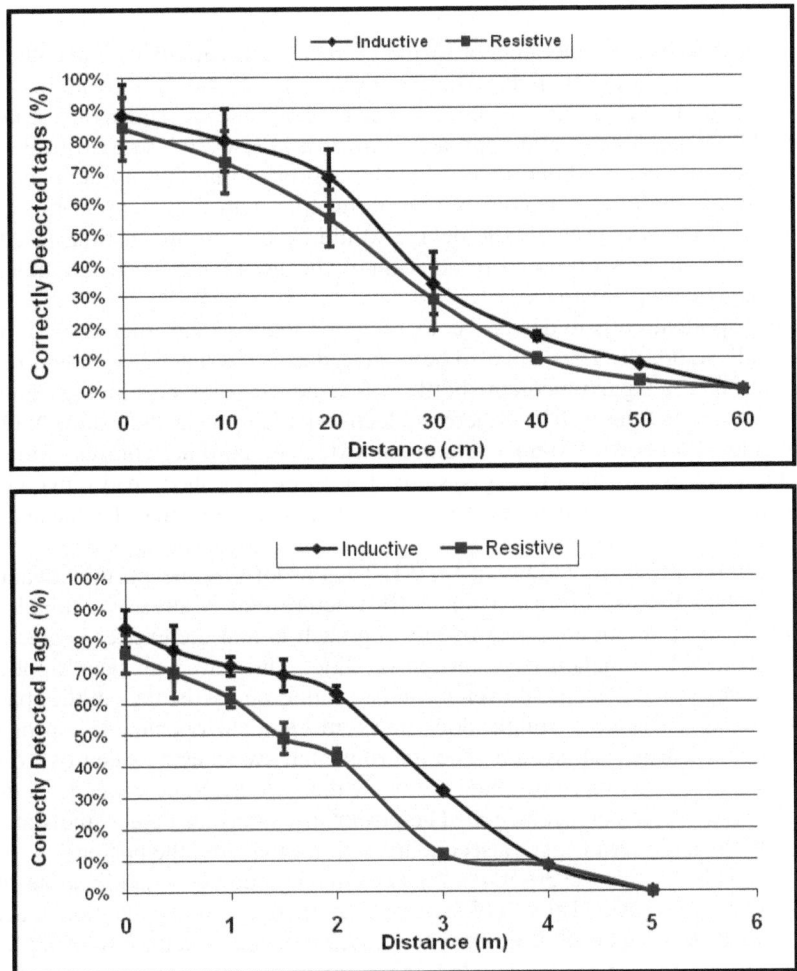

Fig. 11. Average read accuracy of PL-Tags at various locations in a home and commercial office buildings using transients created with an inductive and resistive load. Top: This graph shows the accuracy rate (the percentage of successful detections by the receiver module) for various distance of the tag from a electrical wiring. The position of the receiver module was a fix location 2 meters along the power line from the tag (same room). Bottom: This graph shows the accuracy rate of a tag located at a fixed 20 cm from the electrical wiring, but varying the distance of the plug-in receiver module. The distance is an estimate of the electrical wiring distance between the module and the closes wiring to the tag.

5 Discussion and Future Work

The results of a batteryless tag powered by a transient impulse are encouraging. However, there is still a significant amount of future work that remains. The current detection range is 30-50cm. Despite the short range, there are still some applications and usage scenarios in which this might be feasible. The first is for finding lost items, where tagged items can be queried by a portable device that a person plugs in the wall (similar to our transient generator) or by the nearest electrical device that could generate a strong enough transient pulse. Another use of this approach is the ability to identify who actuated an electrical device in the space if people were to wear the tags as a bracelet. More specifically, the approach shown in [18] would be used to determine which device was actuated and then using the approach outlined in this paper to determine the identity of tag. This would be useful for smart home applications, where the identity of the person using the device would be useful information. In addition, this approach could also be used to determine who uses what electronic devices for energy monitoring applications.

Another application is to detect the absence of a tagged item, such as item missing off of a night stand. PL-Tags can also be used to detect if a tagged book was removed from a shelf or if a tagged medicine bottle was removed from a cabinet (assuming the bookshelf is against the wall and there is electrical wiring near the cabinet). One can also imagine a homeowner simply running an extension cord near places of interest to extend the coverage of the PL-Tag system. The limited or short read range could be used as an advantage when trying to localize where the tag may be located in the environment.

Despite the current limitations of the PL-Tags proof-of-concept, the advantage of this approach is the use of an existing infrastructure that is already available in the space. In addition, a unique aspect of this approach is that the exciter is also a naturally occurring phenomenon over the power line. Thus, we can have applications where human-initiated electrical events can be used to query the tags in the space. For example, we constructed a simple door micro-mechanical open/close sensor that is excited by the nearby wall switch. The micro-switch switches between two different capacitors causing the tag to resonate at one of the two different frequencies depending on the state of the contact switch. The power line interface detects the state of the switch, which can in turn trigger a secondary action (notifying the person).

We acknowledge that the 30-50 cm range limits the number of applications, but extending the range to even 1m would substantially increase the application space. Part of this limitation is because of the near-field nature of the system. Exploring far-field effects by trying to harvest the coupled energy from the transients is a potential way of extending the range. However, we believe the detection range can be extended to 1m by designing a new antenna and employing a weak signal detection scheme for the receiver (similar to those used in powerline communication systems).

Our current approach used minimum components to reduce the complexity, but we can improve the performance by building a tag that has a sharply defined resonate frequency. In particular, one strategy to be considered is to maximize the resonator's Q by using teflon insulated wire and high-Q capacitors and producing the highest practical L/C ratio. The antenna can be enclosed (not necessarily encased) by low-dielectric-constant plastic like ABS or teflon in order to keep some distance between the antenna and other nearby components and objects.

With the current proof-of-concept, our read distance along the power line is also fairly short (3-4 meters). Part of the problem is that the near-field effect has a sharp drop off. So, the use of wavelet transforms instead of a traditional FFT could also be used to help identify the ringing waveform and potentially increase the detection accuracy of weaker signals, which would also help increase the tag read range. A better solution is to employ a weak signal detection scheme. This would work by passing the received signal through two filters. The first filter would be matched to the tag frequency and the second filter at a slightly higher or lower non-tag frequency and then the decay envelopes of the filters would be compared. With no tag present, both envelopes will be similar. With a tag signal, the first filter's envelope will decay more slowly than the second filter's envelope. The filters can be implemented in a DSP chip as autocorrelation filters, where a delayed version of the signal is multiplied by itself, causing reinforcement at a frequency determined by the delay time. This will increase sensitivity at the expense of "smearing" the decay envelopes. A Duffing oscillator will be more sensitive than autocorrelation, however its reference (internal driving) frequency must exactly match the tag frequency.

Although we did not measure it directly, we also found orientation to cause variations in the read accuracy, thus for all the experiments we took care to keep the tag orientation consistent. RFID and other inductively coupled techniques have similar problems. Although the main goal of this paper was to leverage already existing systems, it would be interesting to investigate how we might built a custom transient generator that would "shape" the most effective signal to increase the read distance. A signal can deliberately be generated to excite the tags instead of using transients. Another solution would be to install multiple plug-in modules throughout a home or building that would "repeat" the detection of a tag to other modules. In this case, we would be leveraging the power lines both for interrogating the tags and for communicating back to a central module. Although adding more modules increases the amount of hardware required in the deployment, we believe the potential cost effectiveness and ease of installation are still advantageous.

For simplicity, we mainly focused on a single tag in read range at any given time. However, we did conduct experiments with two tags being inductively coupled simultaneously and were able to identify both of them (Fig. 2 actually shows two ringing waveforms from two different tags). Obviously, in practice there would be many tags that would have to be read simultaneously. This also argues for a modulation scheme or frequency shift keying, which is the focus of our future work. A simple strategy is to chirp the tag. A tag built with capacitors makes a single frequency pulse, where as tags constructed with varactor diodes produce a broader range of sharp frequencies pulses. At the receiver end, the detected chirp will exhibit a decrease in the frequency with time. The chirping may enable decomposing multiple tags appearing in the environment. Another approach is to use a DTMF-like approach with two- or three-frequency tags with frequencies widely spaced to prevent coupling between the coils. Crystals can be used to stabilize a tag's frequency output (by placing the crystal in series with the inductor and capacitor). Finally, another low-power strategy is to use an amplitude modulation. For example, some backscatter RFID tags use a FET across the LC circuit to amplitude modulate the output. PL-Tags can accomplish this by using micropower oscillator chips that would drive the FET.

6 Conclusion

We presented a system, called PL-Tags, for detecting the presence of batteryless tags in a building or home through the power lines. The excitation (or interrogation) and detection of these tags occurs entirely using the powerline infrastructure in a building. The PL-Tags system consists of a single plug-in module that monitors the power line for presence of these tags when they are excited. Although the results of this approach are still preliminary, the primary advantage of this approach is that it requires very little additional infrastructure to be added to home or building. An interesting additional benefit of PL-Tags is that the tags are excited using electrical transient pulses that result from the switching of electrical loads over the power line, which is an existing phenomenon over the power line. We showed how these energy rich transients, which occur by simply turning on a light switch, fan, television, *etc*, can excite these tags and be detected over the power line. This approach highlights another class of batteryfree sensing approaches that leverage already ubiquitous infrastructure in a building, and we intend to expand interest in this line of research. We also discussed potential new ways of improving the performance of this approach and increasing the tag detection range.

References

1. Amirtharajah, R., Chandrakasan, A.P.: Self-Powered Signal Processing Using Vibration-Based Power Generation. IEEE Journal of Solid State Circuits 33(5), 687–695 (1998)
2. BBC News. Ramps Generates Power As Cars Pass (2008),
   http://news.bbc.co.uk/2/hi/uk_news/england/somerset/4535408.stm
3. Bharatula, N.B., Ossevoort, S., Stäger, M., Tröster, G.: Towards Wearable Autonomous Mi-crosystems. In: Ferscha, A., Mattern, F. (eds.) Pervasive 2004. LNCS, vol. 3001, pp. 225–237. Springer, Heidelberg (2004)
4. Deyle, T., Reynolds, M.: PowerPACK: A Wireless Power Distribution System for Wearable Devices. In: The proceedings of ISWC 2008 (2008)
5. Faraday Flashlights (2008), http://www.everlifeflashlight.com/
6. Fogarty, J., Au, C., Hudson, S.E.: Sensing from the Basement: A Feasibility Study of Unobtrusive and Low-Cost Home Activity Recognition. In: the Proc of ACM Symposium on User Interface Software and Technology (UIST 2006) (2006)
7. Gellersen, H., Beigl, M., Krull, H.: The MediaCup: Awareness Technology Embedded in a Everyday Object. In: Gellersen, H.-W. (ed.) HUC 1999. LNCS, vol. 1707, p. 308. Springer, Heidelberg (1999)
8. Howell, E.K.: How Switches Produce Electrical Noise. IEEE Transactions on Electromagnetic Compatibility 21(3), 162–170 (1979)
9. Kishi, M., Nemoto, H., Hamao, T., Yamamoto, M., Sudou, S., Mandai, M., Yamamoto, S.: Micro-thermoelectric modules and their application to wristwatches as an energy source. In: The proceedings of The International Conference on Thermoelectrics (ICT 1999), pp. 301–307 (August 1999)
10. Kurs, K., Karalis, A., Moffatt, R., Joannopoulos, J.D., Fisher, P., Soljacic, M.: Wireless Power Transfer via Strongly Coupled Magnetic Resonances. Science 317(5834), 83–86 (2007)
11. Kymisis, J., Kendall, C., Paradiso, J., Gershenfeld, N.: Parasitic Power Harvesting in Shoes. In: The proceedings of the Second IEEE International Conference on Wearable Computing (ISWC) (October 1998)

12. Ma, H., Paradiso, J.A.: The FindIT Flashlight: Responsive Tagging Based on Optically Triggered Microprocessor Wakeup. In: Borriello, G., Holmquist, L.E. (eds.) UbiComp 2002. LNCS, vol. 2498, pp. 160–167. Springer, Heidelberg (2002)

13. Marubayashi, G.: Noise Measurements of the Residential Power Line. In: The Proceedings of International Symposium on Power Line Communications and Its Applications 1997, pp. 104–108 (1997)

14. Mitcheson, P.D., Green, T.C., Yeatman, E.M., Holmes, A.S.: Analysis of optimized micro-generator architectures for self-powered ubiquitous computers. In: The Adjunct Proceedings of Ubicomp 2002, Goteborg, Sweden (2002)

15. Opasjumruskit, K., Thanthipwan, T., Sathusen, O., Sirinamarattana, P., Gadmanee, E., Poota-rapan, N., Wongkomet, A., Thanachayanont, M.: Thamsirianunt. Self-Powered Wireless Tem-perature Sensors Exploit RFID Technology. IEEE Pervasive Computing Magazine 5(1), 54–61 (2006)

16. Paradiso, J., Feldmeier, M.: A Compact, Wireless, Self-Powered Pushbutton Controller. In: Abowd, G.D., Brumitt, B., Shafer, S. (eds.) UbiComp 2001. LNCS, vol. 2201, p. 299. Springer, Heidelberg (2001)

17. Patel, S.N., Reynolds, M.S., Abowd, G.D.: Detecting Human Movement by Differential Air Pressure Sensing in HVAC System Ductwork: An Exploration in Infrastructure Mediated Sensing. In: Indulska, J., Patterson, D.J., Rodden, T., Ott, M. (eds.) Pervasive 2008. LNCS, vol. 5013, pp. 1–18. Springer, Heidelberg (2008)

18. Patel, S.N., Robertson, T., Kientz, J.A., Reynolds, M.S., Abowd, G.D.: At the Flick of a Switch: Detecting and Classifying Unique Electrical Events on the Residential Power Line. In: Krumm, J., Abowd, G.D., Seneviratne, A., Strang, T. (eds.) UbiComp 2007. LNCS, vol. 4717, pp. 271–288. Springer, Heidelberg (2007)

19. Patel, S.N., Truong, K.N., Abowd, G.D.: PowerLine Positioning: A Practical Sub-Room-Level Indoor Location System for Domestic Use. In: Dourish, P., Friday, A. (eds.) Ubi-Comp 2006. LNCS, vol. 4206, pp. 441–458. Springer, Heidelberg (2006)

20. Smith, J.R., Pauline Powledge, A.S., Mamishev, A., Roy, S.: A wirelessly powered platform for sensing and computation. In: Dourish, P., Friday, A. (eds.) UbiComp 2006. LNCS, vol. 4206, pp. 495–506. Springer, Heidelberg (2006)

21. Starner, T.: Human Powered Wearable Computing. IBM Systems Journal 35(3), 618–629 (1996)

22. Starner, T.: Powerful Change Part 1: Batteries and Possible Alternatives for the Mobile ket. IEEE Pervasive Computing 2(4), 86–88 (2003)

23. Stordeur, M.S.: Low Power Thermoelectric Generator: Self-sufficient energy supply for micro systems. In: The proceedings of the 16th International Conference on Thermoelectrics, pp. 575–577 (1997)

24. Stuntebeck, E.P., Patel, S.N., Robertson, T., Reynolds, M.S., Abowd, G.D.: Wideband power-line positioning for indoor localization. In: The Proceedings of Ubicomp 2008, pp. 94–103 (2008)

25. Yeager, D., Sample, A., Smith, J.: WISP: a passively powered uhf rfid tag with sensing and computation. Rfid Handbook: Applications, Technology, Security, and Privacy (2008)

26. Yun, J., Patel, S.N., Reynolds, M.S., Abowd, G.D.: A quantitative investigation of inertial power harvesting for human-powered devices. In: The Proceedings of Ubicomp 2008, pp. 74–83 (2008)

Geo-fencing: Confining Wi-Fi Coverage to Physical Boundaries

Anmol Sheth[1], Srinivasan Seshan[2], and David Wetherall[1,3]

[1] Intel Research, Seattle
[2] Carnegie Mellon University
[3] University of Washington

Abstract. We present a means of containing Wi-Fi coverage to physical boundaries that are meaningful to users. We call it *geo-fencing*. Our approach is based on directional antennas, and our motivation is to provide wireless access and privacy models that are a natural fit with user expectations. To evaluate geo-fencing, we use measurements from an indoor testbed of Wi-Fi nodes and APs with electronically-steerable directional antennas. We find that by combining directionality, power control and coding across multiple APs, we are able to successfully confine Wi-Fi coverage to clients located within target regions of varying shapes and sizes; we can select between nodes located as close as five feet from each other.

1 Introduction

Local wireless data communications, exemplified by Wi-Fi, are rapidly spreading beyond mobile computers to everyday consumer devices such as cell-phones, personal health monitors and digital cameras. Thus, users may have many wireless devices that they wish to connect to different wireless networks in various settings. Unfortunately, this is burdensome with current usage models because usability is in tension with access and privacy needs.

Current wireless usage requires that the user select which wireless network to use and provide credentials or other key information. The underlying reason for these requirements is that, unlike wired networks, wireless networks lack physical boundaries that are meaningful to users. Their extent is defined by RF propagation, and this creates an inherent disconnect between the users' view of where Wi-Fi service should be available and the actual service area. Since wireless signals go through walls, it is necessary to select which network is intended to provide service in a given location versus other networks that happen to overlap that location. Since wireless signals can be received by parties in nearby locations, it is necessary to encrypt communications to provide confidentiality. This is done with standard mechanisms such as Wi-Fi Protected Access (WPA and WPA2) [1].

We are exploring a different means of using wireless networks based on confining the Wi-Fi service areas to a well-defined physical region. This model, which we call *geo-fencing*, is motivated by the issues described above as well as user

H. Tokuda et al. (Eds.): Pervasive 2009, LNCS 5538, pp. 274–290, 2009.

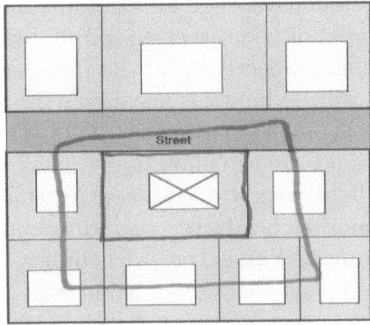

Fig. 1. Representative diagram drawn by a Wi-Fi user from the general public that shows his/her perceived Wi-Fi service area (green) and the smaller desired service area (magenta) for his/her home Wi-Fi network

perceptions of wireless networks. Both can be seen at play in Figure 1, which shows a diagram drawn by a participant in a user study that we conducted [2]. It shows the extent of perceived and desired Wi-Fi coverage for the participants' home network. Interestingly, the perceived coverage follows unrealistic, rectilinear boundaries and shows that users often lack a reasonable understanding of wireless behavior. Of particular interest is the desired coverage, which maps directly to a geographic boundary – the property line. This is an intuitive boundary for Wi-Fi service areas in the same way that rooms are an intuitive boundary for social privacy. We note that this view is similar to the Virtual Walls [3] framework for wireless privacy, which extends the metaphor of physical walls to virtual pervasive environments. Results of a user study suggests that it is easy to understand and use.

With geo-fencing, access is simplified because clients can be authorized to obtain basic connectivity simply by being inside the known region in the same way that, e.g., the users of a conference room are authorized to use the projector in the room. Similarly, access can be revoked as soon as clients leave the region. All of this can be done without involvement of the device owner or the wireless service provider because they do not need to set up cryptographic keys. Privacy is strengthened because information cannot be received by standard clients even in encrypted form beyond the target region. This is valuable because even with encryption, (i.e. WEP/WPA), users may be identified by observing network management traffic that is sent in the clear [4], and knowledge such as applications and even the name of the movie being streamed can be gleaned from side-channels such as packet lengths and timings [5]. Additionally, cryptographic mechanisms are cumbersome to use in relatively unmanaged environments, such as the home and Wi-Fi hotspots intended for a changing customer base. As a result many of these settings forego encryption.

The focus of this paper is to assess how well we can realize the geo-fencing model in practice. Our approach is to use multiple access points (APs) equipped with electronically steerable directional antennas as well as transmit power control. The intuition is that coverage can be controlled by adjusting the orientation and transmit power of the directional antenna to focus signals on the intended area, and by tying connectivity to the intersection of the signals from multiple APs. This is an approximation because it is not possible to precisely confine wireless signals to arbitrary regions in real-world settings without artificial impediments like Faraday cages. Thus the security aspects of our approach should be interpreted as light-weight access control that significantly raises the bar for devices with commodity hardware, including omni-directional antennas, but will not defeat sophisticated attackers with bulky, high-gain antennas. Despite this limitation, we expect that geo-fencing will be sufficient by itself in many deployments, or can serve as the foundation of hierarchical mechanisms that provide defense-in-depth when stronger security is needed.

We report on an experimental study of geo-fencing run on an office testbed of 802.11b/g nodes with 2.4 GHz steerable directional antennas that play the role of APs and clients. We assess several approaches to orient the antennas and choose transmit power levels, from measurement-based fingerprinting to heuristics based on the angle of arrival (AoA) as estimated by the directional antennas. Finding the most effective configuration is a key challenge because the RF environment of an indoor setting, especially with multi-path interference, affects the directionality of the antennas. Nonetheless, the patterns provide sufficient isolation in gain to be able to confine coverage to desired regions of varying shapes and sizes. Our measurements show that with three directional antennas, geo-fencing can isolate regions of different shapes and sizes ranging from a small desk area of 5 feet × 5 feet to regions of large room sizes of 25 feet × 20 feet. Geo-fencing is able to successfully isolate individual clients located 5 feet away from the target client in our 50 feet × 30 feet testbed. For regions defined by a single target client, geo-fencing limits the maximum packet reception rate measured outside the target region to 50% while providing >90% packet reception to the target client. These measurements suggest that geo-fencing can be realized as a novel physical layer mechanism.

The rest of the paper is organized as follows. In Section 2, we present the requirements for the geo-fencing mechanism and our approach. We describe our testbed in Section 3. Section 4 then describes a measurement study of indoor directional antenna behavior. We use these results to sketch a geo-fencing design in Section 5. We evaluate this design using measurements in Section 6. We present related work in Section 7 and conclude in Section 8.

2 Goals and Approach

Geo-fencing enables deployment scenarios that would otherwise be cumbersome to achieve. For example, consider the scenarios where a public library or coffee shop wishes to provide Wi-Fi service to all patrons while they are within the facility, but wishes to deny service when they leave. In this section we list the goals that geo-fencing should meet to enable such scenarios and the approach.

2.1 Goals

Granularity and Region Definition. Providers should be able to confine transmissions to regular shaped regions that range from small room-sized regions to large areas as entire library or cafe. Providers should also be able to change the boundary definitions of the region as need changes.

Region Selectivity. Regions should be well-defined and client devices more than a few feet outside the defined regions should not be able access the network. While 0% packet reception outside of the region and 100% packet reception rate inside the region would be ideal, it cannot be practically achieved. Measurement studies of TCP [6] and UDP based applications, like Skype [7], show that these applications are unusable beyond a link layer loss rate of $> 25\%$. Based on these observations, we desire an absolute link layer packet reception rate threshold of $<70\%$ ($> 30\%$ loss) outside the region and a threshold of $>90\%$ ($< 10\%$ loss) within the region.

Manageable Infrastructure Overhead. We allow geo-fencing to take advantage of a larger number of APs to limit the region coverage more accurately. However, the system should not require an excessive (> 5) number of APs.

Client Compatibility. Our goal is to maintain compatibility with existing mobile client hardware. This means that we assume that only APs have directional capability.

2.2 Approach

While these goals are simple to state, they are challenging to achieve because of the realities of RF propagation. Coverage regions are irregular even with omni-directional antennas. This is because of signal reflections and other RF effects such as scattering that are significant in indoor environments and generally unknown a priori. Directional antennas have not been widely used indoors because their directionality is significantly degraded relative to outdoor settings. Nonetheless, we believe that the indoor use of directional antennas is valuable if their radiation pattern is adapted based on measurements of the environment.

Our approach is to use a distributed set of APs with directional antennas. Figure 2 shows an overview of our approach. AP1 and AP2 are Wi-Fi APs equipped with electronically steerable directional antennas that are configured to form a controlled signal overlap of their radiation patterns in the desired region. The signal overlap is formed by collecting measurements of packet reception rates from the desired region for different antenna configurations. By coding Wi-Fi frames across the two APs, successful packet recovery is restricted only within the region formed by the intersection of the two patterns.

To see why this approach may be possible, consider a simplified view of the capabilities of directional antennas. First, note that wireless radios have a narrow *transition range* ($< 5dBm$) below which no packets are received, and above which packets are received with near certainty. Second, the primary lobe of the antenna pattern can be represented as an isosceles triangle, as seen in Figure 2, and the

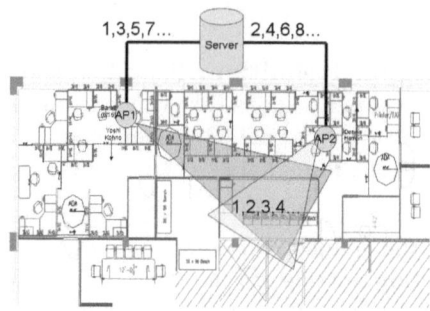

Fig. 2. Geo-fencing approach to confine Wi-Fi coverage to a specified region

secondary lobes of the radiation pattern are negligible. Third, the size of this triangle can be varied using fine grained transmit power control and the triangle can be rotated around its primary vertex with an arbitrary granularity.

Given this view, a minimum of two directional antennas are required to confine coverage to a target area. This can be done by varying the transmit power and beam rotation so that the intersection of the antenna patterns forms a region where the signal from each antenna is above the transition range. For example, a narrowly defined region can be created by overlapping any two vertices, or the edge of one triangle with the vertex of the other. With this technique receivers do not require special hardware.

3   Testbed Description

We use the Phocus Array [8] electronically steerable directional antenna system for our experiments. Figure 3(a) shows the sample directional and omni-directional antenna pattern generated by this system. The directional antenna isolation is measured as the difference in gain between the main lobe and largest secondary lobe of the antenna radiation pattern. The antenna isolation in our system is 20 dB, and thus, the secondary lobes can be ignored so that the antenna pattern is modeled as an isosceles triangle. The antenna pattern can be electronically steered in 360° range with an angular granularity of 22.5°. Thus there are a total of 16 states and each pattern overlaps the adjacent pattern by 22.5°. The beamwidth of an antenna is defined as the angular separation between two identical points on opposite sides of the pattern's peak gain value. The Half Power Beam Width (HPBW) of the antenna system is 45°. Thus, there are only two states on either side of the antenna state with the peak gain which have a gain above the (*peak gain* − 3*dB*) threshold.

An important point to note is that the peak gain of the antenna in directional mode is only 3-4 dB greater than the gain in omni-directional mode. That is, the antenna system does not so much boost gain in the target region as it sharply attenuates gain in the undesired region. This property is well suited for the geo-fencing application as it reduces the extent to which secondary lobes are formed by reflections of the wireless signal.

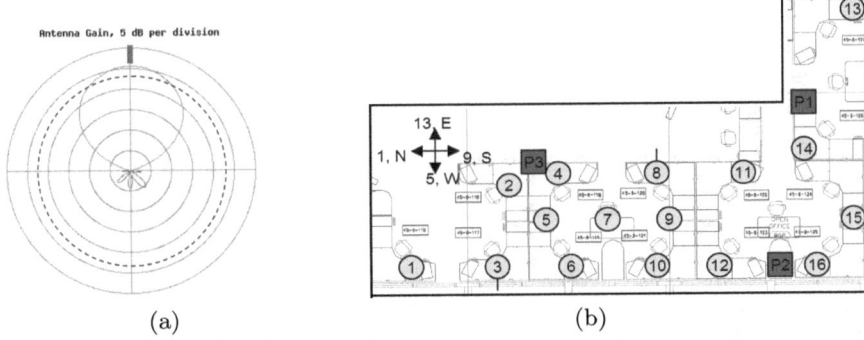

(a)   (b)

Fig. 3. Figure (a) shows the directional and omni-directional antenna patterns. The scale is 5 dB per division with the origin at -5 dB. Figure (b) shows the layout of the indoor testbed with Wi-Fi nodes (circles) and steerable directional antennas (squares).

In addition to APs equipped with the steerable directional antennas, we use single board computers as sensors to gather measurement data. Each Wi-Fi sensor is equipped with an Atheros 802.11a/b/g wireless interface and omni-directional antenna. The stock Madwifi Linux driver was modified to log per packet statistics such as Received Signal Strength (RSS), and to measure the Packet Reception Rate (PRR) across time intervals by measuring gaps in the sequence numbers.

The floor plan for our testbed is shown in Figure 3(b). This testbed was deployed on one floor of size 50 feet × 30 feet in a typical office environment and consisted of 16 sensors (circles) and 3 directional APs (squares). The map also shows the antenna orientation states corresponding to the four primary directions. All three directional APs were oriented facing north. We deployed a sensor in each employee cubicle to provide high spatial resolution for our measurements. The average distance between adjacent sensors was 5 feet. The path between the sensors and APs was obstructed by cubicle walls, cabinets and other hardware equipment and most of the sensors do not have direct line of sight to the APs.

4   Characteristics of Directional Antennas

The key challenge in getting geo-fencing to work is to handle realistic wireless environments. The four primary characteristics of directional antennas that we depend on for geo-fencing are:

- *Antenna isolation:* This allows the antenna to selectively provide coverage between two regions by steering the antenna main lobe away from the undesired region and toward the target region. For geo-fencing to be effective, the antenna isolation should be greater than the transition range — the range below which no packets are received, and above which packets are received with near certainty.

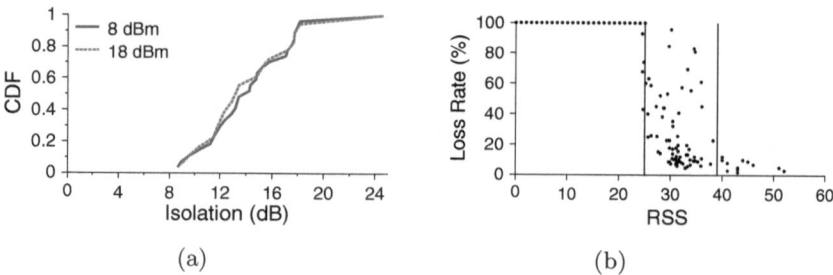

(a)   (b)

Fig. 4. Figure (a) shows distribution of the isolation measured across all the nodes in the testbed. Figure (b) shows the transition region for sensor 5.

- *Beamwidth:* The beamwidth of the directional antenna pattern directly impacts the base of the triangle pattern. With coarse grained control over antenna orientation, a wide beamwidth limits the smallest size of the regions that can be formed.
- *Scaling of antenna pattern:* Transmit power control should allow scaling of the triangle pattern to confine service to regions of different sizes.
- *Stability of antenna pattern:* The antenna patterns and the corresponding PRR should be stable over time and not be sensitive to mobility of people and objects in the environment. This directly impacts region selectivity without requiring frequent realignment of the antenna patterns.

To study these characteristics in our testbed, we gathered measurement data across all 16 sensors while transmitting from the three directional transmitters. The antenna radiation pattern was rotated across all 16 states sequentially and broadcast traffic was generated at each antenna orientation state for 10 seconds. The isolation of the directional antenna is measured at every sensor location by recording the maximum and minimum RSS across the 16 antenna orientation states.

4.1   Directional Antenna Isolation

Previous measurement studies [9] show that the transition range, in absence of external interference and multipath, is 5 dB. Clearly, an isolation of 20 dB of the directional antenna radiation pattern (Figure 3(a)) should be sufficient.

Figure 4(a) shows the distribution of the antenna isolation measured across all the sensors in the testbed for two different transmit power levels at the APs (18 dBm and 8 dBm). The figure shows that there is a wide variation (8 to 24 dB) in the measured isolation across the distributed sensors. The median isolation is reduced from the expected 20 dB to 12 dB. The figure also shows that the distribution of isolation does not change for different transmit power level settings at directional APs as the RSS patterns scale proportionally with the transmit power level.

Figure 4(b) plots the packet loss rate (i.e., $1-$ PRR) measured at sensor 5 during each 10 second measurement against the average RSS of the received

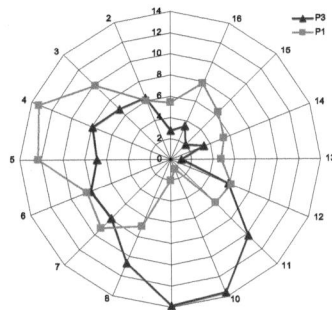

Fig. 5. Antenna radiation pattern of P1 and P3 measured at sensor 11

packets. The two vertical lines demarcate our estimate of the transition range. The transition range is bounded by a loss rate less than 10% to the right of the range, and a loss rate of greater than 90% to the left of the range. In this case, the width of the transition range is 14 dB. The average transition range measured across all the sensors in our testbed was 10 dB.

While the median isolation is only 2 dB higher than the average transition range width, the variability of the isolation achieved from the distributed antennas is sufficient to provide selective coverage between two regions.

4.2   Directional Antenna Beamwidth

Our measurements show that indoor multipath significantly increases the beamwidth of the pattern and also introduces large secondary lobes due to strong reflectors like metal cabinets in the environment. Figure 5 shows the antenna pattern measured by sensor 11 from antenna P1 and P3. The linear axis is in dB with 2 dB per unit. The radial axis represent the 16 possible orientations of the antenna with increments of 22.5° in counter-clockwise direction. For both the directional APs, the peak RSS is measured when the transmitter is pointing its main lobe directly at the receiver. The antenna radiation pattern measured at the receiver show that along with the main lobe, large secondary lobes are also formed due to multipath reflections. The patterns are also specific to the path between the transmitter and receiver as the secondary lobes are different for the two directional transmitters. Across all nodes in the testbed, the number of states that measure a receive gain above the half power threshold ($peak\ gain - 3dB$) ranges from 3 to 15. Thus, antenna orientation in isolation is not sufficient for geo-fencing. In the following section we show how transmit power control can be used to significantly limit the number of antenna states at which successful packet recovery is possible.

4.3   Transmit Power Control

Figure 6(a) shows the effectiveness of using transmit power control to vary the number of states with a high PRR, which directly impacts the angular coverage.

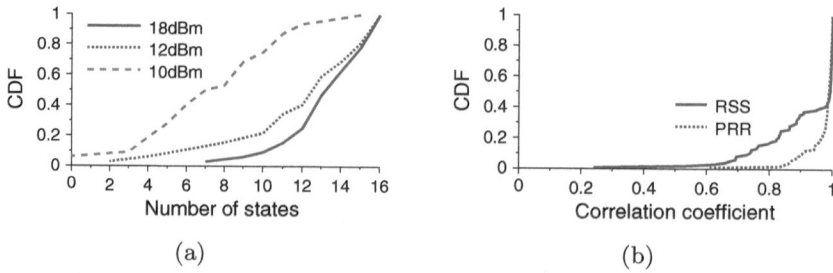

Fig. 6. Figure (a) shows the number of antenna orientation states with PRR > 90% from antenna P2 across all nodes in the testbed. Figure (b) shows distribution of auto-correlation of PRR and RSS patterns across 20 hours. The antenna patterns are stable over long time periods with the median of the distribution close to one.

The figure shows the distribution of the number of antenna states that provide a PRR > 90% from antenna P2 across all sensors in the testbed. We observe that transmit power control significantly reduces the median angular coverage from 12 states at 18dBm to 6 states at 10dBm. Similar distributions were measured for antennas P1 and P3. Thus, despite the wide antenna beamwidths, transmit power control can significantly help in reducing the angular coverage across which packets are received with a high PRR.

4.4   Stability of Antenna Patterns

To understand the temporal stability of the antenna radiation pattern, we measure the antenna's RSS pattern as well as its PRR pattern 10 times over a duration of 20 hours. The transmit power of the three APs was fixed at 12 dBm. To verify that the antenna patterns do not change from one measurement iteration to another, we compute the auto-correlation of the antenna patterns with lag set to one. Hence, for each sensor location we have a set of nine correlation coefficients computed. Figure 6(b) shows the distribution of the correlation coefficients for the PRR patterns and RSS patterns measured at each sensor. The median correlation coefficient is almost one. This shows that the PRR and RSS patterns are stable over long periods of time. This maintains the stability of the geo-fenced region and also reduces the need for frequent antenna alignment.

5   Geo-fencing Technique

Based on our observations, our technique to confine Wi-Fi service to a target region is to align the distributed directional antennas to create controlled overlaps of the transmission patterns and to code packets across the distributed antennas. We define an *antenna configuration* as a combination of the orientation of the main lobe and transmit power level at an AP. The geo-fencing *system configuration* is a combination of antenna configurations across all the distributed APs.

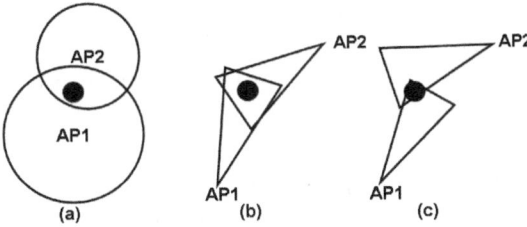

Fig. 7. Overview of three different antenna orientation approaches (a) Omni-directional approach (b) Angle-of-Arrival approach (c) Min-overlap approach.

5.1   Aligning Antenna Patterns

We outline four different approaches for selecting antenna configurations for geo-fencing that we evaluate in the next section. The omni-directional approach (Figure 7(a)) forms the baseline that we use to compare the effectiveness of directional antennas for geo-fencing. The Angle of Arrival (AoA) approach (Figure 7(b)) and the Minimum Overlap heuristic approach (Figure 7(c)) assume triangle-shaped antenna patterns. These approaches require packet reception rate measurements of antenna configurations only from the sensors located within the target region. The dense fingerprinting approach (not pictured) uses measurements from all sensors located inside and outside the target region to optimize the antenna configuration. It provides our ground truth as it accounts for the realistic wireless environment.

Omni-directional approach: The most basic approach is to use omni-directional APs and create controlled overlaps of the antennas patterns by only using transmit power control. Here we find the combination of transmit power levels at the distributed APs which minimizes the maximum PRR at the non-target sensors. The difficulty with this approach is that lack of spatial confinement of the wireless signal and coarse grained transmit power control limits the definition of regions.

Angle-of-Arrival (AoA) approach: A potential approach to select the antenna orientation would be for each AP to form the smallest ideal triangle-shaped antenna pattern and orient the mid-point of the base of the triangle (AoA state) with the target region. Figure 7(b) shows this approach. The mid-point corresponds to the peak gain point of the antenna pattern. The AoA state from an AP to a target region can be approximated by determining the antenna orientation that results in the highest RSS measured by the sensors located within the target region. The smallest antenna pattern can be formed by adjusting the transmit power of the Wi-Fi radio. This approach results in much smaller and controlled overlapping regions than the omni-directional approach.

Minimum Overlap Heuristic approach: For a selected transmit power level, there could be multiple antenna states adjacent (clock-wise and counter clock-wise) to the AoA state that also result in high PRR within the target region. The minimum overlap heuristic, Figure 7(c), aims to minimize the distance of the target region from the boundary of the edge of the triangle pattern, essentially putting the region at the edge or vertex of ideal triangle of coverage. The heuristic approach selects the same transmit power level as the AoA approach, but selects antenna orientation states that are adjacent to the AoA state within a window of ± 2 states.

Dense Fingerprinting approach: As the antenna patterns are irregular and specific to the path between the transmitter and receiver, the dense fingerprinting approach is based on measuring the effective PRR across all the distributed sensors for every antenna configuration. These measurements are collected at a central server which then selects the best system configuration that minimizes the peak PRR outside the intended region.

5.2   Coding Packets Across Distributed APs

To make information available only inside the geo-fenced region, we code packets across the set of APs. Sensors located outside the region may then receive signals from some APs, but cannot decode the overall signal because they are unlikely to receive signals from all the APs. Coding may be done either by transmitting independent packets from different APs, or by using a central point to divide the contents of each packet across the AP transmissions. The former is simple, while the latter provides better containment. For our evaluation, we consider a coding technique based on Shamir's *secret sharing technique* [10]. A client at a given location must then receive packet fragments from all APs to decode the complete packet.

6   Evaluation

The primary metric used to evaluate the effectiveness of geo-fencing on our testbed is Packet Reception Rate (PRR) measured at the sensors within the target region and outside the target region. Our goal is to provide a PRR <70% outside the target region (< 90% PRR with retry limit set to 1) and a PRR >90% (>99% PRR with retry limit set to 1) within the region. In the following subsections, we attempt to answer four key questions:

- *Are multiple directional antennas needed to make geo-fencing work and, if so, how many?* We find that directional antennas provide significantly better confinement than omni-directional antennas. With three directional antennas, we can reduce the maximum PRR outside the target region to 50%, which is significantly lower than the 70% threshold.
- *Can geo-fencing support regions of different shapes and sizes?* We show that while the PRR outside the target region rises slightly with region size, geo-fencing is effective for regions ranging from a small desk area of 5 feet × 5 feet to the size of a large room (20 feet × 20 feet).
- *How effective are the different antenna alignment approaches?* Our evaluation shows that even the most simple approach based on Angle-of-Arrival

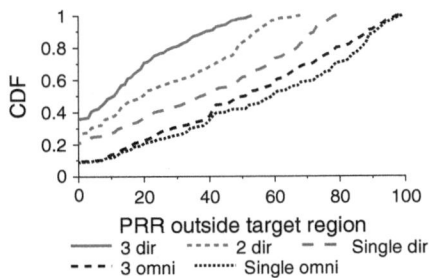

Fig. 8. Isolation achieved with using single/multiple omni/directional antennas.

performs significantly better than the omni-directional approach. The dense fingerprinting based approach, that takes into account measurements from distributed points, provides the best confinement.

- *What type of special hardware does an adversary require to defeat geo-fencing?* Our evaluation shows that geo-fencing denies access to clients with commodity hardware. Clients would need a median omni-directional gain of 8 dB to raise their PRR to 90%, which can only be achieved by bulky high-power antennas. Most mobile clients clients have omni-directional antennas with a gain of 2-3 dB.

To answer these questions, we conducted experiments using the testbed described in Section 3. Traffic was generated at a rate of 1 Mbps UDP CBR broadcast traffic at a fixed modulation rate of 54 Mbps. Packets were coded using the coding scheme described in Section 5.2. Unless otherwise specified, we use the dense fingerprinting approach in all our evaluation and in Section 6.3 we compare the effectiveness of the other approaches.

6.1   Omni-Directional vs. Directional Antennas

We motivate the need to use multiple directional antennas by comparing the isolation achieved with that of omni-directional antennas. The target region is defined by a single sensor. For each sensor, we select the AP and transmit power that achieves greater than 90% PRR at the target sensor but minimizes the maximum PRR measured at all other sensors. We performed this configuration selection for each of the 16 potential sensors, measuring the PRR at all 15 non-target sensors. For both cases, the single best AP and power setting is selected, while for the directional case the best orientation is also considered.

First, we consider one AP. Figure 8 shows the CDF of the PRR measured by all non-target sensors. Even a single directional antenna provide much better confinement of Wi-Fi signals than omni-directional antennas. In the omni-directional case (line "Single omni"), half the sensor locations have a PRR above 70%, while in the directional case (line "Single dir") only approximately 30% of the sensors have a PRR above 70%.

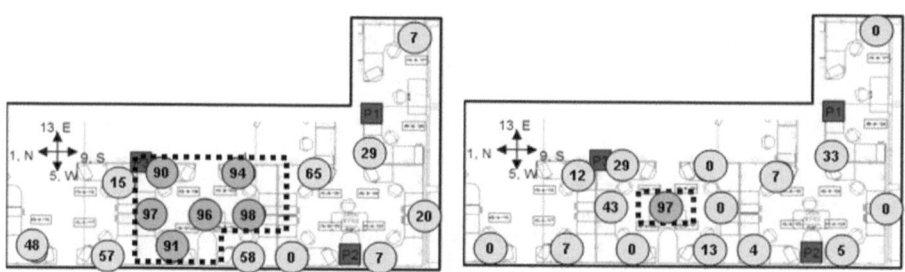

Fig. 9. Target regions of different shapes and sizes can be geo-fenced. Figure shows a target region of the size of a room and a target region of the size of a single desk.

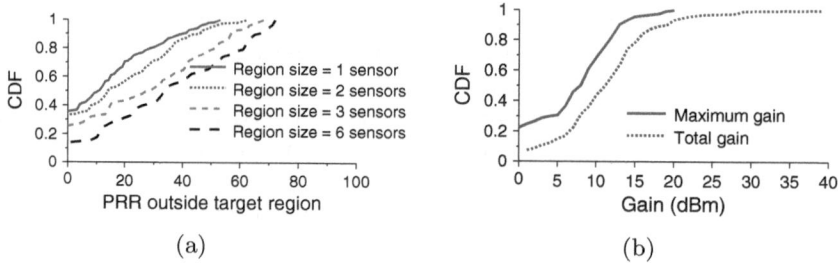

(a)   (b)

Fig. 10. Figure (a) shows the effectiveness of geo-fencing with increasing region size. Figure (b) shows the distribution of the maximum and total gain required for a node outside the region to receive > 90% of the packets.

We now consider multiple APs. Figure 8 shows the limitations of coding network traffic across multiple omni-directional APs (line "3 omni"). It only marginally improves isolation over a single omni-directional antenna. The figure also shows a significant improvement when the number of directional APs are increased. With two APs (line "2 dir"), the maximum PRR measured outside the region reduces to 68%. While this is below the required threshold of 70% required for geo-fencing, adding an extra AP reduces the maximum PRR by almost 25% (line "3 dir"). The median PRR is 11% and the maximum PRR is 52%, making geo-fencing significantly more effective.

6.2   Varying Region Sizes and Shapes

We evaluate the effectiveness of geo-fencing regions of different shapes and sizes by increasing the number of sensors in the target region. Target regions of size n sensors are formed by selecting the closest $n - 1$ sensors for every sensor in the testbed.

To better understand the shape of the target regions formed, Figure 9 shows examples of geo-fencing a large target region and a small target region (5 feet × 5 feet). The figure also shows the PRR measured by the sensors located inside

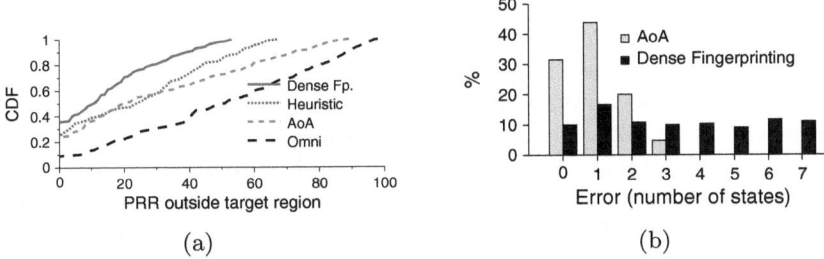

Fig. 11. Figure (a) shows the effectiveness of the different antenna alignment approaches for three antennas. Figure (b) shows the distribution of the deviation of the antenna orientations selected by the two approaches from the actual measured angles.

and outside the target region. From the figure, we observe that for large target regions, sensors located immediately outside the target region do receive a much higher PRR as compared to sensors located farther away from the target region. Geo-fencing is more effective for smaller target regions and the PRR outside the target region is significantly lower. Figure 10(a) shows the CDF of the isolation achieved for target regions specified by 1, 2, 3 and 6 sensors. Geo-fencing is effective even for target regions sizes consisting of 6 sensors (size of 20 feet × 20 feet). The median PRR measured is 34% and the maximum PRR outside the target region is below the 70% threshold.

6.3   Antenna Alignment Approaches

In this section, we evaluate the four different antenna alignment approaches described in Section 5 to understand the tradeoff between the different approaches. Figure 11(a) shows the isolation in PRR measured outside the target region for each approach. Geo-fencing using omni-directional APs cannot confine access to Wi-Fi service and more than 30% of the sensors located outside the target region receive a PRR > 70%. Using directional antennas significantly increases the isolation between target and non-target region. The median of the minimum overlap heuristic approach is the same as the AoA approach in our testbed. However, the heuristic approach significantly reduces the maximum PRR from 89% to 67%, which meets the target PRR for preventing access to Wi-Fi service from from outside the target region.

Among the four approaches, the dense fingerprinting approach provides the maximum isolation. To better understand the isolation achieved by the dense fingerprinting approach, we compare the antenna configuration selected by the AoA and dense fingerprinting approach. While there was not a significant difference in the transmit power level between the two approaches (± 2 dB), the antenna orientation states differed significantly. Figure 11(b) shows the distribution of the deviation of the antenna orientations selected by the two approaches from the actual measured angle between the AP and client. We observe that the antenna orientations selected by the dense fingerprinting approach deviate significantly from the orientation selected by the AoA approach. More than 35%

of the measurements of dense fingerprinting approach deviate by more than five states (90°) from the AoA state. From this distribution, we conclude that the orientation selected by dense fingerprinting approach is not always a part of the primary main lobe. The antenna orientation selected by the dense fingerprinting based approach often aligns the secondary lobes formed due to indoor multipath reflections along the target region. For example, in Figure 5, for target sensor 11 the dense fingerprinting approach may select the smaller secondary lobe formed between states 11 and 12 instead of the wide main lobe formed between states 2 and 6. Thus, the dense fingerprinting based approach achieves better isolation as it accounts for the RF reflectors present in the indoor environment.

6.4   Antenna Gain Requirements

For an adversary located outside the target region to gain access, it would need a high PRR (>90%) from each AP. The direct approach to do this would be to use a single high gain omni-directional antenna. Our measurement analysis show that the additional gain required is significantly higher than the antenna gain of commodity omni-directional antennas that are embedded in devices like laptops and cell phones.

Figure 10(b) shows the distribution of total gain required, defined as the sum of the gains in each direction, and maximum gain required in any one direction by an adversary. The additional required gain is measured by first estimating the width of the transition range at every sensor location. Based on the measurement of the RSS for the particular geo-fenced configuration, we calculate the additional gain required to achieve a PRR >90% for every non-target sensor. For example, for the transition range shown in Figure 4(b), if the average signal strength measured at the sensor for a particular antenna configuration is 25 dB, then it would require an additional 14 dB gain to achieve a signal strength of 37 dB [1] and a high PRR (> 90%).

From the figure we observe that even for a dense deployment, where the sensors are less than 5 feet away from each other, the median of the total gain required is 11 dB. The median of maximum gain in any one direction is measured to be 8 dB. Thus commodity omni-directional antennas embedded in devices like laptops and PDAs, which have a gain of 2-3 dBi, are not sufficient to gain access.

7   Related Work

Wi-Fi localization-based access control: There are commercial [11] as well as academic research prototypes [12,13] that provide location based access control in Wi-Fi networks. Based on the access control policy and the estimated location of the client, the client is either granted or denied access. Although similar in flavor to geo-fencing, these systems do not confine radio signals to the intended service area, and are consequently prone to eavesdropping. Compared

---

[1] Atheros radios report signal strength measurements as the measured signal power level above the preset noise floor of -80 dBm.

to the use of dense arrays of low power Wi-Fi APs [14] for customizing service regions, geo-fencing is expected to provide better control over the boundaries of the coverage area with less infrastructure.

Link-layer security mechanisms: Generally, access control in 802.11 networks is achieved by higher layer cryptographic security techniques such as WEP and WPA/WPA2 [1]. However, these techniques are not suited for hot spot style Wi-Fi deployments which require providing temporary access to clients. The primary limitation for the widespread use of these security mechanisms is tedious key distribution, as evidenced by the number of open APs seen in war driving studies [15].

Directional antennas: Unlike static directional antennas that cannot be electronically steered, steerable directional antennas allow dynamic steering of the antenna orientation [8]. Commercial products, like BeamFlex [16], use these antennas and change the antenna orientation on a per-packet basis to improve coverage and performance in wireless LAN deployments. Most cellular network deployments extend the range of the network by using multiple directional antennas co-located at a central tower [17], where each directional antenna services a sector of 90-180°. In [18], the authors use directional antennas to extend the range of a Wi-Fi link to 100-200 kms. The only other work that we are aware of that uses steerable directional antennas is MobiSteer [19]. MobiSteer aims at improving performance of 802.11 links in the context of communication between a moving vehicle and roadside APs.

8   Conclusion

In this paper we present *geo-fencing* — a novel physical layer mechanism that allows users to define service areas of the Wi-Fi access points to a specified physical region. Geo-fencing uses distributed steerable directional antennas to confine Wi-Fi signals to a specified region in an indoor environment. Geo-fencing confines Wi-Fi service areas by making use of a combination of power control, antenna beam orientation at each AP, and coding of packets across the distributed APs. Our measurements show that with three directional antennas, geo-fencing can isolate regions of different shapes and sizes ranging from a small desk area of 5 feet × 5 feet to regions of large room sizes of 20 feet × 20 feet. Geo-fencing is able to successfully isolate individual clients located 5 feet away from the target client in our 50 feet × 30 feet testbed. For regions defined by a single target client, geo-fencing limits the maximum packet reception rate measured outside the target region to 50% while providing >90% packet reception to the target client.

References

1. Edney, J., Arbaugh, W.A.: Real 802.11 Security: Wi-Fi Protected Access and 802.11i, vol. 1. Addison-Wesley, Reading (2001)

2. Klasnja, P., Consolvo, S., Jung, J., Greenstein, B., LeGrand, L., Powledge, P., Wetherall, D.: When I am on Wi-Fi, I am Fearles: Privacy concerns and practices in everyday Wi-Fi use. In: Proceedings of ACM CHI Conference on Human Factors in Computing Systems (to appear, 2009)
3. Kapadia, A., Henderson, T., Fielding, J.J., Kotz, D.: Virtual walls: Protecting digital privacy in pervasive environments. In: LaMarca, A., Langheinrich, M., Truong, K.N. (eds.) Pervasive 2007. LNCS, vol. 4480, pp. 162–179. Springer, Heidelberg (2007)
4. Pang, J., Greenstein, B., Gummadi, R., Seshan, S., Wetherall, D.: 802.11 user fingerprinting. In: MobiCom 2007: Proceedings of the 13th Annual International Conference on Mobile Computing and Networking (September 2007)
5. Saponas, T., Lester, J., Hartung, C., Agarwal, S., Kohno, T.: Devices that tell on you: privacy trends in consumer ubiquitous computing. In: SS 2007: USENIX Security Symposium, Berkeley, CA, USA, pp. 1–16. USENIX Association (2007)
6. Balakrishnan, H., Padmanabhan, V., Seshan, S., Katz, R.: A comparison of mechanisms for improving TCP performance over wireless links. IEEE/ ACM Transactions on Networking 5(6), 756–769 (1997)
7. Sat, B., Wah, B.: Analysis and evaluation of the skype and google-talk voip systems. In: IEEE International Conference on Multimedia and Expo., pp. 2153–2156. ACM Press, New York (2006)
8. Fidelity Comtech, http://www.fidelity-comtech.com/
9. Charles, R., Ratul, M., Maya, R., David, W., John, Z.: Measurement-based models of delivery and interference in static wireless networks. In: SIGCOMM 2006: Proceedings of the 2006 conference on Applications, technologies, architectures, and protocols for computer communications, pp. 51–62. ACM, New York (2006)
10. Shamir, A.: How to share a secret. Communications of the ACM 22(11), 612–613 (1979)
11. Aruba, http://www.arubanetworks.com/
12. Bahl, P., Padmanabhan, V.: RADAR: An in-building RF-based user location and tracking system. In: INFOCOM (2), pp. 775–784 (2000)
13. Haeberlen, A., Flannery, E., Ladd, A.M., Rudys, A., Wallach, D.S., Kavraki, L.E.: Practical robust localization over large-scale 802.11 wireless networks. In: MobiCom 2004: Proceedings of the 10th annual international conference on Mobile computing and networking, pp. 70–84. ACM, New York (2004)
14. Chandra, R., Padhye, J., Wolman, A., Zill, B.: A location-based management system for enterprise wireless lans. In: Proceedings of the 3rd ACM/USENIX Symposium on Networked Systems Design and Implementation (NSDI), pp. 115–130 (2007)
15. Bittau, A., Handley, M., Lackey, J.: The final nail in wep's coffin. In: Symposium on Security and Privacy, pp. 386–400. IEEE Computer Society, Washington (2006)
16. BeamFlex, http://www.ruckuswireless.com/technology/beamflex.php/
17. Rappaport, T.: Wireless Communications: Principles and Practice, vol. 2, Reading, Massachusetts (2001)
18. Patra, R., Nedevschi, S., Surana, S., Sheth, A., Subramanian, L., Brewer, E.: WiLDNet: Design and implementation of high performance wifi based long distance networks. In: 4th USENIX Symposium on Networked Systems Design and Implementation, pp. 87–100 (2007)
19. Navda, V., Subramanian, A.P., Dhanasekaran, K., Timm-Giel, A., Das, S.: Mobisteer: using steerable beam directional antenna for vehicular network access. In: MobiSys 2007: Proceedings of the 5th international conference on Mobile systems, applications and services, pp. 192–205. ACM, New York (2007)

Securing RFID Systems by Detecting Tag Cloning

Mikko Lehtonen[1], Daniel Ostojic[2], Alexander Ilic[1], and Florian Michahelles[1]

[1] Information Management, ETH Zürich, 8092 Zurich, Switzerland
mlehtonen@ethz.ch, ailic@ethz.ch, fmichahelles@ethz.ch
[2] Pervasive and Artificial Intelligence Research Group, Department of Informatics,
University of Fribourg, Switzerland
daniel.ostojic@unifr.ch

Abstract. Cloning of RFID tags can lead to financial losses in many commercial RFID applications. There are two general strategies to provide security: prevention and detection. The security community and the RFID chip manufacturers are currently focused on the former by making tags hard to clone. This paper focuses on the latter by investigating a method to pinpoint tags with the same ID. This method is suitable for low-cost tags since it makes use of writing a new random number on the tag's memory every time the tag is scanned. A back-end that issues these numbers detects tag cloning attacks as soon as both the genuine and the cloned tag are scanned. This paper describes the method and presents a mathematical model of the level of security and an implementation based on EPC tags. The results suggest that the method provides a potentially effective way to secure RFID systems against tag cloning.

Keywords: Security, clone detection, low-cost, EPC, RFID.

1 Introduction

Radio frequency identification (RFID) is taking its place as a pervasive everyday tool for automatic identification (Auto-ID) of physical objects. Various industries use it to facilitate the handling of physical goods. RFID is also an enabling technology behind the the Internet of Things (IoT) [1]. IoT connects physical objects to networks and databases and makes use of sensors and actuators to enable new levels of measuring and processing accuracy of real-world processes.

RFID is changing the way security is engineered in Auto-ID applications. On the one hand, RFID brings improvements to security vis-a-vis older Auto-ID technologies by providing increased visibility and the possibility to use cryptography [2]. While an object tagged with a non-serialized barcode can be reliably authenticated only with the help of an additional security feature, such as a hologram or special taggants, an RFID tag can enable both identification and authentication of the tagged object. On the other hand, security is needed in many RFID applications. RFID tags are used to grant access to buildings [3], ski resorts [4], and highways [5], as tickets to public transports [6] and Olympic

H. Tokuda et al. (Eds.): Pervasive 2009, LNCS 5538, pp. 291–308, 2009.

games [7], and in mobile payment [8]. Moreover, RFID is being adopted as a product authentication technology to secure supply chains from counterfeit products [9]. In all these applications cloning and impersonation of RFID tags could be financially lucrative for occasional hackers or professional criminals, and severely damaging for the licit companies' revenues and reputation. The potential losses due to security breaches are furthermore amplified by the high level of automation allowed by the technology. Therefore security is not only added value that RFID provides vis-a-vis older Auto-ID technologies – it is also a requirement.

From the point of view of RFID technology, the most challenging security threats in commercial RFID applications are tag cloning and tag impersonation. The research community addresses these threats primarily by trying to make tag cloning hard by using cryptographic tag authentication protocols [2]. The fundamental difficulties of this research revolve around the trade-offs between tag cost, level of security, and performance in terms of reading speed and distance; it is not very hard to protect an RF device from cloning today, but it is extremely challenging to do it using a low-cost barcode-replacing RFID tag. These tags will be deployed in numbers of several millions and the end-user companies have a strong financial incentive to minimize the tag cost and thus the features the tags provide. To illustrate these rigid hardware constraints, according to Sanjay Sarma, the co-founder of the Auto-ID center at MIT, you *can't do anything beyond hashes in passive RFID tags* [10].

Though the research community always provides incremental improvements to the aforementioned trade-offs, there are reasons to believe that low-cost RFID tags cannot be completely protected from cloning in the foreseeable future. Today it takes the computational and physical complexity of approximately a smart card to implement a mobile device that can be considered reasonably secured against most known threats, including side-channel attacks and physical attacks [42]. Low-cost tags are computationally much weaker devices than smart cards, they can use only a fraction of a smart card's energy and power budget, they lack the physical protection, and furthermore even stronger and better protected devices have been cracked. As a result, it is disputable whether it is possible to come up with a truly secure RFID device that addresses all known vulnerabilities without coming up with a device that effectively has the cost and/or performance (i.e. reading speed and distance) of a wireless smart card.

This paper investigates an approach to secure low-cost RFID systems against tag cloning and impersonation based on detection of cloning attacks – an approach that is far from being fully exploited today. Instead of relying on the strength of the weakest and cheapest devices within the system, the tags, this approach relies on the visibility the tags provide. The underlying technical concept is simple and it has already been proposed for ownership transfer and access control [31,32,33] (cf. Section 2). However, it has not been included in review papers (e.g. [2]), and we think that it merits a recognition. Therefore our major contribution is not the idea development itself but innovative application and thorough evaluation of the concept with respect to cloning of RFID tags.

Our focus on low-cost RFID tags stems from two motivations. First, also low-cost tags are used in security-sensitive applications where cloning of tags could lead to big damage. For instance, Pfizer uses low-cost HF and UHF tags as authentication features for their most counterfeited drug product Viagra [9]. Second, if also low-cost tags can be properly secured, RFID could be applied also in security sensitive domains where the cost of cryptographic tags cannot be justified.

This paper is organized as follows. We first provide a structured review of related work in Section 2. We then study the potential of the presented approach by presenting a statistical model of the provided level of security in Section 3, our implementation based on standard off-the-shelf EPC tags in Section 4, and we discuss the pros and cons of the method focusing on anti-counterfeiting and access control applications in Section 5. Section 6 finishes with the conclusions.

1.1 Introduction to RFID

RFID systems include tags that are affixed to objects, interrogators that read and write data on tags, and back-end systems that store and share data. Passive tags get all their power from the reader while more expensive active tags have a battery. The most important standard for networked RFID is overseen by EPCglobal Inc.[1] The focus of the EPC system is on information in databases associated with EPCs. EPC standards are driven by the retail industry and they focus on passive low-cost UHF tags [11]. Moreover, UHF tags are important in logistics applications due to their higher read range compared to LF and HF tags.

While cryptographic RFID tags are currently widely available in the HF band (e.g. Mifare Desfire[2]), today there are no cryptographic tags commercially available in the UHF band. However, the need for security products in the UHF market is emerging and the first implementations exist (e.g. [12,13]).

2 Related Work

In very general terms, security is the process of protecting assets against adversaries' actions and it comprises steps of prevention, detection, and response [15]. In the following we review related work by mapping countermeasures to the three steps in the process of securing an RFID system against tag cloning and impersonation. This resulting overall process is illustrated in Fig. 1.

2.1 Prevention

Prevention is about building barriers that must be broken or bypassed so as to materialize a threat. It constitutes the first level of defense and the most obvious target for adversaries' attacks. A mundane example of preventive security

---

[1] http://www.epcglobalinc.org
[2] http://mifare.net/products/mifare_desfire.asp

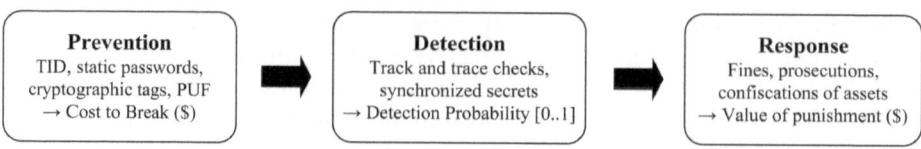

Fig. 1. Process of securing an RFID system against tag cloning and impersonation (the small arrows indicate the outcome and metric of each step)

measures is a lock in a house's front door. Strength of the preventive measures is characterized by their Cost to Break (CtB) that is the minimum effort to find and exploit a vulnerability [16]. Once preventive measures are broken, the exploitation can normally be repeated with a small marginal cost.

Basic preventive measures of standard EPC tags include unique factory programmed, read only, transponder ID (TID) numbers [11] that are somewhat similar to the network card MAC addresses, and password-protected ACCESS and KILL commands (e.g. [18]). The basic measures, however, are vulnerable to eavesdropping and thus they provide only modest protection against tag cloning.

Cryptographic measures include reader-to-tag and tag-to-reader authentication. Several tag-to-reader authentication protocols have been proposed in the literature, usually based on cryptographic primitives like bitwise operations and pseudo-random numbers (e.g., [17,19,20]) or hash-functions (e.g., [22,23,24]). Also different symmetric encryption-based tag authentication protocols exist, for example based on AES algorithm (e.g., [14,12,25]). Asymmetric encryption is currently very challenging on RFID tags but due to advances in elliptic curve cryptography (ECC) it is becoming feasible [26,27]. Moreover, key distribution that is a big future challenge of secure RFID. Another way to authenticate an RFID tag is to use a Physical Unclonable Function (PUF) [28] that is a one way function implemented using minimalistic hardware overhead.

2.2 Detection

Detection is about minimizing the negative effects of materialized threats and increasing the adversaries' probability of getting caught. A video surveillance system is a typical example of detective measures. In some cases detection enables an immediate response that nullifies the negative effects of the materialized threat, and the result is effective prevention of the negative effects. This is analogous to an intrusion detection system that detects the intruder immediately when the intrusion occurs and blocks the intruder before he can do any harm. In other cases there is a delay before detection leads to a response and the materialized threat leads to harmful effects. For instance, this is the case with burglar alarms that do not immediately seize the harm from happening.

In RFID systems, detection-based measures do not require cryptographic operations from the tags but they make use of visibility to detect cloned tags or changes in the tag ownership. The efficiency of a detection based measure is characterized by the probability to detect a threat. In contrast to preventive

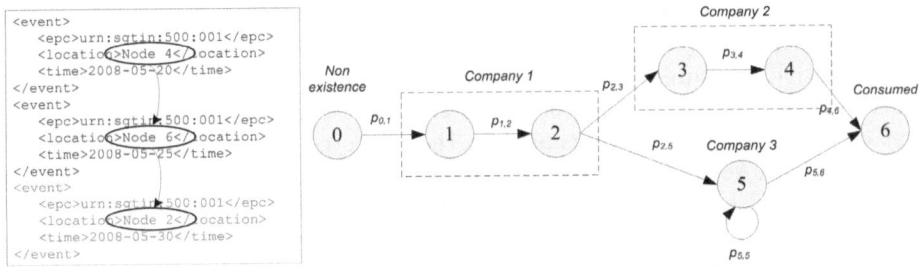

```
<event>
    <epc>urn:sgtin:500:001</epc>
    <location>Node 4</location>
    <time>2008-05-20</time>
</event>
<event>
    <epc>urn:sgtin:500:001</epc>
    <location>Node 6</location>
    <time>2008-05-25</time>
</event>
<event>
    <epc>urn:sgtin:500:001</epc>
    <location>Node 2</location>
    <time>2008-05-30</time>
</event>
```

Fig. 2. Illustration of how cloned tags can be detected from track and trace data (left): since transition from Node 6 to Node 2 ($p_{6,2}$) is not possible according to the supply chain model (right), the last event in Node 2 must be generated by a cloned tag [30]

measures, detective measures can generate *false alarms* where a genuine tag is classified as an impersonator.

Juels [2] noted that serial level identification alone without secure verification of the identities can be a powerful anti-counterfeiting tool. Koh *et al.* [34] made use of this assumption to secure pharmaceutical supply chains by proposing an authentication server that publishes a *white list* of genuine products' ID numbers. Staake *et al.* [29] were among the first to discuss the potential of track and trace based product authentication within the EPC network and they point out some problems that occur when the back-end no longer knows where the genuine object is. Mirowski and Hartnett [3] developed a system that essentially detects cloned RFID tags or other changes in tag ownership in an access control application using intrusion detection methods. To address the problem of limited visibility, Lehtonen *et al.* [30] applied machine learning techniques to automatically detect cloned tags from incomplete location data (cf. Fig. 2).

Ilic *et al.* [31] made use of a similar synchronized secret approach, but the application focus was on ownership transfer and access control. Also Grummt and Ackermann [32] presented the idea behind synchronized secrets approach in an RFID access control application in a scheme called *chosen, temporarily valid secrets*. In addition, Koscher *et al.* [33] describe the same principle in a technical report as a way of increasing the security of ACCESS code based authentication of EPC tags. However, none of the authors discussed and evaluated how the synchronized secrets approach could be applied to address tag cloning attacks.

2.3   Response

Response is what happens after a materialized threat is detected. It comprises of all the actions that minimize the negative effects for the process owner [35] and maximize the negative effects for the adversary in terms of punishments. In commercial RFID applications this can mean, for example, confiscation of the illicit goods, prosecution of the illicit players on contract breaches and illegal activities, and ending business relationships. The lack of effective law enforcement can severely cripple the strength of responsive measures, especially in developing countries. Moreover, small companies have less power to deliver hefty punishments than big companies, making them potentially more lucrative targets.

Responsive measures define the expected value of the punishment and they contribute an important component to the overall *deterrent* effect of security that can be characterized by change in the expected payoff from attempted illicit activities. According to the deterrence theory, the lower the overall payoff including the risk of getting caught, the less willingly and often adversaries attempt to realize the threat. In particular, an asset worth of $100 is safe from rational, risk-neutral, and financially motivated thieves if the cost and risk factor of an attempted theft sum up to more than $100. However, because of asymmetric information, different risk perceptions, irrational decisions, and lack of reliable data, researchers have often failed to find empirical evidence of deterrence decreasing the supply of crime in practice [36].

2.4   Effect of Security

Given the structured view of security, we can now model the overall effect of a system's all security measures on an adversary. Such modeling can be used to evaluate the effect of security on financially motivated thieves, but it is less useful for occasional hackers who are motivated by intellectual challenges, fame, reputation etc. When E denotes the expected net value of an attempted attack for an adversary, CtB the cost to break the preventive measures, P_{det} the probability the an attack is detected by the detective measures, P_{pun} the probability that the adversary is punished if the attack is detected, F the value of the punishment, and L the value of the loot, the process of security affects an adversary's payoff as defined by Equation 1.

$$E = (1 - P_{det})(L - CtB) - P_{det}(P_{pun}F + CtB) \qquad (1)$$

This model bases on Schechter's work on how much security is enough to stop a thief [37] and it shows how both preventive and detective measures can make an adversary's payoffs negative through high CtB or high P_{det}, respectively. In particular, owing to the risk of punishment, a detective measure does not need to have a 100% P_{det} in order to make $E < 0$. This means that a high-enough detection rate is enough to destroy the business model of a thief.

3   Detecting Cloned Tags with Synchronized Secrets

The available methods to secure low-cost RFID tags from cloning are limited. In particular, cryptographic approaches proposed in the literature cannot be used with the existing standard UHF tags since they require changes in the chip's integrated circuit, and existing detective measures do not perform well under limited visibility. The presented method described in this section attempts to partially address these problems. Though the method is simple and it has already been proposed in other RFID applications ([31,32,33]), it has not yet been applied and evaluated to address tag cloning attacks.

3.1 Proposed Method

The presented method makes use of the tags' rewritable memory. In addition to the static object and transponder identifiers (e.g. EPC, TID [11]), the tags store a random number that is changed every time the tag is read. We denote this number a *synchronized secret* since it is unknown to all who do not have access to the tag and it can also be understood as a one-time password. A centralized back-end system issues these numbers and keeps track of which number is written on which tag to detect synchronization errors.

Every time a tag is read, the back-end first verifies the tag's static identifier. If this number is valid, the back-end then compares the tag's synchronized secret to the one stored for that particular tag. If these numbers match, the tag passes the check – otherwise an alarm is triggered. After the check, the back-end generates a new synchronized secret that the reader device writes on the tag. This principle is illustrated in Fig. 3.

If a tag has an outdated synchronized secret, either the tag is genuine but it has not been correctly updated (desynchronization) or someone has purposefully obtained and written an old secret to the genuine tag (sophisticated vandalism), or the genuine tag has been cloned and the cloned tag has been scanned. Since unintentional desynchronization problems can be addressed with acknowledgments and the described form of vandalism appears somewhat unrealistic in today's commercial RFID applications, an outdated synchronized secret is as a strong evidence of a tag cloning attack. If a tag has a valid static identifier but a synchronized secret that has never been issued by the back-end, the tag is likely to be forged.

An outdated synchronized secret alone does not yet prove that a tag is cloned; if the cloned tag is read before the genuine tag after cloning attack occurred, it is the genuine tag that has an outdated synchronized secret. Therefore an outdated synchronized secret is only a proof that tag cloning attack has occurred, but not a proof that a tag is cloned. As a result, the presented method pinpoints the objects with the same identifier but it still needs to be used together with a manual inspection to ascertain which of the objects is not genuine.

To protect the scheme against man-in-the-middle (MITM) attacks and malicious back-ends and readers, the back-end and the readers need a reliable way to prove their authenticity to each other. The protocol itself is agnostic to how this is achieved, and it can be done using for example a trusted reader platform [38] and standard public key infrastructure (PKI).

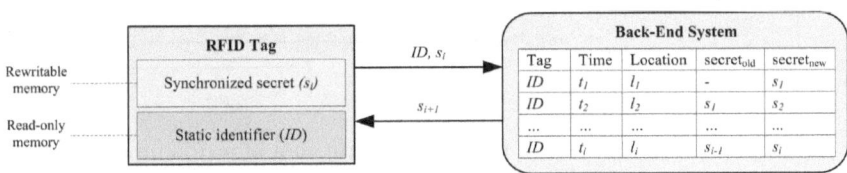

Fig. 3. Illustration of the protocol

In addition to knowing that a cloning attack has occurred, the back-end can pinpoint a time window and a location window where the cloning attack happened. Thus the method makes it also hard to *repudiate* tag cloning to parties who handle the tagged objects. This is a security service that preventive measures do not provide and it can support the responsive actions.

3.2   Level of Security

The level of security of a detection based security measure is characterized by its detection rate (cf. subsection 2.2). In this subsection we evaluate the level of security of the presented method with a statistical model.

We assume a system which consists of a population of tags that have a static identifier and non-volatile memory for the synchronized secret. The tags are repeatedly scanned by readers that are connected to the back-end. The probability that a tag will be scanned sometimes in the future at least once more is constant and denoted by Θ. When a tag is scanned its synchronized secret is updated both on the tag and the back-end as described above in subsection 3.1. The time between these updates for a tag is denoted by a random variable T_{update}. An adversary can copy any tag in the system and inject the cloned tag into the system. The time delay from the copying attack to when the copied tag is scanned is denoted by a random variable T_{attack}. In addition, an adversary can try to guess the value of the synchronized secret.

The system's responses can be statistically analyzed. First, the probability to successfully guess a genuine tag's synchronized secret is $1/(2^N)$, where N denotes the length of the synchronized secret in bits. Even with short sizes, e.g. $N = 32$, guessing the synchronized secret is hard (ca. 2×10^{-9}) and the system can thus be considered secure against guessing attacks[3]. Second, when a copying attack occurs, three mutually exclusive outcomes are possible (cf. Fig. 4):

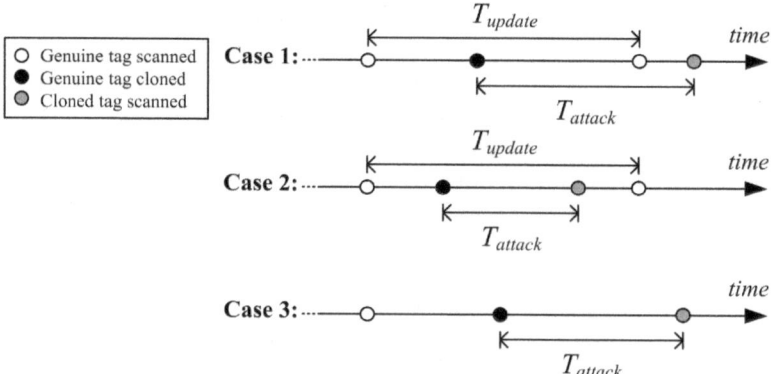

Fig. 4. An illustration of the possible outcomes of a cloning attack

---

[3] N.B.: There is no brute-force attack to uncover this number.

- **Case 1:** The genuine tag is scanned before the copied tag and an alarm is thus triggered when the copied tag is scanned.
- **Case 2:** The copied tag is scanned before the genuine tag and an alarm is thus triggered when the genuine tag is scanned.
- **Case 3:** The genuine tag is not scanned anymore and thus no alarm is triggered for the copied tag.

In Case 1 the cloned tag is detected as soon as it is scanned the first time and the negative effect of the attack can be prevented. In Case 2 the cloned tag passes a check without raising an alarm but the system detects the cloning attack when the genuine tag is scanned. In Case 3 the security fails and the cloning attack goes unnoticed. The system's level of security is characterized by the probability of Case 1 that tells how often threats are prevented, and by the probability of Case 1 or Case 2 that tells how often threats are detected.

$$\text{Prevention rate} = \Pr(\text{Case 1}) \tag{2}$$

$$\text{Detection rate} = \Pr(\text{Case 1} \vee \text{Case 2}) \tag{3}$$

The probability of Case 1 equals the probability that the genuine tag is scanned at least once more, Θ, multiplied by the probability that the genuine tag is scanned before the cloned tag. Let us assume that the time when the cloning attack occurs is independent of when the genuine tag is scanned and uniformly distributed over the time axis, so the average time before the genuine tag is scanned after the copying attack is $T_{update}/2$. We can now estimate the probability of Case 1 as follows:

$$\Pr(\text{Case 1}) = \Theta \cdot \Pr\left(\frac{T_{update}}{2} - T_{attack} < 0\right) \tag{4}$$

Assuming that $T_{update} \sim N(\mu_{update}, \sigma_{update}^2)$ and $T_{attack} \sim N(\mu_{attack}, \sigma_{attack}^2)$, we can estimate the probability of Case 1 using a new random variable $Z = \frac{T_{update}}{2} - T_{attack}$ as follows[4]:

$$\Pr(\text{Case 1}) = \Theta \cdot \Pr(Z < 0) \tag{5}$$

Distribution of Z can be calculated using these rules: if $X \sim N(\nu, \tau^2)$, then $aX \sim N(a\nu, (a\tau)^2)$, and if $Y \sim N(\kappa, \lambda^2)$, then $X + Y \sim N(\nu + \kappa, \tau^2 + \lambda^2)$.

$$Z \sim N\left(\frac{\mu_{update}}{2} - \mu_{attack}, \frac{\sigma_{update}^2}{4} + \sigma_{attack}^2\right) \tag{6}$$

Equation 4 shows that the level of security of the synchronized secrets method depends on the frequency in which the genuine tags are scanned with respect to the time delay of the attack, and on the probability that the genuine tag

---

[4] N.B.: Since T_{update} and T_{attack} cannot be negative, these assumptions yield viable estimates only when the mean values are high and variances low.

is scanned once more. The same finding is confirmed from equations 5 and 6 which show more clearly that, in the case of normally distributed time variables, $\lim_{\mu_{attack}-\mu_{update}\to\infty} \Pr(\text{Case 1}) = \Theta$.

After the last transaction of the genuine tag, a single cloned tag will always go unnoticed (Case 3). We assumed above a statistically average adversary who does not systematically exploit this vulnerability. However, a real-world adversary who knows the system is not likely to behave in this way. Therefore this vulnerability should be patched by flagging tags that are known to have left the system.

4   Implementation

This section presents our experimental implementation of the presented method using UHF tags conforming to the EPC standard [11]. These tags are relatively low-cost (ca. 0.10-0.20 USD), provide only basic functionalities (e.g. 96-bit rewritable identifier, password-protected access, 16-bit pseudo random number generator), and are therefore expected to be employed in large volumes for tracking various kinds of physical objects.

EPC standards define a user memory bank where the synchronized secret can be stored [21]. To illustrate the real hardware constraints of low-cost RFID tags, many existing EPC tags do not have any user memory. To overcome this problem, one can alternatively re-write the 32-bit access-password in the reserved memory bank to store the synchronized secret, or use a part of the EPC memory bank if it is not completely needed for the object identifier.

The protocol between the back-end system, the reader, and the tag is presented in Fig. 5. In the illustration, s^i denotes the current synchronized secret, s^{i+1} the new synchronized secret, RND_{32} a new 32-bit random number, $alarm$ a boolean value whether an alarm is triggered or not, and ack an acknowledgment of a successful update of the synchronized secret. Step 6 is dedicated to establishing a secure connection between the reader and the back-end to mitigate MITM attacks, malicious back-end systems, and to protect the integrity of the back-end.

4.1   Set-Up

We have implemented the presented method using EPC Class-1 Gen-2 tags from UPM Raflatac that use Monza 1A chips manufactured by Impinj. The reader device is A828EU UHF reader from CAEN and it is controlled by a laptop that runs the local client program. The back-end system was implemented as a web server that stores the EPC numbers, synchronized secrets, and time stamps in a MySQL database. The hardware set-up is shown in Fig. 6.

Given that an RFID infrastructure is in place and tags have a modest amount of user memory, the only direct cost of the presented method is the time delay of verifying and updating the synchronized secrets, i.e. steps 4-14 of the protocol

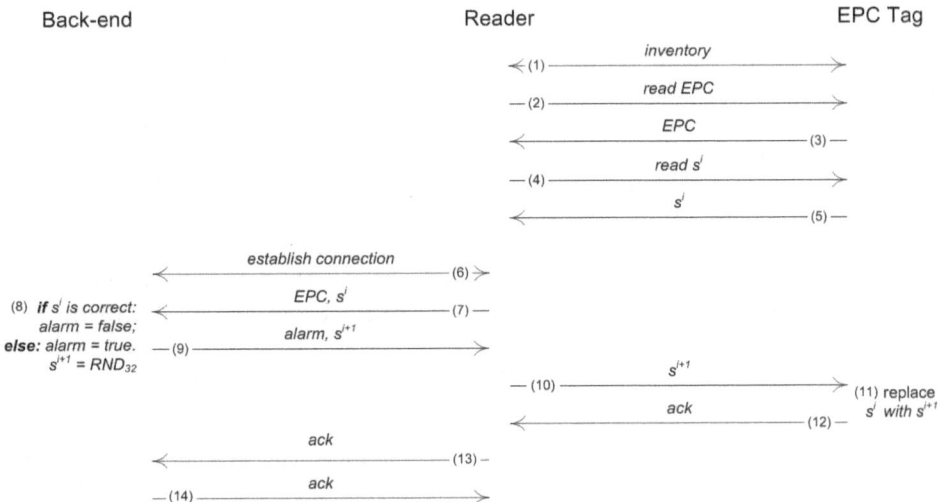

Fig. 5. Implemented protocol

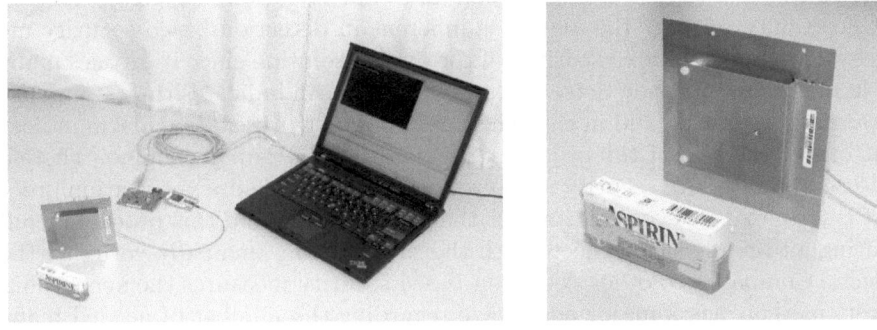

Fig. 6. The hardware set-up

(cf. Fig. 5). We have measured this overhead time from 100 reads where the tagged product faces the antenna in 5 cm distance[5].

4.2   Performance

The average overall processing time of one tag was 864 ms. This includes 128 ms for the inventory command, 181 ms for reading the EPC number, and the remaining 555 ms is the time overhead of the synchronized secrets protocol. The measured average times and standard deviations are presented in Fig. 7. The results show that the time overhead of the protocol increases one tag's processing time approximately by a factor of 300%, after the inventory command. Even

---

[5] Steps 13-14 of the protocol are omitted from the measurements since they do not increase a tag's processing time.

though the time overhead is short in absolute terms, it makes a difference in bulk reading where multiple products are scanned at once. A closer look on the times of different steps reveals that writing a new synchronized secret on the tag is only a slightly slower than reading a secret from the tag, and that the biggest variance is experienced within the back-end access (steps 6-9).

The performance depends on implementation and has potential for improvement through optimization of reader and back-end software. In addition, variance in web server latency makes the time overhead hard to predict. Despite these limitations, this simple experiment provides evidence that the overhead time can limit the usability of the presented method in time-constrained bulk reading.

5   Discussion

The challenge of the system designers is to engineer the systems to resist not only the occasional hackers, but also the *law of greed* which says that whenever there exist a possibility to gain from unintended or illegal use, sooner or later someone will do it. Since RFID is primarily used for object identification, the first step of protection is to make sure the objects are what they claim they are. This translates into addressing cloning and impersonation of tags.

Uncertainty relating the alarms is inherent in detection-based security measures and an important cost driver of the overall solution since it invokes manual work; in typical intrusion detection systems an alarm indicates that an intrusion *might* have happened and in the synchronized secrets method an alarm indicates that one of the one of the objects with the same ID is not genuine. Therefore end-users of detective security measures need to implement a verification process that is triggered by every alarm. For the presented method this process includes locating all the physical objects with the same ID and manually verifying these objects. Compared to other detection-based security measures the synchronized secrets method has a major advantage regarding the number of needed manual verification; since an alarm in the synchronized secrets method always indicates a cloning attack – given that desynchronization problems are addressed – the

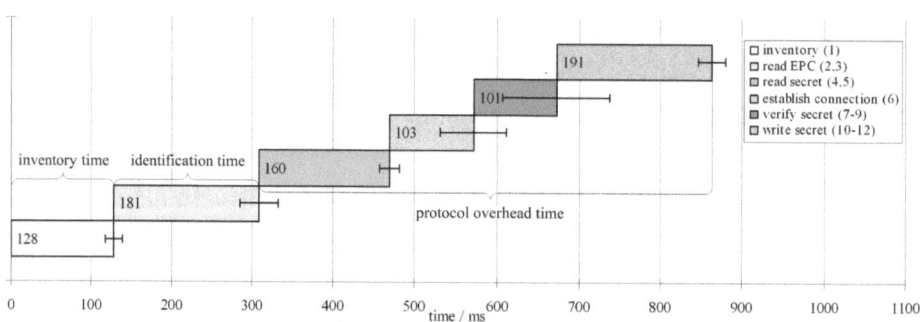

Fig. 7. Measured average times and standard deviations (error bars) of different steps (numbers in brackets) in the implemented protocol

method does not generate any pure false positives. In track and trace based methods, however, alarms can also be generated by any irregular supply chain events such as reverse logistics. This advantage is illustrated with a numeric example in subsection 5.2 below.

The synchronized secrets method does not require sharing of track and trace data, which is a benefit for companies that find this information too sensitive for disclosing. However, if there are large delays between the scans, the synchronized secrets method can trigger an alarm for the cloned tag only after a large delay. In some applications this delay cannot be allowed since it could mean, for example, that a counterfeit medicine has already been consumed. In track and trace based clone detection methods the alarm is triggered – if it is triggered – primarily right after the cloned tag is scanned, and thus similar delays are less likely to occur.

One physical back-end system is unlikely to be scalable enough to run the synchronized secrets protocol for the large numbers of objects that will be tagged. Fortunately, this kind of scalability is also not needed. The back-end can be distributed to virtually an unlimited number of servers by having, for example, one back-end server per product family, per product type, per geographical region, or per a subset of certain kinds of products. This can be implemented either with static lists that map EPC numbers to different back-end systems and that is known by readers, or with the help of EPC Object Naming Service (ONS) or Discovery Services (DS) that provide one logical central point for queries about information and services related to a product [39]. Moreover, the scalability requirements of the presented method are the same as in any RFID system where the back-end knows the current location/status of the items. Additional network requirements of the presented method include strong authentication between the reader devices and the back-end to secure the protocol against MITM attacks.

All EPC tags are potentially vulnerable to tampering of the tag data which can be used as a Denial of Service (DoS) attack against the presented method. This DoS vulnerability can be mitigated with the access passwords of EPC tags [11] by having the reader retrieve the access password and unlock the tag after identification (cf. step 2 in Fig. 5), and lock the tag again after updating the synchronized secret. Moreover, write and read protection of the user memory where the synchronized secret is stored can be used to as a complementing security measure to prevent tag cloning and tampering. In addition, the use of synchronized secrets opens a door for a new DoS attack that makes a genuine tag cause an alarm even when there are no cloned tags in the system; an adversary that is located near to an authorized reader can eavesdrop the static ID number and the synchronized secret of a genuine tag and impersonate this tag to an authorized reader before the genuine tag is scanned. As a result, the genuine tag will raise an alarm next time it is scanned. This results into an unnecessary manual inspection of the genuine tag (which will reveal the time and location of the impersonation attack). This DoS attack is possible only when adversaries have access to an authorized reader device, which is typically not the case in supply chain applications such as anti-counterfeiting. Furthermore, the time and location of this DoS attack are registered, while there are also simpler attacks

that achieve the same outcome without leaving any such trace, namely physical
or electromagnetic destruction of the tags.

5.1   Anti-counterfeiting

The presented method, complemented by flagging of all sold or consumed products, makes injection of counterfeit products into protected supply chains very
difficult; counterfeit products that do not have RFID tags or that have RFID
tags with invalid ID numbers are revealed as fakes, and counterfeit products with
cloned RFID tags cause a desynchronization that the back-end detects (Case 1
or Case 2). In particular, there is nothing that an adversary can do to a cloned
tag that would prevent the system from detecting the cloning attack, given that
the genuine and cloned tags will be scanned. In addition to protecting the system from tag cloning, the presented method also provides a proof of when the
tags are cloned. This helps further in pinpointing the illicit players and problematic locations. Since the readers and products are located in the premises of
the supply chain partners, the risk of above mentioned DoS attacks is low. As
a result, the presented method provides a considerable increase in security compared to standard EPC/RFID-enabled supply chains where tag cloning attack
is not addressed.

5.2   Access Control

Level of security of the presented method depends on how often the tags are
scanned and on how much time the adversary needs to conduct the cloning and
impersonation attack. We study the scan rates of genuine tags based on a public
access control data set [43]. This data set is an activity record of proximity
cards within an access control system that controls the access to parts of a
building. The probability that a tag was scanned again within this data set

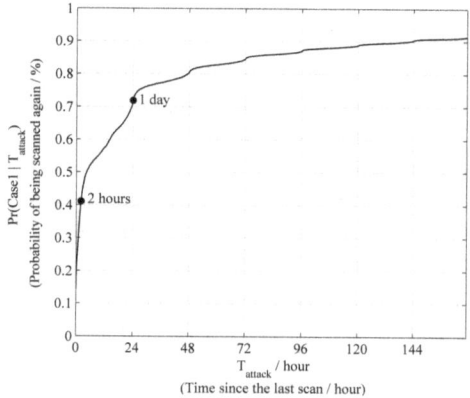

Fig. 8. Time delay between consecutive reads in an access control data set ([43])

is presented in Fig. 8 as a function of time delay from the previous scan. This value equals the probability that a arbitrarily injected cloned tag raises an alarm (Case 1) given the attack delay. For example, an adversary who clones a genuine tag when it is scanned and injects the tag 2 or 24 hours after cloning has a 41% or a 72% chance of raising an alarm upon impersonation, respectively. The overall probability of a tag being scanned again, Θ, was 99.15%, which corresponds to the detection rate (Equation 3). The findings suggest that only very few cloning attacks would potentially go completely unnoticed in the studied application, and that an adversary needs to conduct the impersonation attack within a few hours after tag cloning to have a relative good chance of not raising an alarm.

Last, we compare the performance of the synchronized secrets method to that of Deckard, a system that was designed to detect cloned tags within the aforementioned data set based on statistical anomalies [3]. In average, Deckard was able to detect 76% of cloned tags with an 8% false alarm rate from simulated attack scenarios within the aforementioned data set. Assuming that 1% of transactions are generated by cloned tags, this means that for each alarm triggered by a cloned tag there are approximately 11 false alarms triggered by genuine tags. As a result the probability that a tag that triggers an alarm is really a cloned one is only 8.4%. Within the synchronized secrets method, however, each alarm indicates a cloning attack and the probability that a tag that triggers an alarm is really a cloned one is 50%, compared to only 8.4% of Deckard. In addition, an alarm would be triggered to 99.15% of cloned tags, compared to 76% of Deckard. This numeric example illustrates the improved reliability of the synchronized secrets method compared to another detection-based RFID security measure.

6   Conclusions

Detecting cloned RFID tags appears attractive for securing commercial RFID applications since it does not require more expensive and energy thirsty cryptographic tags. This paper presents a synchronized secrets method to detect cloning attacks and to pinpoint the different tags with the same ID. The presented method requires only a small amount of rewritable memory from the tag but it provides a considerable increase to the level of security for systems that use unprotected tags. A major benefit of the presented measure is that it can be used with existing standard low-cost RFID tags, such as EPC Gen-2, and it can be applied in all RFID applications where the tags are repeatedly scanned. The additional cost factor of the presented method is manual verifications needed to ascertain which of the tags (objects) with the same ID number is the cloned one, but we show that the number of needed verifications for the presented method is considerably smaller than for comparable detective security measures. Overall, the presented method has the potential to make harmful injection of cloned tags into RFID systems considerably harder using only a minimal hardware overhead.

Acknowledgment

This work is partly funded by the Auto-ID Labs and by the European Commission within the Sixth Framework Programme (2002-2006) projects BRIDGE (Building Radiofrequency IDentification solutions for the Global Environment), IP Nr. IST-FP6-033546. The authors would like to thank the following persons for their comments and help: Michael Fercu and Luke Mirowski.

References

1. Fleisch, E., Mattern, F.: Das Internet der Dinge: Ubiquitous Computing Und RFID in Der Praxis: Visionen, Technologien, Anwendungen, Handlungsanleitungen. Springer, Berlin (2005)
2. Juels, A.: RFID security and privacy: A research survey. IEEE Journal of Selected Areas of Communication 24(2), 381–894 (2006)
3. Mirowski, L., Hartnett., J.: Deckard: A System to Detect Change of RFID Tag Ownership. International Journal of Computer Science and Network Security 7(7) (2007)
4. Michahelles, F., Flörkemeier, C., Lehtonen, M., Hinske, S.: An RFID-tag in Every Ski Item-Level Tagging in the Ski Industry. In: Pervasive Technology Applied - Real-World Experiences with RFID and Sensor Networks, Proceedings of the Pervasive 2006 Workshops, Dublin (2006)
5. Swedberg, C.: RFID Drives Highway Traffic Reports. RFID Journal (2004)
6. IDTechEx: Oyster Transport for London TfL, card UK (2007)
7. RFID News: Olympic tickets to carry wealth of personal info. (2008)
8. Texas Instruments: ExxonMobil Speedpass (2008)
9. Bacheldor, B.: Pfizer Prepares for Viagra E-Pedigree Trial. RFID Journal (Feburary 2007)
10. Sarja, S.: Introductory Talk: Some issues related to RFID and Security. In: Keynote Speech in Workshop on RFID Security 2006, Graz (2006)
11. EPCglobal Inc.: Class-1 Generation-2 UHF RFID Conformance Requirements Specification v. 1.0.2 (2005)
12. Feldhofer, M., Aigner, M., Dominikus, S.: An Application of RFID Tags using Secure Symmetric Authentication. In: 1st International Workshop on Privacy and Trust in Pervasive and Ubiquitous Computing, pp. 43–49 (2005)
13. Plos, T., Hutter, M., Feldhofer, M.: Evaluation of Side-Channel Preprocessing Techniques on Cryptographic-Enabled HF and UHF RFID-Tag Prototypes. In: Workshop on RFID Security 2008, Budapest (July 2008)
14. Dominikus, S., Oswald, E., Feldhofer, M.: Symmetric authentication for RFID systems in practice. In: ECRYPT Workshop on RFID and Lightweight Crypto, Graz (2005)
15. Schneier, B.: Beyond Fear. Thinking Sensibly of Security in an Uncertain World. Copernicus Books, New York (2003)
16. Schechter, S.E.: Quantitatively differentiating system security. In: The First Workshop on Economics and Information Security, Berkeley (2002)
17. Juels, A.: Minimalist cryptography for low-cost RFID tag. In: Blundo, C., Cimato, S. (eds.) SCN 2004. LNCS, vol. 3352, pp. 149–164. Springer, Heidelberg (2005)
18. Juels, A.: Strengthening EPC Tags Against Cloning. In: Jakobsson, M., Poovendran, R. (eds.) Proceedings of the 2005 ACM Workshop on Wireless Security, pp. 67–76. ACM Press, Cologne (2005)

19. Vajda, I., Buttyán, L.: Lightweight authentication protocols for low-cost RFID tags. In: Workshop on Security in Ubiquitous Computing, Ubicomp 2003 (2003)
20. Tsudik, G.: YA-TRAP: Yet another trivial RFID authentication protocol. In: IEEE International Conference on Pervasive Computing and Communications, pp. 640–643 (2006)
21. EPCglobal Inc.: Class-1 Generation-2 UHF RFID Protocol for Communications at 860 MHz - 960 MHz v. 1.1.0 (2005)
22. Yang, J., Park, J., Lee, H., Ren, K., Kim, K.: Mutual authentication protocol for low-cost RFID. In: ECRYPT Workshop on RFID and Lightweight Crypto, Graz (2005)
23. Dimitriou, T.: A lightweight RFID protocol to protect against traceability and cloning attacks. In: IEEE Conference on Security and Privacy for Emerging Areas in Communication Networks SecureComm., Athens, Greece (2005)
24. Avoine, G., Oechslin, P.: A scalable and provably secure hash based RFID protocol. In: IEEE International Workshop on Pervasive Computing and Communication Security, pp. 110–114 (2005)
25. Bailey, D., Juels, A.: Shoehorning Security into the EPC Tag Standard. In: De Prisco, R., Yung, M. (eds.) SCN 2006. LNCS, vol. 4116, pp. 303–320. Springer, Heidelberg (2006)
26. Wolkerstorfer, J.: Is Elliptic-Curve Cryptography Suitable to Secure RFID Tags? In: ECRYPT Workshop on RFID and Lightweight Crypto, Graz (2005)
27. Batina, L., Guajardo, J., Kerins, T., Mentens, N., Tuyls, P., Verbauwhede, I.: An Elliptic Curve Processor Suitable For RFID-Tags. Cryptology ePrint Archive, Report 2006/227 (2006)
28. Devadas, S., Suh, E., Paral, S., Sowell, R., Ziola, T., Khandelwal, V.: Design and Implementation of PUF-Based "Unclonable" RFID ICs for Anti-Counterfeiting and Security Applications. In: IEEE International Conference on RFID 2008, pp. 58–64 (2008)
29. Staake, T., Thiesse, F., Fleisch, E.: Extending the EPC Network – The Potential of RFID in Anti-Counterfeiting. In: Symposium on Applied Computing, New York, pp. 1607–1612 (2005)
30. Lehtonen, M., Michahelles, F.: Fleisch, E.: Probabilistic Approach for Location-Based Authentication. In: 1st International Workshop on Security for Spontaneous Interaction IWSSI 2007, 9th International Conference on Ubiquitous Computing (2007)
31. Ilic, A., Michahelles, F., Fleisch, E.: The Dual Ownership Model: Using Organizational Relationships for Access Control in Safety Supply Chains. In: IEEE International Symposium on Ubisafe Computing (2007)
32. Grummt, E., Ackermann, R.: Proof of Possession: Using RFID for large-scale Authorization Management. In: Mühlhäuser, M., Ferscha, A., Aitenbichler, E. (eds.) Constructing Ambient Intelligence, AmI-07 Workshops Proceedings. Communications in Computer and Information Science, pp. 174–182 (2008)
33. Koscher, K., Juels, A., Kohno, T., Brajkovic, V.: EPC RFID Tags in Security Applications: Passport Cards, Enhanced Drivers Licenses, and Beyond (2008) (Manuscript)
34. Koh, R., Schuster, E., Chackrabarti, I., Bellman, A.: Securing the Pharmaceutical Supply Chain. Auto-ID Labs White Paper (2003)
35. Mitropoulos, S., Patsos, D., Douligeris, C.: On Incident Handling and Response: A state-of-the-art approach. Computers and Security 25(5), 351–370 (2006)
36. Cameron, S.: The Economics of Crime Deterrence: A Survey of Theory and Evidence. Kyklos International Review for Social Sciences 41(2), 301–323 (1988)

37. Schechter, S.E., Smith, M.: How Much Security is Enough to Stop a Thief? The Economics of Outsider Theft via Computer Systems and Networks. In: Seventh International Financial Cryptography Conference, Guadeloupe (2003)
38. Soppera, A., Burbridge, T., Broekhuizen, V.: A Trusted RFID Reader for Multi-Party Services. EU RFID Convocation (2007)
39. EPCglobal Inc.: EPCglobal Architecture Framework Version 1.0 (2005)
40. Wang, J., Li, H., Yu, F.: Design of Secure and Low-cost RFID Tag Baseband. In: International Conference on Wireless Communications, Networking and Mobile Computing, pp. 2066–2069 (2007)
41. Sandhu, R.: Good-Enough Security: Toward a Pragmatic Business-Driven Discipline. IEEE Internet Computing 7(1), 66–68 (2003)
42. Weingart, S.: Physical Security Devices for Computer Subsystems: A Survey of Attacks and Defenses. In: Workshop on Cryptographic Hardware and Embedded Systems, Massachusetts, pp. 302–317 (2000)
43. Mirowski, L., Hartnett, J., Williams, R., Gray, T.: A RFID Proximity Card Data Set. Tech. Report University of Tasmania (2008),
    http://eprints.utas.edu.au/6903/1/a_rfid_proximity_card_data_set.pdf

Towards Ontology-Based Formal Verification Methods for Context Aware Systems*

Hedda R. Schmidtke and Woontack Woo**

GIST U-VR Lab., 500-712 Gwangju, South Korea
{schmidtk,wwoo}@gist.ac.kr

Abstract. Pervasive computing systems work within, and rely on, a model of the environment they operate in. In this respect, pervasive computing systems differ from other distributed and mobile computing systems, and require new verification methods. A range of methods and tools exist for verifying distributed and mobile concurrent systems, and for checking consistency of ontology-based context models. As a tool for verifying current pervasive computing systems both are not optimal, since the former cover mainly tree-based location models, whereas the latter are not able to address the dynamic aspects of computing systems. We propose to formally describe pervasive computing systems as distributed concurrent systems operating on the background of a mereotopological context model.

Keywords: context modelling, mereotopology, program verification, ontologies.

1 Introduction

Pervasive computing systems can be understood as distributed and mobile concurrent computing systems that are able to react flexibly to changes in their physical environment [30]. A model of the environment is therefore a fundamental part of a pervasive computing system, and research on context modelling methodology has led to novel data structures and ontologies for representing numerous aspects of context, such as location, time, social structure, computational structure, and generally the physical properties of the environment. However, current approaches to verification of pervasive computing systems [4, 27] based on Ambient Calculus [6] or the theory of Bigraphs [24] are focussed on tree-based location models, which are inappropriate to represent overlapping contexts, continuous domains of context, and continuous change. Ranganathan and Campbell [27] concluded that program verification and verification of context models are complementary tasks. However, important interactions exist and properties at the interface, where a process queries the context model or uses it to communicate information to other processes, should be verifiable [27].

---

* This research is supported by the UCN Project, the MIC 21st Century Frontier R&D Program in Korea.
** Corresponding author.

H. Tokuda et al. (Eds.): Pervasive 2009, LNCS 5538, pp. 309–326, 2009.
© Springer-Verlag Berlin Heidelberg 2009

We propose a method for verification of pervasive computing systems with complex context models. We follow the suggestion of Ranganathan and Campbell [27] to separate verification of the context model from verification of the control structure. We argue that the semantics of context models can be described adequately within the theory of mereotopology [37], which is used widely in the area of formal ontology for describing domains such as space, time, partonomies, and taxonomies. We show that mereotopology is suitable to describe lattices, hierarchical structures that are not trees, and continuous domains, such as ranges of sensor values and uncertainty regions around GPS coordinates. On this background, we outline how querying of a context model and communication via activation of contexts in the context model can be added as new primitives to the semantics of a programming language. We thus separate the context model from the state of a program, so that constraints on both parts can be verified separately. The resulting simple example language is already expressive enough to allow specification of relevant distributed and mobile concurrent algorithms that operate with hybrid location models and are triggered by sensors.

The article consists of three main sections. In the next section (Sect. 2), we analyse related works on the representation of context from the areas of ubiquitous computing, formal verification methods for context-aware computing, and from the areas of knowledge representation and ontologies. Then, we outline a simple variant of the theory of mereotopology and describe how the theory applies to the problem of context modelling (Sect. 3). We show in Sect. 4 how this language can be combined with a classical CSP-style programming language.

2   Related Works

Ubiquitous computing became possible after distributed systems and mobile computing, in particular, the idea of wireless ad hoc networks, had been developed [30]. Accordingly, location is one of the best understood and most important parameters of context. The active map of Schilit and Theimer [31] organises locations of users and objects in a containment hierarchy of names for regions organised as a tree. Hybrid location models [20, 22] in addition allow for adding coordinate information into the hierarchical structure. The hierarchical structure is realised in [20, 31] with a tree data structure, and in [22] with a more general lattice structure that in contrast to a tree structure also allows that regions share sub-regions, that is, that regions can overlap. However, Leonhardt [22] demands that the leaf nodes of the hierarchical structure, the so-called zones, may not overlap; the zones thus provide a partitioning of space. A lattice structure based on partitions has also been suggested by Ye et al. [40]. Schmidtke and Woo [34] show that a partitioning of space can lead to problems: it imposes restrictions on the representation, leads to inflated hierarchies, and does not allow to properly reflect uncertainty of location information. We discuss in Sect. 3 how lattice-based location models can be described within the formal framework of mereotopology.

A range of approaches for verifying mobile systems that contain a notion of location has been suggested in the area of program verification, in particular approaches building upon *pi-calculus* [17], and the theories of *bigraphs* [4, 24], and *ambients* [6]. However, the notion of location or domains employed in these approaches is limited to tree-hierarchies [6] or allows for general graph structures [17]. Milner [24] suggests to combine a hierarchical tree-structure of localities with a global, unrestricted network structure. From the perspectives of knowledge representation [34] and context modelling [40], however, a tree structure is not sufficient for specifying location: in tree-based hierarchies, overlapping regions are not allowed and have to be split, and movement of an agent from one region to a neighbouring region is always discrete. Consequently, Ranganathan and Campbell [27] argue that proof of correctness with ambients [6] needs to be complemented with a proof of correctness of the context modelling parts of a pervasive computing environments, for which they use an ontology described in first-order logic [28]. However, it is not clear how the location model in an ambient specification should be made consistent with the more refined location model in an ontology [27].

We follow the suggestion in [27] and propose a coupled semantics of context modelling and programs, so that inconsistencies and unnecessary complexity can be avoided. Instead of using one of the above mentioned verification methods, which come with an in-built location model, we start from a simple textbook variant [2] of CSP as a widely familiar program verification language and complement it with a context modelling language [35] that allows for the description of contemporary lattice-based location models. The location model in our approach is maintained exclusively by the context model, or *context knowledge base*. The knowledge base is verified via a mereotopological, first-order logic theory of context aspects, that is, as an ontology, but not as a computing system. Our aim is to characterise a formal system that describes a broad range of existing context-aware pervasive computing systems in both their context modelling and context adaptation functionality.

With respect to context adaptation, two types of control structures for processing of contextual information have been distinguished [18]: *triggering* of program code is needed to start or activate applications in response to changes of context; and *branching* of program code is needed to let applications produce different behaviour depending on content of the context model at a given time. Dey [10] realised adaptation with a type-system and a condition-based publish-subscribe mechanism. We show in Sect. 4.2 how such conditions [10] and context dependent *branching* and *triggering* [18] of processes can be formalised.

Ontologies have been suggested as the key technology for adding semantics to applications. Ontologies for space and time have received considerable interest in the areas of formal ontologies and qualitative reasoning. One of the predominant approaches in these areas is mereotopology [37], with successful applications in robotics, geographical information science, the biosciences, and even for motion tracking. Surprisingly, it has not yet received much attention in pervasive computing or ambient intelligence [8, 16]. In contrast to point-set topology, which characterises neighbourhoods in a continuous domain, such as spatial regions or temporal intervals, based on sets of extension-less points, mereotopology starts from extended portions of the world [37], that is, from the neighbourhoods or contexts themselves. Being independent from the number of dimensions of a domain, mereotopology cannot only be applied to describe spatial contexts but also temporal contexts and sensor value ranges, and can also be used to model discrete domains as well as concept hierarchies and collections [5].

The primitive relations in mereotopology are that of part-hood ($\sqsubseteq$) and connection (C) between regions, replacing the set-theoretic notion of membership ($\in$) between points and sets of points. The theory dates back to works on point-free geometry by Whitehead, Tarski, and Clarke [37]. It has been applied for formal ontologies in information systems by Randell et al. [26] and many others; a recent overview has been given by Varzi [37]. The idea to move away from point-sets to more meaningful primitive extended entities has not only been successful for the spatial domain but also for the temporal domain [1]. Moreover, results of Galton [14] suggest that concepts of qualitative reasoning, which make the mereotopological calculi attractive for applications in artificial intelligence [9], can be generalised, so as to cover the broad range of sensors employed for establishing context awareness [32]. We illustrate (Sect. 3.2) that concepts of mereotopology can be used to describe continuous quantitative domains of sensor values, in particular when uncertainty [7, 33] or privacy requirements [11, 29] are important. In these cases, sensory information may not be given by exact values or coordinates, but only in the form of intervals or uncertainty regions.

An important aspect in pervasive computing is how to identify and address specific entities (objects, places, users, services, times etc.) in the environment. One way to do this is to assign unique identifiers to every entity, for instance by attaching bar codes or RFID, and to give a web presence to people, places, and things. This approach poses considerable challenges for privacy, since users and their smart objects cannot remain anonymous, and it can also lead to inefficiency [3] caused by unnecessarily lengthy descriptions. A more context-oriented approach, which respects the privacy principle of locality [21] better, is to connect smart objects based on their current context [3]. Our model allows to formally capture such novel address methods. In general, a process in our model activates a context, whose activation is then transmitted either upward or downward to all processes that are registered with super-contexts or sub-contexts, respectively (Sect. 4). We can thus model contemporary methods of inter-process communication, such as *semantic triggering* of a processes when a change of state is detected, and *semantic broadcasts*, e.g., broadcasts over a certain spatial area, to a certain community, or to an otherwise anonymous recipient who has certain properties. Communication via IDs can be modelled realistically as a broadcast to a very specific context; one could then reason about whether a certain protocol can ensure formally that only a unique listener is addressed.

It has often been stated [28, 36, 38, 39] that modelling context with OWL, the de facto standard for ontologies, would not be possible. A main difficulty is that OWL (version 1.0) does not support to describe a relation as a partial ordering relation, which would be required to characterise location hierarchies such as that of Ye et al. [40], other containment hierarchies, and ordering of sensor values, since reflexivity and antisymmetry of relations cannot be described in the description logic $\mathcal{SHOIQ}$ underlying OWL-DL. The solution chosen by many approaches to ontology-based context modelling is to use expressive logical languages, such as F-Logic [36] or first order logic [28, 38], which come at the price of less efficient reasoning. An alternative is the context-oriented logical language (for brevity called CL in this article) proposed in [35], which we use in this article as a context modelling language (Sect. 4). Although in several ways less expressive than OWL, CL supports hierarchical reasoning over multiple domains. The syntax is similar to that of first order logic, and it can easily be embedded into the first order logic needed to characterise mereotopological structures. CL is thus compatible with general first order logic ontology frameworks containing mereotopological notions, such as SUMO [25].

3   Mereotopology as a Theory for Context Modelling

In this section, we describe a mereotopological theory that captures notions of context found in the literature on context modelling and location modelling for mobile, ubiquitous, and pervasive computing. The general theory is given in Sect. 3.1 in the form of a domain-independent first-order logic axiomatisation. For illustration purposes, examples from the spatial domain and location modelling are used in this section. Domains beyond location are the topic of Sect. 3.2, where we introduce a more general first order language for context modelling and illustrate its applicability with multi-domain examples of context modelling. In Sect. 4 we outline the more tractable language CL [35], which serves as a sub-language for describing context and querying context models in a simple context-aware programming language.

3.1   The Language of Mereotopology

The framework of mereotopology in this section mainly follows the explications given by Varzi [37]. However, our focus is less on philosophical questions of ontology than on specifying properties underlying contemporary context models.[1] We assume the standard basic language of first order predicate logic with identity ($=$) as given.[2]

---

[1] We also deviate in syntax: where Varzi [37] – like many other authors in the field of mereotopology – uses prefix notation, we use mainly infix notation, for instance: we write $x \sqsubseteq y$ instead of Pxy for expressing that x is part of y.

[2] For increasing readability of formulae we introduce rules for saving brackets. The following precedence of logical connectives $\neg, \wedge, \vee, \rightarrow, \leftrightarrow$ and term connectives $\sim, \sqcap, \sqcup$ is assumed. Quantifiers $\forall$ and $\exists$ are to be read as having maximal scope, that is, until the first bracket closing an opening bracket before the quantifier or until the end of the formula. We highlight the boundaries of atomic formulae, such as $[x_1 \sqsubseteq x_2]$, with square brackets, so that complex terms and complex formulae are separated clearly. Sentences are numbered with respect to their function: axioms are indicated with the prefix A, definitions with D and examples with E.

Parts. The fundamental notion of mereotopology is the mereological primitive relation of *part-hood* ($\sqsubseteq$). In location modelling, the relation $\sqsubseteq$ corresponds to spatial containment and the entities that are ordered by $\sqsubseteq$ can be understood as spatial contexts or regions, whether obtained via symbolic location sensing or coordinate location sensing.

The relation $\sqsubseteq$ is characterised axiomatically as a partial ordering relation, that is, as reflexive, transitive, and antisymmetric. Reflexivity states that each x is a part of itself (A1). Transitivity demands that if x_1 is part of x_2, and x_2 is part of x_3, then x_1 is also part of x_3 (A2). Antisymmetry establishes the ordering: if x_1 is part of x_2, and x_2 is part of x_1, then they must be identical (A3).

$$\forall x : [x \sqsubseteq x] \tag{A1}$$

$$\forall x_1, x_2, x_3 : [x_1 \sqsubseteq x_2] \wedge [x_2 \sqsubseteq x_3] \rightarrow [x_1 \sqsubseteq x_3] \tag{A2}$$

$$\forall x_1, x_2 : [x_1 \sqsubseteq x_2] \wedge [x_2 \sqsubseteq x_1] \rightarrow [x_1 = x_2] \tag{A3}$$

$$[x_1 \sqsubset x_2] \overset{\text{def}}{\Leftrightarrow} [x_1 \sqsubseteq x_2] \wedge \neg[x_2 \sqsubseteq x_1] \tag{D1}$$

From the relation $\sqsubseteq$, we can define a strict variant: a *proper part* x_1 of an entity x_2 (relation symbol: $\sqsubset$) is a part that does not contain x_2 as a part (D1).

Overlap and Underlap. We further define two partial functions: for two entities x_1 and x_2, the smallest entity that contains both is called the *sum* (D2), and the largest entity which they both contain is called here the *intersection*[3] (D3).

$$[x_1 \sqcup x_2 = x] \overset{\text{def}}{\Leftrightarrow} [x_1 \sqsubseteq x] \wedge [x_2 \sqsubseteq x] \wedge \tag{D2}$$
$$\forall y : [y \sqsubseteq x] \wedge [x_1 \sqsubseteq y] \wedge [x_2 \sqsubseteq y] \rightarrow [y = x]$$

$$[x_1 \sqcap x_2 = x] \overset{\text{def}}{\Leftrightarrow} [x \sqsubseteq x_1] \wedge [x \sqsubseteq x_2] \wedge \tag{D3}$$
$$\forall y : [x \sqsubseteq y] \wedge [y \sqsubseteq x_2] \wedge [y \sqsubseteq x_1] \rightarrow [y = x]$$

$$\forall x : [\bot \sqsubseteq x] \wedge [x \sqsubseteq \top] \tag{A4}$$

With respect to common sense spatial intuition, these two functions are partial functions, since there can be no intersection between regions that do not share a part, and there can be no sum of two regions if there is no region that contains them both. It is one of the benefits of mereotopology from a philosophical point of view that it does not require arbitrary sums and intersections to exist. However for a lattice structure such as demanded for the hybrid location models [22, 40], the existence of arbitrary sums and intersections of regions has to be ensured. A simple way to achieve this is to introduce an empty region $\bot$ that is part of every region and a region $\top$ extending over the complete domain (A4).

With these special regions introduced, the relations of overlap and underlap can be defined: two regions *overlap* if and only if they share a part that is not

---

[3] We use the term intersection here as a mnemonic and reference to spatial intuition. The mathematically more appropriate term used by Varzi [37] is *product*.

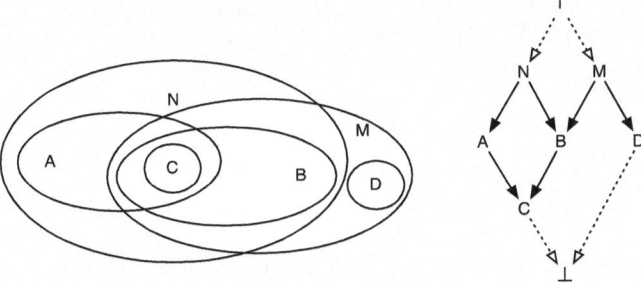

Fig. 1. Overlap and underlap in a location hierarchy: A and B have C as a common part and are both part of the region N

identical to the empty region (D4); and two regions *underlap* if and only if they are both part of a region that is not identical to the whole domain (D5).

$$[x_1 \, O \, x_2] \overset{\text{def}}{\Leftrightarrow} \exists x : \neg[x = \bot] \wedge [x \sqsubseteq x_1] \wedge [x \sqsubseteq x_2] \tag{D4}$$

$$[x_1 \, Y \, x_2] \overset{\text{def}}{\Leftrightarrow} \exists x : \neg[x = \top] \wedge [x_1 \sqsubseteq x] \wedge [x_2 \sqsubseteq x] \tag{D5}$$

$$[\sim x = y] \overset{\text{def}}{\Leftrightarrow} [x \sqcup y = \top] \wedge [x \sqcap y = \bot] \tag{D6}$$

Additionally, we can define a complement operation ($\sim$): the region y is the complement of x if and only if the sum of x and y is identical to the whole domain $\top$, and the intersection is identical to the empty region $\bot$ (D6).

In location models, the relation of overlap can be used to span accessibility graphs: if x overlaps y, then it is possible to go from x to y [40], or to enter y from x [28]. For a finite set of regions in a location model, the graph of the overlap relation already allows us to do path planning: if the hallway overlaps with Alice's office and Bob's office, then we can go from Alice's office to Bob's office via the hallway. The notions of overlap and underlap are crucial for modelling such notions of reachability in a context hierarchy (Fig. 1).

Connection. The above specification is sufficient to characterise containment hierarchies and we can already specify two regions as being connected by overlapping in a shared sub-region. For location modelling [34, 40] however, it can be more convenient to use a weaker form of connection that does not require the modeller to be committed to asserting that there is an overlap region, for instance: we might specify two roads as being connected without being committed to asserting that there is an overlap region; or we might want to contrast viable and relevant *overlap*, such as a monitored doorway region shared by a hallway and an office, with *external connection* as sharing of boundary parts, such as two offices separated by a shared wall.

However, topological notions such as boundary, interior, or closure cannot yet be expressed. At first glance, these notions seem to require a point-set perspective on regions. However, a purely region-based characterisation can be obtained if a relation of *connection* (C) between regions is introduced. Connection is characterised as a reflexive and symmetric relation: every non-empty region is connected to itself (A5); and if x_1 is connected to x_2, then also x_2 is connected to x_1 (A6):

$$\forall x : \neg[x = \bot] \to [x \,\mathrm{C}\, x] \tag{A5}$$

$$\forall x_1, x_2 : [x_1 \,\mathrm{C}\, x_2] \to [x_2 \,\mathrm{C}\, x_1] \tag{A6}$$

$$\forall x_1, x_2 : [x_1 \sqsubseteq x_2] \to \forall y : [y \,\mathrm{C}\, x_1] \to [y \,\mathrm{C}\, x_2] \tag{A7}$$

Additionally, we need a bridging principle [37] between the mereological parts ($\sqsubseteq$) and the topological parts (C) of the theory. Following Varzi [37], we demand that C is monotonous with respect to $\sqsubseteq$: if x_1 is part of x_2, then any region y connected to x_1 is also connected to x_2 (A7). We can then define the well-known RCC relations [8, 26]: *external connection* (*EC*) is connection without overlap (D7); a *non-tangential proper part* (*NTPP*) is a part x of y that is only connected to regions z that overlap y (D8); and a *tangential proper part* (*TPP*) is a proper part that is not *NTPP* (D9).

$$[x \,\mathrm{EC}\, y] \overset{def}{\Leftrightarrow} [x \,\mathrm{C}\, y] \wedge \neg[x \,\mathrm{O}\, y] \tag{D7}$$

$$[x \,\mathrm{NTPP}\, y] \overset{def}{\Leftrightarrow} [x \sqsubset y] \wedge \forall z : [x \,\mathrm{C}\, z] \to [y \,\mathrm{O}\, z] \tag{D8}$$

$$[x \,\mathrm{TPP}\, y] \overset{def}{\Leftrightarrow} [x \sqsubset y] \wedge \neg[x \,\mathrm{NTPP}\, y] \tag{D9}$$

It is clear that O fulfils all three axioms of C, but if we set C to be identical to O, then the definitions of *EC* and *TPP* become empty and *NTPP* becomes identical to $\sqsubset$. In current location models, overlap and connection are not yet distinguished.

The idea of a distinction between overlap, as sharing of parts, and connection, as sharing of something that is not necessarily a relevant element of the domain, has been generalised in theories of granularity [12, 34]. As context models can easily become very large, modelling granularity is crucial to ensure scalability of hierarchical context models, cf. [33] for an approach applicable to mereotopologically specified context models.

3.2   Beyond Location

The above discussion mentioned mainly location models, and the notion of context is conceived broader in pervasive computing [32]. However, key properties for context modelling in many other domains are similar hierarchical structures and a related notion of connection. Mereotopology is a general domain-independent mathematical theory that can describe such structures. We demonstrate this claim by presenting hierarchical structures in five other domains.

Following Jang et al. [19], a representation of context should be expressive enough to answer at least the six questions: *who* caused a certain interaction with *what where when how* and *why*. Following this mnemonic, we can identify five domains with corresponding relations, besides spatial containment and spatial overlap, which are in the following individuated with symbols $\sqsubseteq_{where}$ and O_{where}.

We need to model at least: *times* with a relation of temporal containment $\sqsubseteq_{when}$ of intervals, *groups of agents* with a containment relation between groups, *classes of objects* with a taxonomic subclass relation $\sqsubseteq_{what}$, *states*, including measured or inferred states of the environment, structured by an embedded logical entailment relation $\sqsubseteq_{how}$, and *events*, including externally available commands and actions of a system, partially ordered by a causation relation $\sqsubseteq_{why}$.

The contexts we mostly want to reason about have a meaningful extension in more than one of the six dimensions; a conference, for instance, has a time, location, and participants. We employ combined relations of general sub-contexts (D10) and partial sub-contexts (D11):

$$[x \sqsubseteq_\forall y] \overset{def}{\Leftrightarrow} [x \sqsubseteq_{when} y] \wedge [x \sqsubseteq_{where} y] \wedge [x \sqsubseteq_{who} y] \wedge \quad \text{(D10)}$$
$$[x \sqsubseteq_{what} y] \wedge [x \sqsubseteq_{how} y] \wedge [x \sqsubseteq_{why} y]$$

$$[x \sqsubseteq_\exists y] \overset{def}{\Leftrightarrow} [x \sqsubseteq_{when} y] \vee [x \sqsubseteq_{where} y] \vee [x \sqsubseteq_{who} y] \vee \quad \text{(D11)}$$
$$[x \sqsubseteq_{what} y] \vee [x \sqsubseteq_{how} y] \vee [x \sqsubseteq_{why} y]$$

$$[x =_m y] \overset{def}{\Leftrightarrow} [x \sqsubseteq_m y] \wedge [y \sqsubseteq_m x] \quad \text{(D12)}$$

The axioms of Sect. 3.1 hold for all six relations $\sqsubseteq_m$ if we replace identity $(=)$ with appropriate equivalence relations $=_m$ (D12). Note that our limitation to a fixed set of relations allows us to avoid distinctions between different types of entities in the language and facilitates reification. The only types of entities we employ are *contexts* as portions of the world around us that can have an extension in one or more of the six dimensions.[4]

Temporal Containment and Ordering. The extension from spatial containment to temporal containment is straightforward if we consider time as a one-dimensional space. We can express with a relation of temporal containment that one context is during the time of another ($\sqsubseteq_{when}$), that contexts are synchronous ($=_{when}$), or temporally overlapping (O_{when}). Basic time conflicts can thus be represented:

$$[\text{MeetingA} \sqsubseteq_{when} \text{Aug/1/2008}] \wedge [\text{Aug/1/2008} \sqsubseteq_{when} \text{ConfB}] \wedge \quad \text{(E1)}$$
$$[\text{MeetingA} \sqsubseteq_{where} \text{CityA}] \wedge [\text{ConfB} \sqsubseteq_{where} \text{CityB}] \wedge \neg[\text{CityA} \, O_{where} \, \text{CityB}]$$
$$\forall c : [c \sqsubseteq_\forall \text{ConfB}] \rightarrow \neg[c \sqsubseteq_\forall \text{MeetingA}] \quad \text{(E2)}$$

If a calendar contains an entry for a meeting A on the day *Aug/1/2008*, and this day is during the time of a conference B; and A and B take place in two different, not overlapping city regions (E1), then we can conclude that any context c that is a proper sub-context of B cannot be a proper sub-context of A (E2).

---

[4] With a more classical AI representation format, we can conceive, for instance, $[a \sqsubseteq_{when} b] \wedge [a \sqsubseteq_{where} b]$ to be an abbreviation of $[\text{time}(a) \sqsubseteq_{when} \text{time}(b)] \wedge [\text{loc}(a) \sqsubseteq_{where} \text{loc}(b)]$, where the relations $\sqsubseteq_{when}$ and $\sqsubseteq_{where}$ are relations restricted to the types of times and regions, respectively. However, the operators $\sqcap$ and $\sqcup$ would also have to be replaced by functions $\sqcap_m$ and $\sqcup_m$ yielding the correct types.

In comparison to the interval-based calculus of time proposed by Allen [1], our notion of intervals includes not only convex intervals but also arbitrary sums of intervals, such as generalised intervals [23] and periodically recurring times, since we allow arbitrary sums and intersections. The directedness of time cannot be represented with temporal containment alone; but, given the relation $\sqsubseteq_{\text{why}}$ denoting causation, partial temporal ordering can be derived (E3):

$$[x \text{ before } y] \leftrightarrow \neg[x \ O_{\text{when}} \ y] \wedge [x \sqsubseteq_{\text{why}} y] \tag{E3}$$

The relation $\sqsubseteq_{\text{why}}$ orders the domain of events, such as items in a plan or other program-like structures provided by users as required for anticipating situations and producing pro-active behaviour.[5] However, additional axioms would be needed to reason about relations between temporal structure and causation, in particular, a distinction between event tokens and event types, cf. Galton [15].

Logical Relations between States. In context modelling, important notions of states include status information, such as "on vacation," and sensory data, such as data from physiological or temperature sensors "temperature is higher than 25°C." States can be understood as reified propositions of an embedded logic, similar to a monadic predicate logic. A relation $\sqsubseteq_{\text{how}}$, which has the properties of a mereological containment relation, is the relation of implication between such formulae. For states, the operators $\sqcap$ and $\sqcup$ can be understood as conjunction and disjunction operators of the reified logic. An expression such as (E4) can be used for specifying the state of a context called *Measured*, reduced to the aspect of temperature, as being in the interval [25,30]. A corresponding formula in monadic predicate logic would be (E5).

$$[\text{Measured} \sqcap \text{Celsius} \sqsubseteq_{\text{how}} {\geq}25 \sqcap {\leq}30] \tag{E4}$$
$$\forall x : \text{Measured}(x) \wedge \text{Celsius}(x) \rightarrow {\geq}25(x) \wedge {\leq} 30(x) \tag{E5}$$

We specify the computation of numerical comparisons, such as ≥ 25 in Sect. 4.2. By combining states and times, we can represent a state *Temp* changing over time:

$$\forall x : [x =_{\text{when}} t_1] \rightarrow [x \sqcap \text{Temp} \sqsubseteq_{\text{how}} \text{Cold}] \tag{E6}$$
$$\forall x : [x =_{\text{when}} t_2] \rightarrow [x \sqcap \text{Temp} \sqsubseteq_{\text{how}} \text{Warm}] \tag{E7}$$

Taxonomic Knowledge. A mereological axiomatisation of taxonomic knowledge has been discussed in detail by Bittner et al. [5]. Appropriate hierarchical relations for $\sqsubseteq_{\text{who}}$ and $\sqsubseteq_{\text{what}}$ are *group inclusion* on groups of agents and the *subclass* relation on classes of objects. With this interpretation, the properties of partial ordering relations, as stated above, are intuitively plausible. An example

---

[5] A candidate relation for $\sqsubseteq_{\text{why}}$ that fulfils the requirements of a partial ordering relation would, for instance, be the reflexive and transitive hull of the transition relation between programs in the verification of concurrent programs (cf. the relation $\longrightarrow^*$ in Sect. 4.2).

for transitivity of $\sqsubseteq_{who}$, for instance, is given in (E8): a group of users *Admin* included in the group of users *Staff* is also included in any group that includes the latter, such as *NotificationRecipient*. Knowing that Bob is in the group of administrators, we know that he is eligible to receiving a notification (E9).

$$[\text{Admin} \sqsubseteq_{who} \text{Staff}] \wedge [\text{Staff} \sqsubseteq_{who} \text{NotificationRecipient}] \tag{E8}$$
$$\rightarrow [\text{Admin} \sqsubseteq_{who} \text{NotificationRecipient}]$$

$$[\text{Bob} \sqsubseteq_{who} \text{Admin}] \rightarrow [\text{Bob} \sqsubseteq_{who} \text{NotificationRecipient}] \tag{E9}$$

In our mereological framework, a single agent, such as Bob, or a single object in an interaction is always interpreted as a group of one agent or object. That is, Bob is interpreted not by a token (userID, name, etc.) corresponding to the user Bob but by a group containing only one individual. This may seem as a counterintuitive by-effect of the mereological axiomatisation. However, this property has distinct advantages, for instance for obfuscation: if we do not distinguish between individuals and groups, every application needs in principle to be enabled to handle coarsened information [29]. Also, application objects are addressed by class not by ID, so that interoperability and greater flexibility of applications can be ensured easier.

4   Programming Dynamic Behaviour in a Context

We describe the syntax (Sect. 4.1) and semantics (Sect. 4.2) of a simple context-aware programming language for specifying pervasive computing systems that operate on the background of a context model. We use a widely known simple textbook variant [2] of an imperative concurrent programming language, so as to illustrate our approach with a particularly familiar example. Our aim is to present a methodology of how the semantics of a given language can be augmented with context-awareness.

For adding context-awareness to programs, a mechanism is needed with which the context-aware program can access the context model. We assume in the following that such an interface is provided by an ontology language in which queries to a knowledge base can be formulated. However, we do not restrict how the knowledge base is realised, whether as a database [18], a graph-based data structure [22], or as a logical reasoning mechanism [28, 35].

As a representation language, we use the context-oriented logical language (CL) proposed by [35], since it is most similar to the first order logical language used above for reasoning *about* context-models. The main difference between the two languages is that CL does not allow quantification over variables. We give an abstract grammar for CL derived from the definitions in [35] and extended it with numerical expressions and context terms for numerical comparison.

4.1   Syntax

We assume two predefined data types: a simple number type (*number*) and a data type of alphanumeric character strings (*alphanum*) for names of atomic context terms. CL allows for arbitrarily complex context terms (D15). Numerical expressions (N) are numbers, numerical variables and arithmetic expressions constructed from these (D13).

$$N \stackrel{def}{=} number \mid NVar \mid -N \mid N+N \mid N-N \mid \qquad \text{(D13)}$$
$$N * N \mid N \div N \mid N \bmod N$$

$$NumCT \stackrel{def}{=} \leq N \mid \geq N \mid \#N \mid <N \mid >N \qquad \text{(D14)}$$

$$CT \stackrel{def}{=} alphanum \mid NumCT \mid CTVar \mid \qquad \text{(D15)}$$
$$CT \sqcap CT \mid CT \sqcup CT \mid \sim CT$$

$$CLF \stackrel{def}{=} [CT \sqsubseteq_m CT] \mid CLF \wedge CLF \mid CLF \vee CLF \mid \qquad \text{(D16)}$$
$$\neg CLF \mid CLF \rightarrow CLF \mid \textbf{true} \mid \textbf{false}$$

Numerical comparison in the style of Dey [10] can be described with *numerical context terms* (*NumCT*) that consist of a numerical expression prefixed with one of the symbols $\#, \leq, \geq, <, >$ (D14). Intuitively, numerical context terms denote portions of a numerical domain: the interval $(5, 7]$ can be expressed with the term $>5 \sqcap \leq 7$ as the intersection of the intervals $(5, \infty)$ and $(-\infty, 7]$. The context term $\#5$ denotes the interval $[5, 5]$. In order for these expressions to receive the intended numerical meaning, we need to augment the semantics of CL terms with numerical computations (Sect. 4.2).

Context terms can be related with respect to one or more aspects of the six categories. For simplicity, we include only the fundamental six containment relations $\sqsubseteq_m$, where $m \in \{when, where, who, what, how, why\}$, and consider the formulae $[c =_m d]$, $[c \, O_m \, d]$ to be abbreviations of the more complex formulae by which they are defined above. CL allows not only atomic formulae, but also for arbitrarily complex combinations constructed with the propositional connectives $\wedge, \vee, \rightarrow, \neg$ and two special formulae **true** and **false** (D16).

We characterise a simple imperative language for modelling distributed concurrent processes and supplement it with an additional type for context terms.

$$S \stackrel{def}{=} \textbf{skip} \mid CTVar := CT \mid NVar := N \mid S \, ; S \mid \qquad \text{(D17)}$$
$$\textbf{do } NChoice \textbf{ od} \mid \textbf{if } NChoice \textbf{ fi}$$

$$NChoice \stackrel{def}{=} CLF \rhd S \mid NChoice \, \square \, NChoice \qquad \text{(D18)}$$

$$P \stackrel{def}{=} S \, ; \textbf{do } GChoice \textbf{ od} \qquad \text{(D19)}$$

$$GChoice \stackrel{def}{=} CLF \, ; IO \rhd S \mid GChoice \, \square \, GChoice \qquad \text{(D20)}$$

$$IO \stackrel{def}{=} CT \, ? \, CTVar \mid \uparrow CT \mid \downarrow CT \qquad \text{(D21)}$$

$$PSyst \stackrel{def}{=} P \mid PSyst \| PSyst \qquad \text{(D22)}$$

We allow assignment to context term variables and numerical variables in programs (D17). In non-deterministic choice (D18) and guarded commands (D20), Boolean expressions are replaced with CL-formulae. A pervasive computing system (D22) consists of one or more processes (D19) running in parallel. The most prominent difference is the set of IO-commands (D21): the command $ct\,?\,x$ listens for activation of a context, the commands $\uparrow CT$ and $\downarrow CT$ send an activation *upward* through the hierarchical context knowledge base, to all more general contexts, or downward, to more specific contexts.

We can now, in a very simple manner, describe context-aware algorithms. Example E10 shows a template for a sensor-like component that consists simply of a loop that activates a context reflecting the current temperature by sending the sensed value v in a way that accuracy (± 1) and unit (*Celsius*) are reflected. Example E11 is a corresponding template for an actuator, as a process executing some action, such as turning a heater on, in response to being activated by a context *TooCold*.

$$\langle \text{sense } v\rangle;\ \textbf{do true};\ \uparrow(\geqslant(v-1)\sqcap\leqslant(v+1)\sqcap\text{Celsius})\ \triangleright\ \langle \text{sense } v\rangle\ \textbf{od}\qquad\text{(E10)}$$

$$\textbf{skip};\ \textbf{do true};\ \text{TooCold}\,?\,x\triangleright\langle\text{execute action}\rangle\ \textbf{od}\qquad\qquad\text{(E11)}$$

$$\gamma=\{[\leqslant17\sqcap\text{Celsius}\sqsubseteq_{\text{how}}\text{TooCold}]\}\qquad\qquad\qquad\text{(E12)}$$

With a knowledge base γ specifying that an environment is too cold when the temperature is lower than 17°C (E12), the process (E11) should be triggered if $v < 16$ holds after $\langle\text{sense } v\rangle$. We can prove such properties if a semantics is given.

4.2   Semantics

We describe an operational semantics for the language [2]. The basic notion in the semantics is a *transition* relation $\longrightarrow\,\subseteq\,\Gamma\times\Gamma$ between *configurations* Γ of a computing system. The main idea is that the transition relation describes how one step of execution of a program in a state moves the program on and changes the state. The relation $\longrightarrow^*$ is the transitive and reflexive closure of $\longrightarrow$.

A configuration in our framework is represented as a triple $\langle S,\sigma,\gamma\rangle$, where S is a program, process, or pervasive system to be run, σ is a state, and γ a context knowledge base. A terminal configuration $\langle E,\sigma,\gamma\rangle$ is a configuration, in which the program to be run is the empty program E, where $E\,;S$ and $S\,;E$ are the same as S. A state σ is a substitution function, which can be thought of as a mathematical realisation of a list of variable bindings. For instance, applying $\sigma = [x:0\mid y:2]$ to the numerical expression $(x+y)$ results in the numerical expression $(x+y)\sigma = (0+2)$, whereas applying $\sigma = [x:0\mid y:2]$ to the context term $\leqslant(x+y)$ results in the context term expression $\leqslant(x+y)\sigma = \leqslant(0+2)$.

We define an interpretation function $I_n : N\to\mathbb{R}$ representing evaluation of numerical expressions and an interpretation function I_c that assigns a value to numerical context terms based on I_n. For $x\in number$, $I_n(x)$ trivially maps x to the number in $\mathbb{R}$ that it represents. The compound numerical expressions map to their corresponding evaluations (D23). To give a meaning to the numerical context terms we characterise them as corresponding to the orderings $<,\leq,>,$ $\geq$ on the domain $\mathbb{R}$ (A8-A10).

$$I_n(x + y) = I_n(x) + I_n(y) \qquad I_n(x - y) = I_n(x) - I_n(y) \qquad \text{(D23)}$$
$$I_n(x * y) = I_n(x) * I_n(y) \qquad I_n(x \div y) = I_n(x) \div I_n(y)$$
$$I_n(-x) = -(I_n(x)) \qquad I_n(x \bmod y) = I_n(x) \bmod I_n(y)$$
$$\forall x, y \in N : [\leq\!x \sqsubseteq_\forall \leq\!y] \leftrightarrow I_n(x) \leq I_n(y) \qquad \text{(A8)}$$
$$\forall x, y \in N : [\geq\!x \sqsubseteq_\forall \geq\!y] \leftrightarrow I_n(x) \geq I_n(y) \qquad \text{(A9)}$$
$$\forall x \in N : [>\!x = \sim \leq\!x] \wedge [<\!x = \sim \geq\!x] \wedge [\#x = \leq\!x \sqcap \geq\!x] \qquad \text{(A10)}$$

The transition relation can be characterised using axiom schemata and inference rules. The rules for assignment of numerical variables (A11) and context variables (A12) rely on the above interpretation functions. The standard constructs **skip** (A13) and sequential composition (A14) receive standard semantics extended with the parameter γ representing the knowledge base, which can change independently from the program.

$$\langle v := e, \sigma, \gamma \rangle \longrightarrow \langle E, [\sigma \mid v : I_n(e)\sigma], \gamma' \rangle, \text{ if } v \in NVar \text{ and } e \in N, \qquad \text{(A11)}$$
$$\langle v := e, \sigma, \gamma \rangle \longrightarrow \langle E, [\sigma \mid v : I_c(e)\sigma], \gamma' \rangle, \text{ if } v \in CTVar \text{ and } e \in CT \qquad \text{(A12)}$$
$$\langle \mathbf{skip}, \sigma, \gamma \rangle \longrightarrow \langle E, \sigma, \gamma' \rangle \qquad \text{(A13)}$$
$$\frac{\langle s_0, \sigma, \gamma \rangle \longrightarrow \langle s_0', \sigma', \gamma' \rangle}{\langle s_0\,;\,s_1, \sigma, \gamma \rangle \longrightarrow \langle s_0'\,;\,s_1, \sigma', \gamma' \rangle} \qquad \text{(A14)}$$

The CL formulae in non-deterministic choice (A15) and loops (A16) are evaluated with respect to the knowledge base γ.

$$\langle \mathbf{if}\ \phi_0 \rhd s_0 \,\square \ldots \square\, \phi_n \rhd s_n\ \mathbf{fi}, \sigma, \gamma \rangle \longrightarrow \langle s_i, \sigma, \gamma' \rangle, \text{ if } \gamma\sigma \models \phi_i \text{ for any } \phi_i \quad \text{(A15)}$$
$$\langle \mathbf{if}\ \phi_0 \rhd s_0 \,\square \ldots \square\, \phi_n \rhd s_n\ \mathbf{fi}, \sigma, \gamma \rangle \longrightarrow \langle E, \mathbf{fail}, \gamma' \rangle, \text{ if } \gamma\sigma \not\models \phi_i \text{ for all } \phi_i$$
$$\langle \mathbf{do}\ \phi_0 \rhd s_0 \,\square \ldots \square\, \phi_n \rhd s_n\ \mathbf{od}, \sigma, \gamma \rangle \longrightarrow$$
$$\qquad \langle s_i\,;\mathbf{do}\ \phi_0 \rhd s_0 \,\square \ldots \square\, \phi_n \rhd s_n\ \mathbf{od}, \sigma, \gamma' \rangle, \text{ if } \gamma\sigma \models \phi_i \text{ for some } \phi_i \quad \text{(A16)}$$
$$\langle \mathbf{do}\ \phi_0 \rhd s_0 \,\square \ldots \square\, \phi_n \rhd s_n\ \mathbf{od}, \sigma, \gamma \rangle \longrightarrow \langle E, \sigma, \gamma' \rangle, \text{ if } \gamma\sigma \not\models \phi_i \text{ for all } \phi_i$$

All input and output is related via the knowledge base within whose scope a process is working. The command $c\,?\,v$ registers a process with the context c. The process then waits for a context term related to c to be activated by the knowledge base and stores the activated context it received in a context variable v. The command $\uparrow t$ activates all processes registered to contexts c that are in some aspect higher in the hierarchy $[t \sqsubseteq_\exists c]$ (A17); likewise, $\downarrow t$ activates processes registered to contexts below the context t (A18). Upward and downward activation are non-blocking operations that can always be executed and do not change the state σ (A19, A20).

$$\frac{\langle \uparrow t, \sigma, \gamma \rangle \longrightarrow \langle E, \sigma, \gamma' \rangle}{\langle c\,?\,v, \sigma, \gamma \rangle \longrightarrow \langle E, [\sigma \mid v : I_c(t)\sigma], \gamma' \rangle}, \text{ if } \gamma \models [t \sqsubseteq_\exists c], v \in CTVar \qquad (A17)$$

$$\frac{\langle \downarrow t, \sigma, \gamma \rangle \longrightarrow \langle E, \sigma, \gamma' \rangle}{\langle c\,?\,v, \sigma, \gamma \rangle \longrightarrow \langle E, [\sigma \mid v : I_c(t)\sigma], \gamma' \rangle}, \text{ if } \gamma \models [c \sqsubseteq_\exists t], v \in CTVar \qquad (A18)$$

$$\langle \uparrow c, \sigma, \gamma \rangle \longrightarrow \langle E, \sigma, \gamma' \rangle, \text{ where } c \in CT \qquad (A19)$$

$$\langle \downarrow c, \sigma.\gamma \rangle \longrightarrow \langle E, \sigma, \gamma' \rangle, \text{ where } c \in CT \qquad (A20)$$

We can then specify communication between parallel processes via the knowledge base in a standard way. If none of the conditions of the process can be fulfilled, the process finishes (A21). If one of the conditions ϕ_i holds and the IO-command ioc_i that it guards can be executed, the body s_i is executed (A22). For simplicity, parallel composition is characterised in the same way as sequential composition, except that the order of execution is irrelevant (A23).

$$\langle \textbf{do } \phi_0 \,;\, ioc_0 \rhd s_0 \,\square \dots \square\, \phi_n \,;\, ioc_n \rhd s_n \textbf{od}, \sigma, \gamma \rangle \longrightarrow \langle E, \sigma, \gamma' \rangle,$$
$$\text{if } \gamma\sigma \not\models \phi_i \text{ for all } \phi_i \qquad (A21)$$

$$\frac{\langle ioc_i, \sigma, \gamma \rangle \longrightarrow \langle E, \sigma', \gamma' \rangle}{\langle \textbf{do } \phi_0 \,;\, ioc_0 \rhd s_0 \,\square \dots \square\, \phi_n \,;\, ioc_n \rhd s_n \textbf{od}, \sigma, \gamma \rangle \longrightarrow}, \qquad (A22)$$
$$\langle s_i \,;\, \textbf{do } \phi_0 \,;\, ioc_0 \rhd s_0 \,\square \dots \square\, \phi_n \,;\, ioc_n \rhd s_n \textbf{od}, \sigma', \gamma' \rangle$$
$$\text{if } \gamma\sigma \models \phi_i \text{ for some } \phi_i$$

$$\frac{\langle p_0, \sigma, \gamma \rangle \longrightarrow \langle p_0', \sigma', \gamma' \rangle}{\langle p_0 \| p_1, \sigma, \gamma \rangle \longrightarrow \langle p_0' \| p_1, \sigma', \gamma' \rangle} \text{ and } \frac{\langle p_1, \sigma, \gamma \rangle \longrightarrow \langle p_1', \sigma', \gamma' \rangle}{\langle p_0 \| p_1, \sigma, \gamma \rangle \longrightarrow \langle p_0 \| p_1', \sigma', \gamma' \rangle} \qquad (A23)$$

We can now prove that given a knowledge base γ that contains (E12) and a system consisting of the two processes (E10) and (E11), the $\langle$execute action$\rangle$ is reached if $v < 16$ holds for the sensed value v. The crucial step in the proof is that of activation of a context term. The sensor activates all contexts above the term $\geq(v-1) \sqcap \leq(v+1) \sqcap$ Celsius. The mereotopological axiom for transitivity (A2) together with the definition of $\sqcap$ (D3) and the axioms for the interpretation of numerical context terms (A8-A10) then entail $\geq(v-1) \sqcap \leq(v+1) \sqcap$ Celsius $\sqsubseteq_{\text{how}}$ $\leq 17 \sqcap$ Celsius. We can infer that the context $\leq 17 \sqcap$ Celsius is activated, and it follows that $\geq(v-1) \sqcap \leq(v+1) \sqcap$ Celsius $\sqsubseteq_{\text{how}}$ TooCold holds (A2). From the knowledge base γ (E12), we finally obtain that the context TooCold is also activated. By (A17) and (A22), the process (E11) can therefore proceed to the $\langle$execute action$\rangle$ statement.

The parameter γ allows us to model changes in the external world and the context knowledge base independent from changes of the computational state σ. The facts that the physical context of a program can change during execution and that this change highly influences the program are the primary difference between context aware computing systems and classical distributed computing systems. Consequently, we need a proof theory that allows us to reason not only about changes evoked by the program, but also about changes evoked in the external world. A user carrying a mobile device leaving a room or a room becoming warmer are not the result of a computation process but of processes external to the computational system. Computational processes might be able to influence physical processes: a user might be alarmed with a signal to leave a room, a room may warm up after a heater turned on. However, a theory of computational processes should not be burdened with modelling the processes in the physical reality together with computational processes: the user might not be able to hear the signal because he is wearing a noise protection device, the heater might be out of order. In the simple example, γ remained unchanged over the whole execution. For more realistic scenarios and to obtain robust pervasive computing systems, we need to evaluate under which stability assumptions proper execution can be guaranteed. Tractable theories of physical reality, as investigated in the field of knowledge representation within the areas of *naive physics* and *qualitative reasoning* [13, 14], can be used to formulate and efficiently reason about such stability conditions.

5   Conclusions

We described context-aware computing as a proper extension of distributed computing with ontology reasoning. Our description is faithful to principles employed in existing context-aware computing frameworks, so that it can be used to evaluate and develop pervasive computing systems with a wide range of frameworks. The proposed theory integrates developments from two areas providing formal models relevant for pervasive computing: the area of formal ontologies for information systems and the area of formal verification of programming languages.

For formally specifying context models, we proposed to apply mereotopological theories. We showed that key ideas of mereotopology, such as hierarchical organisation and overlap are meaningful for many aspects of context relevant in pervasive computing, such as for modelling uncertainty, ensuring privacy through obfuscation, and context-based address methods beyond unique ids.

Our approach allows to describe context-aware algorithms and pervasive computing systems on a high level of abstraction. The state of a system, which can be directly influenced by a computational process, is cleanly separated from external physical processes reflected in an external knowledge base to which the process is connected. We showed for the familiar example of a CSP-style programming language, how a language can be extended with high-level concepts of context-awareness. However, the method is general enough to be applicable also to other programming languages.

References

[1] Allen, J.: Towards a general theory of action and time. Artificial Intelligence 23, 123–154 (1984)
[2] Apt, K.R., Olderog, E.-R.: Verification of Sequential and Concurrent Programs. Springer, Heidelberg (1991)
[3] Beigl, M., Zimmer, T., Decker, C.: A location model for communicating and processing of context. Personal and Ubiquitous Computing 6(5/6), 341–357 (2002)

[4] Birkedal, L., Debois, S., Elsborg, E., Hildebrandt, T.T., Niss, H.: Bigraphical models of context-aware systems. In: Aceto, L., Ingólfsdóttir, A. (eds.) FOSSACS 2006. LNCS, vol. 3921, pp. 187–201. Springer, Heidelberg (2006)

[5] Bittner, T., Donnelly, M., Smith, B.: Individuals, universals, collections: On the foundational relations of ontology. In: Varzi, A., Vieu, L. (eds.) Third Conference on Formal Ontology in Information Systems. IOS Press, Amsterdam (2004)

[6] Cardelli, L., Gordon, A.D.: Mobile ambients. Theoretical Computer Science 240(1), 177–213 (2000)

[7] Chalmers, D., Dulay, N., Sloman, M.: Towards reasoning about context in the presence of uncertainty. In: Workshop on Advanced Context Modelling, Reasoning and Management, Nottingham, UK (2004)

[8] Chen, H., Perich, F., Finin, T., Joshi, A.: SOUPA: Standard ontology for ubiquitous and pervasive applications. In: International Conference on Mobile and Ubiquitous Systems: Networking and Services (2004)

[9] Cohn, A.G., Hazarika, S.M.: Qualitative spatial representation and reasoning: An overview. Fundamenta Informaticae 46(1-2), 1–29 (2001)

[10] Dey, A.K.: Providing Architectural Support for Building Context-Aware Applications. PhD thesis, Georgia Institute of Technology (2000)

[11] Duckham, M., Kulik, L.: A formal model of obfuscation and negotiation for location privacy. In: Gellersen, H.-W., Want, R., Schmidt, A. (eds.) Pervasive 2005. LNCS, vol. 3468, pp. 152–170. Springer, Heidelberg (2005)

[12] Euzenat, J.: Granularity in relational formalisms - with application to time and space representation. Computational Intelligence 17(3), 703–737 (2001)

[13] Forbus, K.D.: Qualitative process theory. Artificial Intelligence 24(1-3), 85–168 (1984)

[14] Galton, A.: Qualitative Spatial Change. Oxford University Press, Oxford (2000)

[15] Galton, A.: Operators vs. arguments: the ins and outs of reification. Synthese 150, 415–441 (2006)

[16] Gottfried, B., Guesgen, H.W., Hübner, S.: Spatiotemporal reasoning for smart homes. In: Augusto, J.C., Nugent, C.D. (eds.) Designing Smart Homes, pp. 16–34 (2006)

[17] Hennessy, M.: A Distributed Pi-Calculus. Cambridge University Press, Cambridge (2007)

[18] Henricksen, K., Indulska, J.: Developing context-aware pervasive computing applications: Models and approach. Pervasive and Mobile Computing 2, 37–64 (2006)

[19] Jang, S., Ko, E.-J., Woo, W.: Unified user-centric context: Who, where, when, what, how and why. In: Ko, H., Krüger, A., Lee, S.-G., Woo, W. (eds.) Personalized Context Modeling and Management for UbiComp Applications, vol. 149, pp. 26–34 (2005) CEUR-WS

[20] Jiang, C., Steenkiste, P.: A hybrid location model with a computable location identifier for ubiquitous computing. In: Borriello, G., Holmquist, L.E. (eds.) UbiComp 2002. LNCS, vol. 2498, pp. 246–263. Springer, Heidelberg (2002)

[21] Langheinrich, M.: Privacy by design - principles of privacy-aware ubiquitous systems. In: Abowd, G.D., Brumitt, B., Shafer, S. (eds.) UbiComp 2001. LNCS, vol. 2201, pp. 273–291. Springer, Heidelberg (2001)

[22] Leonhardt, U.: Supporting Location Awareness in Open Distributed Systems. PhD thesis, Imperial College, London, UK (1998)

[23] Ligozat, G.: Generalized intervals: A guided tour. In: Proceedings of the ECAI 1998 Workshop on Spatial and Temporal Reasoning, Brighton, UK (1998)

[24] Milner, R.: Bigraphs and their algebra. Electronic Notes on Theoretical Computer Science 209, 5–19 (2008)

[25] Pease, A., Niles, I., Li, J.: The suggested upper merged ontology: A large ontology for the semantic web and its applications. In: AAAI 2002 Workshop on Ontologies and the Semantic Web (2002)

[26] Randell, D., Cui, Z., Cohn, A.: A spatial logic based on region and connection. In: Knowledge Representation and Reasoning, pp. 165–176. Morgan Kaufmann, San Francisco (1992)

[27] Ranganathan, A., Campbell, R.H.: Provably correct pervasive computing environments. In: PerCom, pp. 160–169 (2008)

[28] Ranganathan, A., McGrath, R.E., Campbell, R.H., Mickunas, M.D.: Use of ontologies in a pervasive computing environment. The Knowledge Engineering Review 18(3), 209–220 (2003)

[29] Rashid, U., Schmidtke, H.R., Woo, W.: Managing disclosure of personal health information in smart home healthcare. In: Stephanidis, C. (ed.) International Conference on Universal Access in Human-Computer Interaction, Held as Part of HCI International, pp. 188–197. Springer, Heidelberg (2007)

[30] Satyanarayanan, M.: Pervasive computing: Vision and challenges. IEEE Personal Communications, 10–17 (2001)

[31] Schilit, B.N., Theimer, M.M.: Disseminating active map information to mobile hosts. IEEE Network 8(5), 22–32 (1994)

[32] Schmidt, A., Beigl, M., Gellersen, H.-W.: There is more to context than location. Computers and Graphics 23(6), 893–901 (1999)

[33] Schmidtke, H.R., Woo, W.: A formal characterization of vagueness and granularity for context-aware mobile and ubiquitous computing. In: Youn, H.Y., Kim, M., Morikawa, H. (eds.) UCS 2006. LNCS, vol. 4239, pp. 144–157. Springer, Heidelberg (2006)

[34] Schmidtke, H.R., Woo, W.: A size-based qualitative approach to the representation of spatial granularity. In: Veloso, M.M. (ed.) Twentieth International Joint Conference on Artificial Intelligence, pp. 563–568 (2007)

[35] Schmidtke, H.R., Hong, D., Woo, W.: Reasoning about models of context: A context-oriented logical language for knowledge-based context-aware applications. Revue d'Intelligence Artificielle 22(5), 589–608 (2008)

[36] Strang, T., Linnhoff-Popien, C., Frank, K.: CoOL: A context ontology language to enable contextual interoperability. In: Stefani, J.-B., Demeure, I., Hagimont, D. (eds.) DAIS 2003. LNCS, vol. 2893, pp. 236–247. Springer, Heidelberg (2003)

[37] Varzi, A.C.: Spatial reasoning and ontology: Parts, wholes, and locations. In: Aiello, M., Pratt-Hartmann, I., van Benthem, J. (eds.) Handbook of Spatial Logics, pp. 945–1038. Springer, Heidelberg (2007)

[38] Wang, X.H., Zhang, D.Q., Gu, T., Pung, H.K.: Ontology based context modeling and reasoning using owl. In: PerCom Workshops, pp. 18–22. IEEE Computer Society Press, Los Alamitos (2004)

[39] Ye, J., Coyle, L., Dobson, S., Nixon, P.: Ontology-based models in pervasive computing systems. The Knowledge Engineering Review 22, 315–347 (2007)

[40] Ye, J., Coyle, L., Dobson, S., Nixon, P.: A unified semantics space model. In: Hightower, J., Schiele, B., Strang, T. (eds.) LoCA 2007. LNCS, vol. 4718, pp. 103–120. Springer, Heidelberg (2007)

Situvis: A Visual Tool for Modeling a User's Behaviour Patterns in a Pervasive Environment

Adrian K. Clear, Ross Shannon, Thomas Holland,
Aaron Quigley, Simon Dobson, and Paddy Nixon

Systems Research Group
UCD Dublin, Ireland
`adrian.clear@ucd.ie`

Abstract. One of the key challenges faced when developing context-aware pervasive systems is to capture the set of inputs that we want a system to adapt to. Arbitrarily specifying ranges of sensor values to respond to will lead to incompleteness of the specification, and may also result in conflicts, when multiple incompatible adaptations may be triggered by a single user action. We posit that the ideal approach combines the use of past traces of real, annotated context data with the ability for a system designer or user to go in and interactively modify the specification of the set of inputs a particular adaptation should be responsive to. We introduce Situvis, an interactive visualisation tool we have developed which assists users and developers of context-aware pervasive systems by visually representing the conditions that need to be present for a situation to be triggered in terms of the real-world context that is being recorded, and allows the user to visually inspect these properties, evaluate their correctness, and change them as required. This tool provides the means to understand the scope of any adaptation defined in the system, and intuitively resolve conflicts inherent in the specification.

1 Introduction

Context-aware pervasive systems are designed to support a user's goals by making adaptations to their behaviours in response to the user's activities or circumstances. The accuracy and utility of these adaptations is predicated on the system's ability to capture and recognise these circumstances as they occur. We system designers characterise these adaptation opportunities by collecting context data from multiple heterogeneous sensors, which may be networked physical instruments in the environment (measuring factors like temperature, humidity or noise volume), software sensors retrieving information from the web or various data feeds, or wearable sensors measuring factors like acceleration or object use. These context data are voluminous, highly multivariate, and constantly being updated as new readings are recorded.

Situations are high-level abstractions of context data, which free the user from having to deal with raw context and allow more expressive adaptations [1]. We define situations in terms of context that has been encapsulated to a level of

H. Tokuda et al. (Eds.): Pervasive 2009, LNCS 5538, pp. 327–341, 2009.

understanding appropriate for a developer specifying a situation (e.g., symbolic locations), rather than the raw sensor readings (e.g., 3D coordinates). Situations are straightforward for both system designers and system users to work with, as they symbolically define commonly-experienced occurrences such as a user "taking a coffee break", or being "in a research meeting", without requiring the user to understand any of the dozens of distinct sensor readings that may have gone into making up these situations. Situations are thus a natural view of a context-aware system, while the individual pieces of context are each "a measurable component of a given situation" [2].

Thomson et al. observe that there are two approaches to situation determination: specification-based and learning-based approaches [3]. The specification-based approach suffers from complexity. As the context information available to a context-aware system at any moment is so extensive, dynamic and highly dimensional, it is a significant challenge for a system observer to ascribe significance to changes in the data or identify emergent trends, much less capture the transient situations that are occurring amid the churn of the data.

On the other hand, learning-based approaches require training data and interpretation. Many situations a user finds themselves in are subjective and hence require a degree of personalisation. Here, we propose a hybrid of these user-driven and data-driven approaches that utilises minimal annotated samples to frame a situation specification, combined with a novel visualisation that simplifies the manual process of fine-tuning.

Situvis is our scalable visualisation tool for illustrating and evaluating the makeup of situations in a context-aware system. By incorporating real situation traces and annotations as ground truth, Situvis assists system developers in constructing and evaluating sound and complete situation specifications by essentially bootstrapping the manual process, affording the developer a better understanding of the situation space, and the reliability of modeling with situations based on real, recorded sensor data. It is a framework that allows developers to understand, at a high level, how their system will behave given certain inputs.

The following section provides some details of other approaches to the recognition of context abstractions, a formal description of situation specifications and a review of some challenges faced when working with context and situations. We then describe the details of the Situvis tool, including a demonstration of its utility, followed by a discussion of its properties.

2   Background

2.1   Activity Recognition

Techniques for activity recognition use machine learning techniques—both supervised and unsupervised—to infer high-level activities from low-level sensor data. Logan et al. present a long-term experiment of activity recognition in a home setting using a semi-supervised technique [4]. Like the majority of activity recognition, the focus is on concepts that can be described and recognised by body movements and object use. 104 hours of video data was manually annotated following the data collection. Many activities, such as dishwashing and meal preparation, were accurately classified to a high degree. However, the study showed that even with this large amount of data and annotation, some activities, such as drying dishes, could not be learned effectively due to lack of training data caused by their infrequent occurrence, even over a 104-hour period.

Krause et al. describe an approach to learning context-dependent personal preferences using machine learning techniques to refine the behaviour of Sensay, a context-aware mobile phone [5]. The behaviour modifications, such as changing the state of the ringer volume from loud to silent, are known in advance. The task is to find the user's "state" (or contextual circumstances) that corresponds to them modifying the behaviour of their phone so that in the future it can be done automatically. Machine learning of personalised states is favoured over manual specification of general states as a result of a study showing that states and desired phone behaviour differed among individuals. Because the behaviour modifications are known, this method requires no supervision. Essentially, the behaviour modifications serve as labels for the recorded sensor values.

Recognising higher-level abstractions. Recent work has aimed to recognise high-level abstractions of context called routines [6]. Routines are structured as compositions of several activities that may be influenced by time, location and the individual performing them. Examples include "commuting" or "working". In contrast to activities, routines cannot be identified through their local physical structure alone: they consist of variable patterns of multiple activities; they range over longer periods of time; and they often vary significantly between instances. Moreover, they are subjective. As a result, the authors chose topic maps as an alternative approach for recognition. Topic maps are a family of probabilistic models often used by the text processing community and enable the recognition of daily routines as a composition of activity patterns.

We too aim to be able to recognise high-level abstractions, and our approach is designed to achieve this with minimal annotation. Situations and routines are similar in that they are subjective, making long periods of annotation unscalable; and they require more factors to recognise them than simply body posture, body movement, or object use. Therefore, accurate situation determination cannot rely completely on data-driven techniques. Situations are generally short term and hence are logically more complex than routines—they can be partially described in terms of individual activities but they are not lengthy nor activity-rich enough to be represented as the most probable activities that are occurring over a long time window. As a result, we are taking a hybrid approach to recognition that includes a short ground truth collection period followed by manual fine-tuning by a domain expert.

2.2 Situation Specifications

Based on the extensive literature on the subject of modeling context for adaptive systems [1,2,7,8,9,10], we can make some observations: the incoming sources of context into a pervasive application are viewed as a finite number of variables: either nominal or categorical values, e.g., activity levels {idle, active, highly active ... }; or quantitative ordinal values which may be defined over some known interval, e.g., noise level in decibels {0, 140}.

Location information will typically arrive as individual values for an object's x, y and z coordinates in a space, and may be recorded by numerous disparate positioning systems, but is modeled as a higher-level abstraction to make it easier to reason with. Previously conducted research allows component x, y and z coordinates to be composed into a symbolic representation, given some domain information [11], and so we can work with locations as readable as "Simon's office" or "Coffee Area". Our visualisation tool works equally well with simple quantitative data or these higher-order categorised data.

Situations are high-level abstractions that serve as a suitable model with which to develop context-aware systems, because they are intuitive concepts for both designers and users to think in. In order for a computing system to be able to recognise situations, they must first be specified in some way. Theory on the semantics of situation specification can be seen in the work of Henricksen [1] and Loke [12]. Based on this work, we also model situations using declarative languages, which can simply be plugged-in to our tool.

Situation specifications are boolean expressions (or assertions)—they are either true or false, denoting occurrence and non-occurrence, respectively. Assertions may be composed using the logical operators AND ($\wedge$), OR ($\vee$), and NOT ($\neg$), resulting in richer expressions. Domain-specific functions can also be defined to enrich specification semantics (e.g., a distance operator could return a numerical value of the distance between two locations).

We can thus define a situation specification as a concatenation of one or more assertions about contexts, which leads us to the following formal definition:

A situation specification consists of one or more assertions about context that are conjoined using the logical operators AND ($\wedge$), OR ($\vee$), and NOT ($\neg$). Assertions may comprise further domain-specific expressions on context, given that the required semantics are available.

2.3   Interactive Machine Learning

Existing work has applied the coupling of data- and user-driven processes to carry out difficult tasks. In particular, the general Interactive Machine Learning (IML) model consists of iterations of classical machine learning followed by refinement through interactive use. In the Crayons project by Fails and Olsen [13], users can build classifiers for image-based perceptual user interfaces using a novel IML model that involves iterative user interaction in order to minimise the feature set and build a decision tree. Moreover, Dey's *a CAPpella* is a prototyping environment, aimed at end-users, for context-aware applications [14]. It uses a *programming by demonstration* approach, through a combination of machine-learning and user interaction, to allow end-users to build complex context-aware applications without having to write any code.

2.4   Visualisation of Context Data

The field of visual analytics uses interactive visual interfaces to aid end-users in analysing and understanding large and complex multivariate data sets. Interactive visualisation tools help the viewer perform visual data analysis tasks: exploring patterns and highlighting and defining filters over interesting data. For example, Andrienko et al. developed a toolset for analysing and reasoning about movement data (e.g., GPS coordinates). Following some preprocessing steps, the data can be clustered according to different properties, such as start and end points of trips, or similar behaviour over time [15]. Such properties can be portrayed using different types of visualisations to increase user understanding.

There exist myriad visualisation techniques, from time-series to multi-dimensional scatter plot methods, which can be adapted to the exploration of multidimensional context data. Our focus here is not only on the exploration of such context data, but also the scope of the higher order situations, their specification, and data cases which fall outside the set boundaries. The Table Lens, a focus+context visualisation, supports the interactive exploration of many data values in a semi-familiar spreadsheet format [16]. In practice, due to the distortion techniques employed, users can see 100 times as many data items within the same screen space as compared with a standard spreadsheet layout. Rather than showing the detailed numeric values in each cell, a single row of pixels, relating to the value in the cell, is shown instead. The Table Lens affords users the ability to easily study quantitative data sets, but categorical values are not well supported.

3   Parallel Coordinates

Parallel Coordinate Visualisations (PCVs) are a standard two-dimensional technique ideally suited to large, multivariate data sets [17]. The technique excels at visually clustering cases that share similar attribute values across a number of independent discrete or continuous dimensions, as they can be visually identified through the distribution of case lines within the visualisation [18]. The user can see the full range of the data's many dimensions, and the relative frequencies at which values on each axis are recorded. These features are visible in Figure 1, which shows context data from our user study, which we will describe in the next section.

PCVs give users a global view of trends in the data while allowing direct interaction to filter the data set as desired. A set of parallel vertical axes are drawn, which correspond to attributes of the readings in the system. Then, a set of n-dimensional tuples are drawn as a set of *polylines* which intersect each axis at a certain point, corresponding to the value recorded for that attribute. Discrete and quantitative axes can be presented in the same view.

As all the polylines are being drawn within the same area, the technique scales well to large data sets with arbitrary numbers of attributes, presenting a

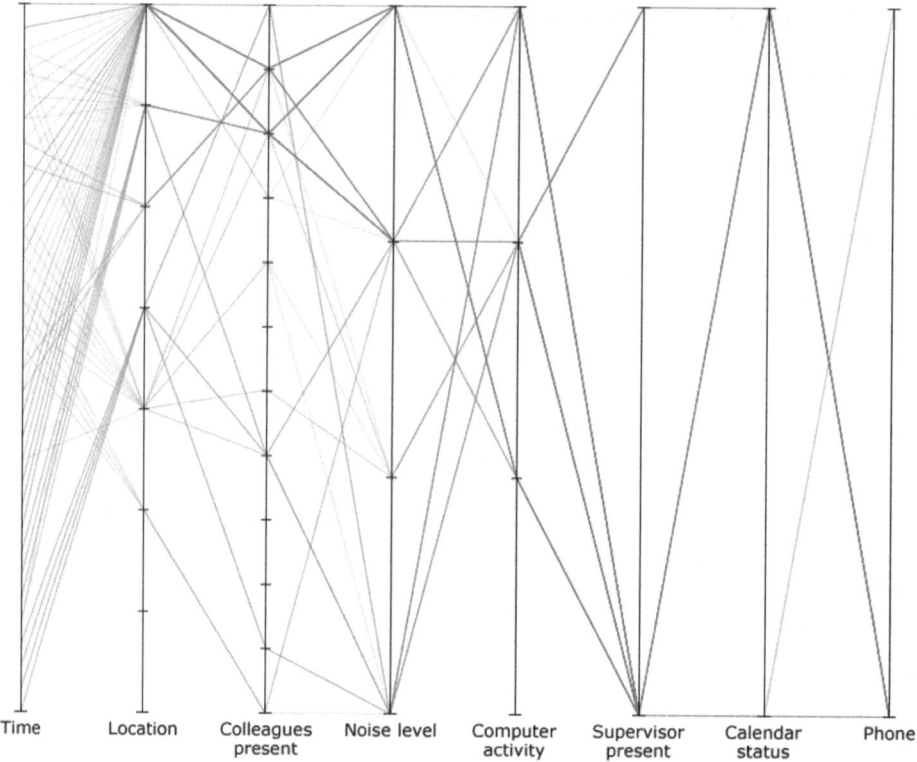

Time   Location   Colleagues   Noise level   Computer   Supervisor   Calendar   Phone
                    present   activity   present   status

Fig. 1. Part of the main Situvis window showing our Parallel Coordinates Visualisation. This is a view of 96 overlaid context traces with 8 data dimensions gathered over three days of summer. Strong correlations can be seen between the days recorded: the subject spent the majority of all three days at their desk (the first value on the "Location" axis), with some deviations due to coffee breaks or visits to their supervisor's office at irregular times.

compact view of the entire data set. Axes can be easily appended or removed from the visualisation as required by the dimensions of the data.

As Parallel Coordinates have a tendency to become crowded as the size of the data set grows larger, techniques have been designed to cluster or elide sub-sets of the data to allow the dominant patterns to be seen [19]. Direct interaction to filter and highlight sections of the data encourages experimentation to discover additional information, as seen in Figure 2.

Hierarchical clustering [20] uses colour to visually distinguish cases that share a certain range of values into a number of sets, increasing the readability of the diagram. We use a similar technique to group case lines that are assigned to a certain situation, colour-coding these as a group. Different situations can be colour-coded so that the interplay of the context traces that correspond to them can be easily seen. We will illustrate this ability in the next section.

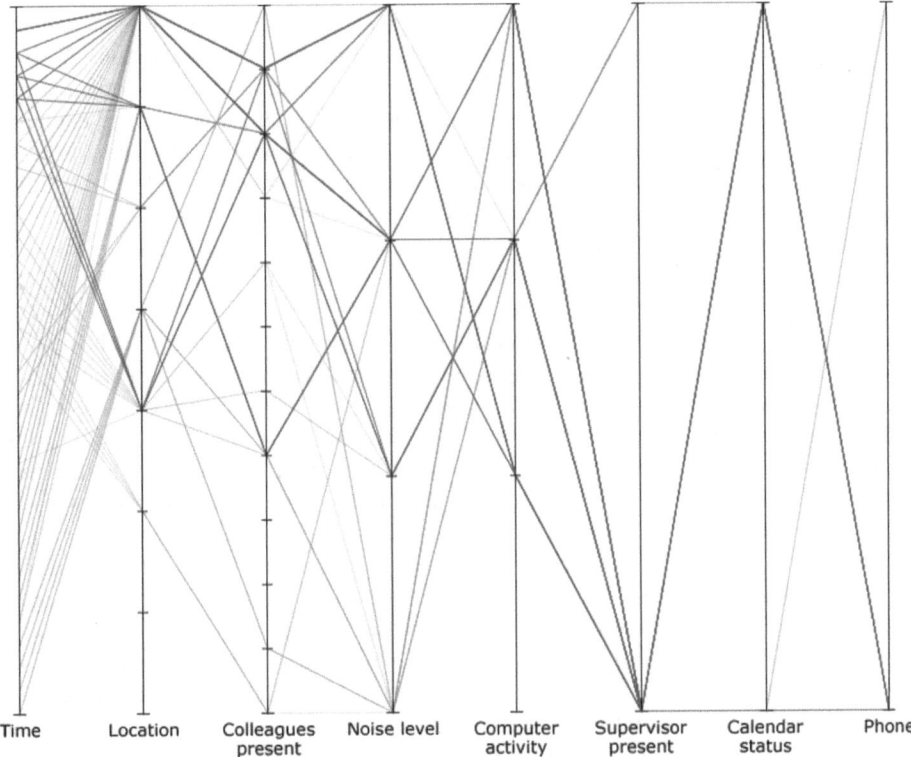

Time   Location   Colleagues   Noise level   Computer   Supervisor   Calendar   Phone
                       present   activity   present   status

Fig. 2. Here the user has "brushed" over a set of case lines (those that correspond to times before 11am) by right clicking and dragging a line across them between the first and second axes. This highlights these polylines throughout the diagram, allowing the patterns that occurred among these times to be seen. This same operation can be performed on any axis to select any subset of the polylines.

4   Evaluating Situations with Situvis

4.1   Description and Case-Study

The goal of Situvis is to combine data-driven and user-driven techniques for situation determination without relying on machine learning to make sense of the data. The tool displays all of the situation trace data, along with annotations where available, in a single view; and allows the user to clearly and easily highlight the situation traces associated with an annotation label. The user can then adjust the resulting set of ranged intervals over the context to create a more complete and accurate situation specification.

Situvis is built using Processing [21], a Java-based visualisation framework that supports rapid prototyping of visualisation techniques.[1] Each context

---

[1] Situvis is freely-available software, which you are encouraged to download from our website at http://situvis.com.

dimension is represented in Situvis as a vertical axis, and each axis contains a set of points that correspond to permitted values for the dimension. A situation trace is represented as a polyline—a line drawn starting at the leftmost axis and continuing rightwards to the next adjacent and so on, intersecting each axis at the point that represents the value that the context has in that situation trace. For example if, in a given situation, a user's computer activity level is "idle", and their location is "canteen", and these two axes are adjacent, then a line will be drawn between those two points. Each situation trace is plotted on the axes and the result is a view of all of the situations, significant and insignificant, that occurred in the system over a period of time.

To carry out our case-study, we required real context data with which we could characterise situations. We chose to gather context data and situation annotations manually over two three-day periods. While the capabilities exist to collect these context data automatically, for this first trial we chose to collect the data through manual journaling, so that we did not need to factor in issues with the aggregation, uncertainty or provenance of the context data. As mentioned previously, we assume the data is at an appropriate level of abstraction to begin with.

We had a single trial participant record their context every fifteen minutes (between 10am–6pm) for three consecutive weekdays, on two distinct occasions. The first occasion was in summer and the second was in autumn. The journaling gap between these two data sets is designed to capture adjustments in the routines and descriptions of situations that the trial participant found himself in. The captured context consists of time, location, noise-level, number of colleagues present, their supervisor's presence (true or false), their phone use (either taking a call or not), calendar data (being busy or having no appointments), and computer activity. For simplicity, the noise-level was recorded on a 4-point scale of quiet, conversation, chatty, and noisy. Likewise, computer activity level was scaled as idle for an hour or more, idle for less than an hour, active, and highly active. We defined six symbolic locations: meeting room, canteen, sports center, supervisor's office, subject's desk, and a lecture theatre. Figure 1 shows a view of the Situvis tool with all of the traces from the first three consecutive days of collection plotted together in one view.

The participant also annotated what, if any, situation he was in at the time of data capture. These annotations are used in Situvis to identify situations that require specification in the system, and to provide some ground truth to initiate their specification.

4.2   Specifying Situations with Context

Situation specifications are structured according to the definition we discussed in Section 2.2. Situvis enables a developer to select all occurrences of a given annotated situation, and add further cases to this definition using interactive brushing of polylines as in Figure 2, or by dragging a range indicator on the left of the axis to expand or contract the range of values covered by this specification. The user can evaluate existing situation specifications overlaid against actual trace data and see where they need to be modified.

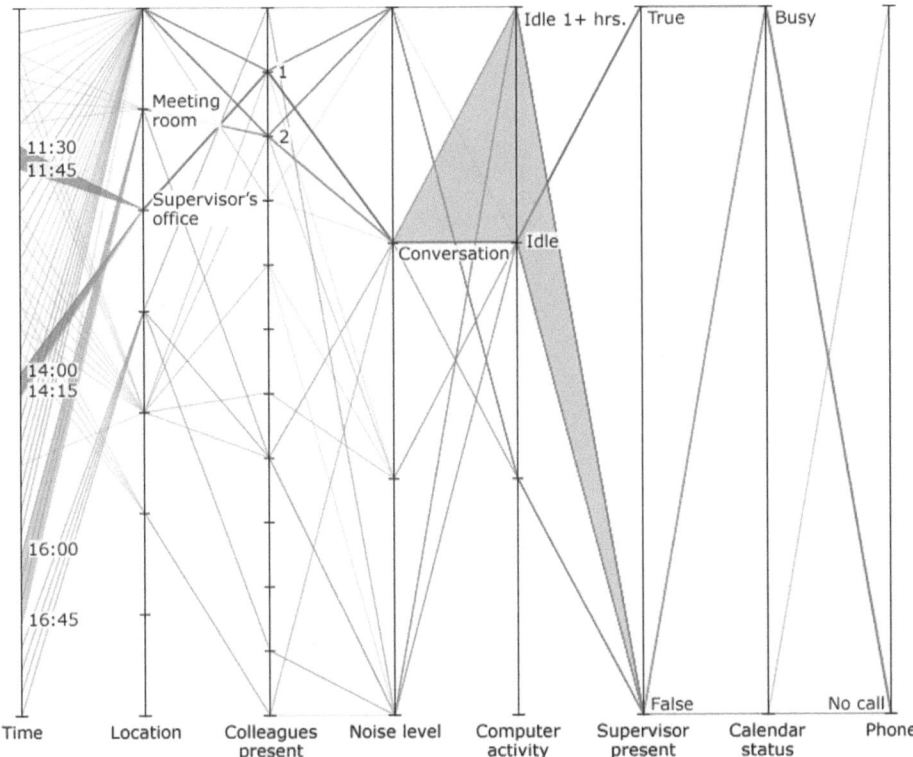

Fig. 3. A view of the Situvis tool with our initial summer data set. The highlighted traces were annotated as a "meeting" situation. These situations occurred at many different times throughout the day in two different locations, with a range of values for the other contexts. Labels have been added to the axis for clarity. They are normally shown when the user hovers over the axis.

An example of this process can be seen in Figure 3 and 4. The trial subject annotated multiple occurrences of a "Meeting" situation [2]. By selecting these traces, it is evident what context dimensions characterise them. We can see that "Time" and "Supervisor presence" are not useful due to the multiple split lines on their axes, and so are ineffective when defining constraints. The specification is clear from the other dimensions, however, and could be expressed as:

{Location = (Meeting room $\vee$
                      Supervisor's office)} $\wedge$
{1 $\leq$ Colleagues present $\leq$ 2} $\wedge$
{Noise-level = conversation} $\wedge$
{Computer activity $\geq$ idle} $\wedge$

---

[2] "Meeting" is a generalisation of two situations that the participant labelled, namely, "Meeting with colleagues" and "Meeting with supervisor". These are assigned separate colours in the figures.

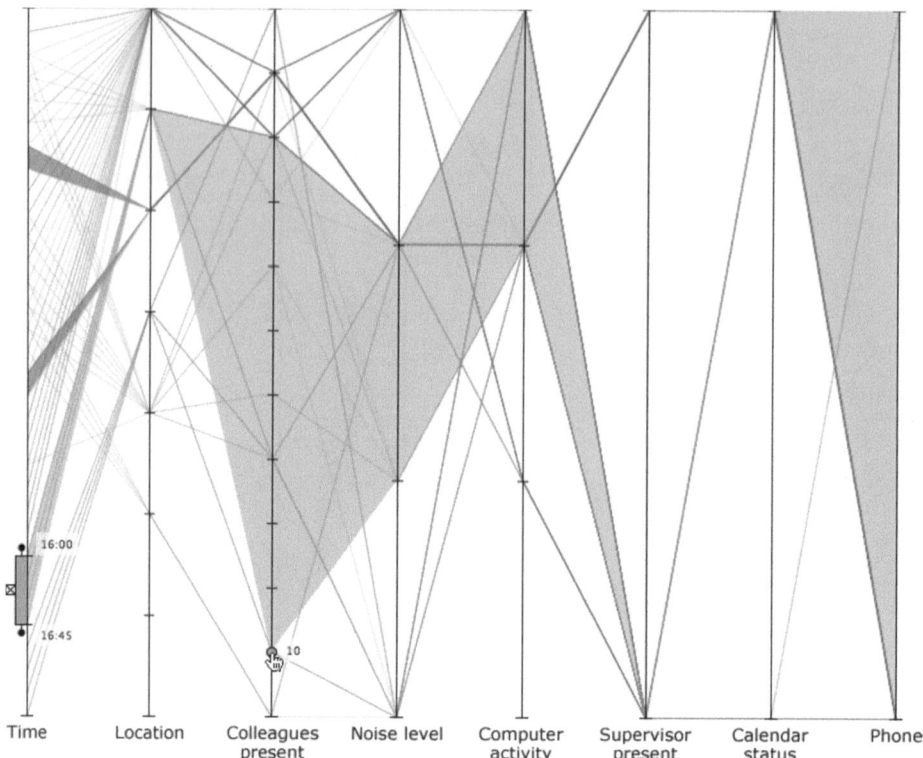

Fig. 4. The user can interactively expand or contract the situation definition along any of the axes. In this case, they have chosen to modify the situation specification to allow for more colleagues to be present, the noise level to be greater and the possibility of talking on the phone.

{Calendar status = busy} ∧
{Phone use = none}

None of these values alone can characterise "Meeting", as the trace data illustrates. Furthermore, each dimension may not always be available. Situvis allows one to identify combinations of dimensions which, when taken together can provide a good estimation of the situation. For example, "Location" taken with "Colleagues present" is a good indication of "Meeting". This can also give system developers an insight into which sensors in their system are the most useful, and which types of sensors they should invest in to gain the most added benefit in terms of the expressiveness of their system.

4.3   Situation Evolution

When existing specifications are overlaid on the trace polylines, the developer can see where they are too strong or weak. Constraints that are too strong will cause the system to sometimes fail in determining when that situation is occurring.

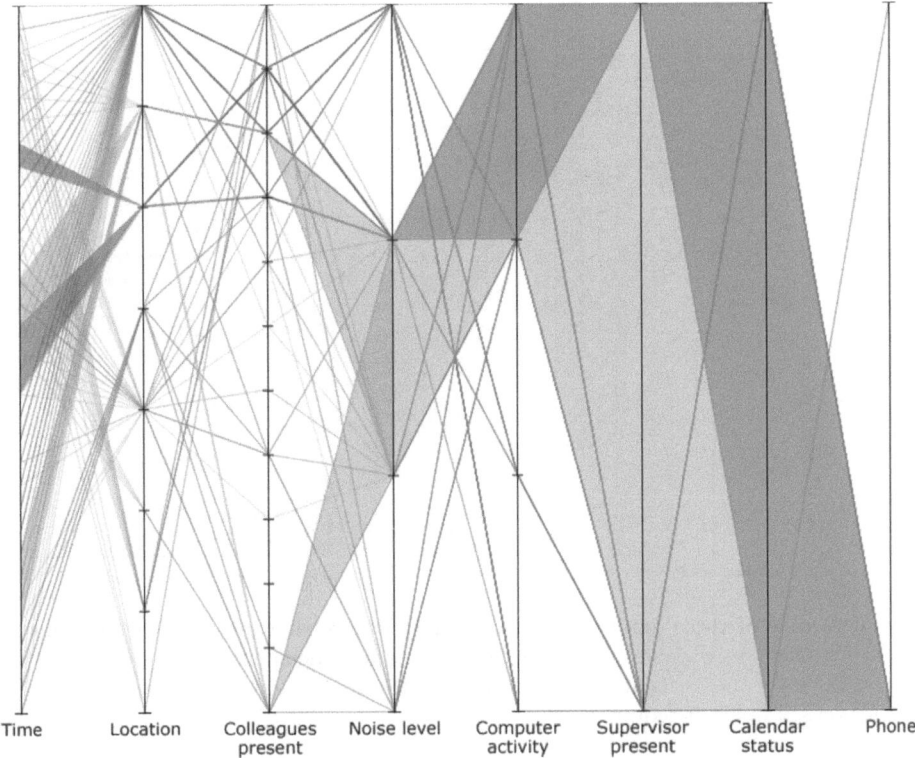

Time   Location   Colleagues   Noise level   Computer   Supervisor   Calendar   Phone
                    present   activity   present   status

Fig. 5. After additional data collection, the case lines annotated as "Meeting" exhibit a different pattern

Constraints that are too weak may be wrongly interpreted as an occurrence of the specified situation, when in fact a different situation is occurring. By overlaying our specification on top of the polylines, it will be obvious where constraints need to be strengthened, weakened or even excluded altogether. Situvis enables a developer to drag the boundaries of specifications to change the polylines that they cover, essentially changing the constraints of the situation.

When the overlaid situation encompasses traces that are not relevant, the user can strengthen the constraints by narrowing the range of values covered by this situation specification (the shaded area in the diagram). Similarly, the user can weaken constraints to include traces that happen to fall outside the existing specification by widening the specification, as we have done in Figure 4.

As more trace data is added and annotated, the constraints that we have defined for "Meeting" may be shown to be too strong. This is what we found to be the case in Figure 5, which contains the traces for both our initial summer data set, as well as the additional days from the autumn data set, for a total of 48 hours of context traces. What we see is that in most cases, the previously apparent patterns are strengthened, as essentially they have recurred. Comparing Figure 3 and Figure 5, we can see that the annotated data that the

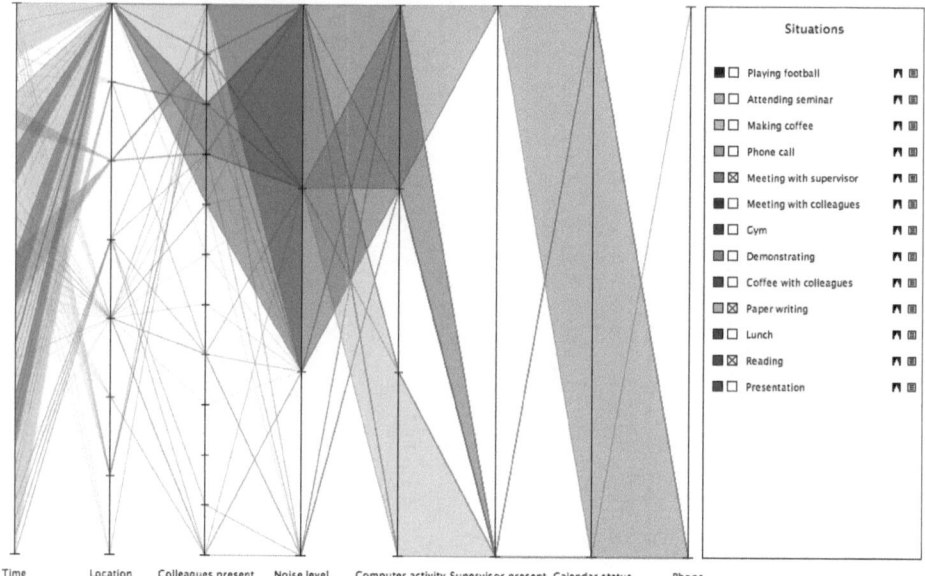

Fig. 6. A view of three distinct situations. Here we are showing the specifications for a meeting with supervisors, paper writing time, and time spent reading. The dissimilarities between these situations are clear from the tool, and the specifications can be further teased apart if required.

user has defined as corresponding to meetings results in a different situation specification.

4.4   Situation Evaluation

Context-aware adaptive systems are very sensitive to incompatible behaviours. These are behaviours that conflict, either due to device restrictions, such as access to a public display, or due to user experiences, such as activating music playback while a meeting is taking place. Situations are closely tied to behaviours—they define envelopes in which behaviour occurs. As a result, their specifications are directly responsible for adherence to compatibility requirements. By harnessing this factor, we can address another key aspect of situation evaluation.

Conceptually relating situations to each other from a behaviour compatibility standpoint is an overwhelming task for a developer. We recognise that there are two situation relationships that may lead to incompatibility:

subsumption if a subsumes b, and b occurs, then a will certainly occur.
overlap if a overlaps b, then a and b may co-occur.

Our tool allows multiple situation specifications to each be coloured distinctly. When two or more situations are shown together, the overlap in their constituent contexts is clear, as well as the extent of their dissimilarities. This view allows the developer to alter constraints where necessary, while the overlap and subsumption relationships are refreshed and displayed on-the-fly. A screenshot of this scenario is seen in Figure 6, which also shows the specification selection panel on the right-hand side. This area allows the user to toggle specifications on and off, so that they can be compared and manipulated.

5   Conclusions and Future Work

We have presented Situvis, a tool that uses a Parallel Coordinate Visualisation to illustrate situation traces and specifications. We have shown, using a case-study, the utility of Situvis in the situation specification and evaluation processes. Situvis presents a developer with a reference point for situation specification and evaluation through the display of actual trace data and situation annotations. The relevance of the underlying context to a specification is made clear, and contrasting situation traces can be used as a guide for specification.

Context-aware systems are dynamic—sensors, users and habits are constantly changing. Hence, we cannot expect situation specifications to remain static. It must be possible to re-evaluate them accordingly. Situvis allows developers to visually overlay specifications on traces, and tailor their constraints as a result. Unlike traditional methods, Situvis clearly depicts cases where constraints are too strong or too weak. Machine-learning techniques would require extra time-consuming training periods for the re-evaluation process, whereas Situvis provides the option of collecting a minimal amount of annotated data to initiate the manual process.

By visually analysing the overlap of situation specifications within their system, the developer can identify where multiple situations require similar context values to be activated. Such overlaps may imply problems in the situation specifications, as conflicting behaviours may be triggered by conceptually similar situations. Thus, the developer can compare situations against others, and change the situation's specifications to become stronger or weaker as necessary.

A feature missing from the current version of the Situvis tool is explicit support for probabilities in situation specifications. In many context-aware applications, robust probabilistic inference is a requirement to handle the naturally fuzzy data in the system. We are considering the addition of an overlay which will allow users to set up a probability distribution, though this requires a more in-depth study of the treatment of uncertainty in situations.

Some context dimensions are not easily represented on a line. In particular, Location within buildings, with its domain relations like subsumption, is difficult to represent in two dimensions. We are researching techniques to flatten hierarchies for a more intuitive representation of this context.

Situvis could also be used by users of the context-aware system as a gateway to user programming: helping them to unroll the cause of a situation activation, so that they can gain insight into why the system began to behave as it did.

Acknowledgments. This work is partially funded under The Embark Initiative of the Irish Research Council for Science, Engineering and Technology, and by Science Foundation Ireland under grant number 03/CE2/I303-1, "LERO: the Irish Software Engineering Research Centre."

References

1. Henricksen, K.: A Framework for Context-Aware Pervasive Computing Applications. PhD thesis, The School of Information Technology and Electrical Engineering, University of Queensland (September 2003)
2. Knox, S., Clear, A.K., Shannon, R., Coyle, L., Dobson, S., Quigley, A., Nixon, P.: Towards Scatterbox: a context-aware message forwarding platform. In: Fourth International Workshop on Modeling and Reasoning in Context in conjunction with Context 2007, Roskilde, Denmark, pp. 13–24 (August 2007)
3. Thomson, G., Stevenson, G., Terzis, S., Nixon, P.: A self-managing infrastructure for ad-hoc situation determination. In: Smart Homes and Beyond - ICOST2006 4th International Conference On Smart Homes and Health Telematics. Assistive Technology Research Series, pp. 157–164. IOS Press, Amsterdam (2006)
4. Logan, B., Healey, J., Philipose, M., Tapia, E.M., Intille, S.: A Long-Term Evaluation of Sensing Modalities for Activity Recognition. In: Krumm, J., Abowd, G.D., Seneviratne, A., Strang, T. (eds.) UbiComp 2007. LNCS, vol. 4717, pp. 483–500. Springer, Heidelberg (2007)
5. Krause, A., Smailagic, A., Siewiorek, D.P.: Context-aware mobile computing: Learning context-dependent personal preferences from a wearable sensor array. IEEE Transactions on Mobile Computing 5(2), 113–127 (2006)
6. Huỳnh, T., Fritz, M., Schiele, B.: Discovery of activity patterns using topic models. In: UbiComp 2008: Ubiquitous Computing, 10th International Conference, Seoul, South Korea, pp. 1–10 (September 2008)
7. Clear, A.K., Knox, S., Ye, J., Coyle, L., Dobson, S., Nixon, P.: Integrating multiple contexts and ontologies in a pervasive computing framework. In: C&O 2006: ECAI 2006 Workshop on Contexts and Ontologies: Theory, Practice and Applications, Riva Del Garda, Italy, pp. 20–25 (August 2006)
8. Coutaz, J., Rey, G.: Foundations for a theory of contextors. In: CADUI: 4th International Conference on Computer-Aided Design of User Interfaces, pp. 13–34. Kluwer, Valenciennes (2002)
9. Dey, A.K.: Understanding and using context. Personal Ubiquitous Computing 5(1), 4–7 (2001)
10. Coutaz, J., Crowley, J., Dobson, S., Garlan, D.: Context is key. Communications of the ACM 48(3), 49–53 (2005)
11. Ye, J., Coyle, L., Dobson, S., Nixon, P.: A unified semantics space model. In: Hightower, J., Schiele, B., Strang, T. (eds.) LoCA 2007. LNCS, vol. 4718, pp. 103–120. Springer, Heidelberg (2007)
12. Loke, S.W.: Representing and reasoning with situations for context-aware pervasive computing: a logic programming perspective. The Knowledge Engineering Review 19(3), 213–233 (2004)
13. Fails, J., Olsen, D.: A design tool for camera-based interaction. In: CHI 2003: Proceedings of the SIGCHI conference on Human factors in computing systems, pp. 449–456. ACM Press, New York (2003)

14. Dey, A.K., Hamid, R., Beckmann, C., Li, I., Hsu, D.: A cappella: programming by demonstration of context-aware applications. In: CHI 2004: Proceedings of the SIGCHI conference on Human factors in computing systems, pp. 33–40. ACM Press, New York (2004)
15. Andrienko, G., Andrienko, N., Wrobel, S.: Visual analytics tools for analysis of movement data. SIGKDD Explorations Newsletter: Special issue on visual analytics 9(2), 38–46 (2007)
16. Tenev, T., Rao, R.: Managing multiple focal levels in table lens. In: Infovis 1997: IEEE Symposium on Information Visualization, p. 59. IEEE Computer Society Press, Los Alamitos (1997)
17. Inselberg, A., Dimsdale, B.: Parallel coordinates: a tool for visualizing multi-dimensional geometry. In: VIS 1990: Proceedings of the 1st conference on Visualization 1990, pp. 361–378. IEEE Computer Society Press, Los Alamitos (1990)
18. Card, S., Mackinlay, J., Schneiderman, B.: Readings in Information Visualization: Using Vision to Think. Morgan Kaufmann, San Francisco (1999)
19. Artero, A., de Oliveira, M., Levkowitz, H.: Uncovering clusters in crowded parallel coordinates visualizations. In: IEEE Symposium on Information Visualization, pp. 81–88 (2004)
20. Fua, Y.H., Ward, M.O., Rundensteiner, E.A.: Hierarchical parallel coordinates for exploration of large datasets. In: VIS 1999: Proceedings of the conference on Visualization 1999, pp. 43–50. IEEE Computer Society Press, Los Alamitos (1999)
21. Reas, C., Fry, B.: Processing: a learning environment for creating interactive web graphics. In: SIGGRAPH 2003: ACM SIGGRAPH 2003 Sketches & Applications, p. 1. ACM Press, New York (2003)

Methodologies for Continuous Cellular Tower Data Analysis

Nathan Eagle[1,2], John A. Quinn[3], and Aaron Clauset[2]

[1] Massachusetts Institute of Technology, 20 Mass Ave, Cambridge, 02139
[2] The Santa Fe Institute, 1399 Hyde Park Rd, Santa Fe, NM 87501
[3] Makerere University, Kampala, Uganda
nathan@mit.edu, john.quinn@ed.ac.uk, aaronc@santafe.edu

Abstract. This paper presents novel methodologies for the analysis of continuous cellular tower data from 215 randomly sampled subjects in a major urban city. We demonstrate the potential of existing community detection methodologies to identify salient locations based on the network generated by tower transitions. The tower groupings from these unsupervised clustering techniques are subsequently validated using data from Bluetooth beacons placed in the homes of the subjects. We then use these inferred locations as states within several dynamic Bayesian networks (DBNs) to predict dwell times within locations and each subject's subsequent movements with over 90% accuracy. We also introduce the X-Factor model, a DBN with a latent variable corresponding to abnormal behavior. By calculating the entropy of the learned X-Factor model parameters, we find there are individuals across demographics who have a wide range of routine in their daily behavior. We conclude with a description of extensions for this model, such as incorporating contextual and temporal variables already being logged by the phones.

1 Introduction

Every one of the approximately 4 billion mobile phones in use today have continuous access to information about proximate cellular towers. We believe these continuous cellular tower data streams can provide valuable insight into a user's behavior. Here we introduce a novel method of segmenting, validating and modeling this data. A major contribution of this paper involves the application and design of community structure algorithms that are appropriate for the identification of location clusters relevant to a user's life. We show that using temporal data from cellular towers, information every phone has access to, a simple generative model can be used to infer these salient locations and anticipate subsequent movements.

There has recently been a significant amount of research quantifying and modeling human behavior using data from mobile phones. We will highlight a selection of the literature on GSM trace analysis and subsequently discuss recent work on location segmentation and movement prediction from GPS data.

Mobile phones are continuously, passively monitoring signals from proximate cellular towers. However, due to power constraints, a mobile phone typically

H. Tokuda et al. (Eds.): Pervasive 2009, LNCS 5538, pp. 342–353, 2009.

does not continuously send back similar signals alerting the nearby towers of its particular location. While there has been recent work on analysis of call data records (CDR) from mobile phone operators [1,2], this data only provide estimates of locations when the phone is in use. Additionally, the only method of obtaining continuous cellular tower data without working with an operator is by installing a logging application on the mobile phone itself.

There have been a variety of projects that have involved installing a mobile phone application that logs visible cellular towers and Bluetooth devices on a set of subjects phones including HIIT's Context project, MIT's Reality Mining project [3] and the PlaceLab [4,5] research at Intel Research. Additionally, other research projects have demonstrated the utility of cellular tower data for a broad spectrum of applications ranging from contextual image tagging [6] to inferring the mobility and location of an individual [7,8,9]. Generally this logging software records between one to four of the cellular towers with the highest signal strength, however, recent research suggests it is possible to localize a handset down to 2.5 meter accuracies if the number of detected towers is dramatically increased [10].

Dynamic Bayesian Networks (DBNs) have been widely used for quantifying and predicting human behavior. For analysis of human movement, typically these models involve location coordinates that are much more precise than cellular tower data, such as GPS data. These models are trained on general human movement [11] or more specific data such as transportation routes [12].

As opposed to the previous work above, our dataset comes from randomly sampled individuals in a large US metropolitan city. We introduce several segmentation algorithms taken from the community structure literature and apply them to networks of cellular towers. Coupling bluetooth beacon data placed in the homes of each subject with the tower data, we validate the output of the community structure algorithms with the community of towers co-present with the beacon exposures. We then describe several DBNs that use the inferred locations clusters as states to parametrize and predict subsequent movements. One such DBN we use for behavioral modeling includes a latent variable, the X-Factor, corresponding to a binary switch indicative of "normal" or "abnormal" behavior. We compare the entropy of the learned X-Factor parameters across different demographics and conclude with ideas for extensions to these models as future work.

2   Methods

2.1   Data Description

Our data was generated from the phones of 215 subjects from a major US city. After providing informed consent, these subjects were given phones that logged the ID of the four cellular towers with the strongest signal strength every 30 seconds. Additionally, the phones conducted Bluetooth scans every minute. Bluetooth beacons were deployed in the homes of each subject; as the beacons are detected only if the phone is within 10 meters of the beacon, detection implies the subject is at home. The data was compressed on the mobile phone and uploaded to a central server after each day.

In contrast to previous datasets, every subject in our study was randomly sampled from a particular city. By offering a smartphone and free service, over 80 percent of the randomly selected individuals agreed to participate in the study. The demographic information we have about the subjects is evenly distributed among ethnic groups and income levels, accurately reflecting the distribution that makes up the city's inhabitants. No longer constrained to the study of academics or researchers, our data represents one of the first comprehensive behavioral depictions of the inhabitants within a major urban city.

2.2   Segmentation via Community Structure

Each phone records the four towers with the strongest signal at 30 second intervals. This data can therefore be represented as a cellular tower network (CTN) where each node is a unique cellular tower, an edge exists between two nodes if both towers co-occur in the same record, and each edge is weighted with the total amount of time (over all records) the pair co-occurred. A CTN is generated for each of the subjects, which includes every tower logged by the phone during the 5-month period. The nodes in the CTN that have the highest total edge weight (the node's "strength") correspond to the towers that are most often visible to the phone. Further, a group of nodes with a large amount of weight within the group, and less weight to other nodes, should correspond to a "location" where the user spends a significant amount of time. Figure 2 shows a 32-tower subgraph of one CTN, segmented into five such locations.

To allow for a meaningful comparison, we use three qualitatively different heuristics for clustering nodes into locations.

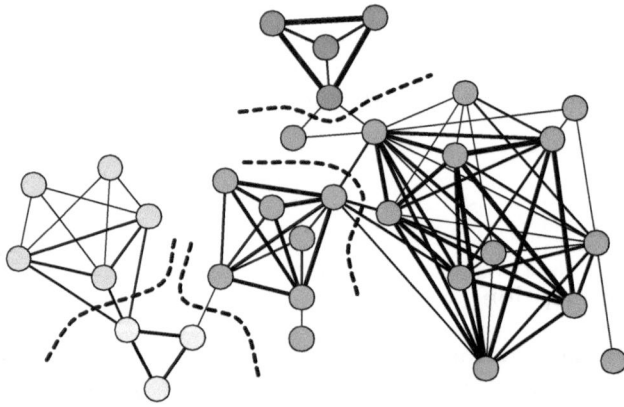

Fig. 1. A 32-tower subgraph of one of our cellular tower networks, segmented into five "locations," clusters of nodes in which towers frequently co-occur in the phone's records

Ncut. The first segmentation algorithm depends on Shi and Malik's *normalized cut* (Ncut) criterion [13], which, like many cut criteria, is NP-hard to optimize. Our implementation uses a spectral approach to find a bisection of the graph that minimizes the size of the normalized cut. Applied recursively, a graph can be split into a specified number of dense clusters. Although originally developed to segment images, the Ncut method can naturally be applied to networks.

Q-Modularity. The second method, drawn from the large literature on detecting "communities" in complex networks [14], depends on Newman and Girvan's popular *modularity* measure Q [15], which measures the density of clusters relative to a simple, randomized null model.

$$Q = \sum_{s=1}^{m} \left[\frac{l_s}{L} - \left(\frac{d_s}{2L} \right)^2 \right] \tag{1}$$

where l_s is the number of edges between the nodes within cluster s, L is the total number of edges in the network, and d_s is the sum of degrees of the nodes in cluster s. While finding the segmentation that maximizes Q is NP-complete, there has been a significant amount of work towards this goal. Although also NP-hard to maximize, we use Clauset *et al.*'s greedy optimizer [16], which has been shown to perform reasonably well on real-world data.

Threshold Groups. The third method is a simple-minded heuristic: we first identify the nodes in the upper decile of "strength," and then perform a breadth-first search on the induced subgraph. Each connected component in this subgraph is labeled as a unique location, and all remaining nodes in the original graph placed in an additional group.

Although all based on somewhat similar principles, in practice these methods produce dramatically different segmentations of our CTNs. This is in part because the first algorithm requires as input the number of segments to be found, unlike the other two.

2.3   Inference via Bluetooth Beacons

One objective measure of these clusterings is to use independent information derived from the Bluetooth beacons, installed in the homes of each subject in the study. Every minute the phone scans for visible Bluetooth devices and if a beacon is within 10 meters of the phone, it is logged as proximate. Creating training data from the set of cellular towers detected at the same time as the bluetooth beacons, we have used several methodologies to infer if a subject is at home given a particular set of visible cellular towers.

Bayesian Posteriors. It is possible to calculate the posterior probability a subject is home, $P(L_{home})$, conditioned on the four towers currently detected by the phone, T_{abcd}, using the likelihood, the marginal and the prior probability of being at home (based on the beacon data).

$$P(L_{home}|T_{abcd}) = \frac{P(T_{abcd}|L_{home})P(L_{home})}{P(T_{abcd})} \qquad (2)$$

Gaussian Processes. While the naive Bayesian model above works well in many cases, simply using the ratio of tower counts co-present with the Bluetooth beacon tends to fail if the phone regularly moves beyond ten meters of the beacon while still staying inside the home. Instead of normalizing by total number of times each tower is detected, it is possible to obtain additional accuracy by incorporating the signal strengths from the detected towers. There are many models for signal strength of a single cellular tower, t. $p_t(s_t|\mathbf{l})$, one such model uses training data to estimate Gaussian distributions over functions modeling signal propagation from cellular towers [17]. In our case, the training data comes from the signals of towers detected at the same time as the Bluetooth beacon in the subject's home, and the inference is binary (home or not home); however, these models are easily extendable for more broad localization.

Deviations in Tower Signal Distributions. The two models above generate a probability of being at home associated with a single sample of detected towers (ie: the four tower IDs and their respective signal strengths). However, during the times when a subject is stationary, the phone continuously collects samples of the detected towers' signal strengths. These samples can form 'fingerprint' distributions of the expected signal strengths associated with that particular location. It is possible to detect deviations within these distributions of signal strengths using a pairwise analysis of variance (ANOVA) with the Bonferroni adjustment to correct for different sample sizes. Training the home distributions on the times when the beacon is visible (or if there are no beacons, on times when the subject is likely home such as 2-4am), an ANOVA comparing this home distribution with a distribution of recent tower signal strengths makes it possible to identify if the subject is truly at home, or is at a next-door neighbor's house. In previous work, such tower probability density functions have successfully localized a phone down to the office-level [3].

2.4   Prediction via Dynamic Bayesian Networks

The clusters of towers identified above can be incorporated as states of a dynamical model. Given a sequence of locations visited by a subject, we can learn patterns in their behaviour and calculate the probability of them moving to different future locations. We start with a baseline dynamical model and introduce additional observed and latent variables in order to model the situation more accurately.

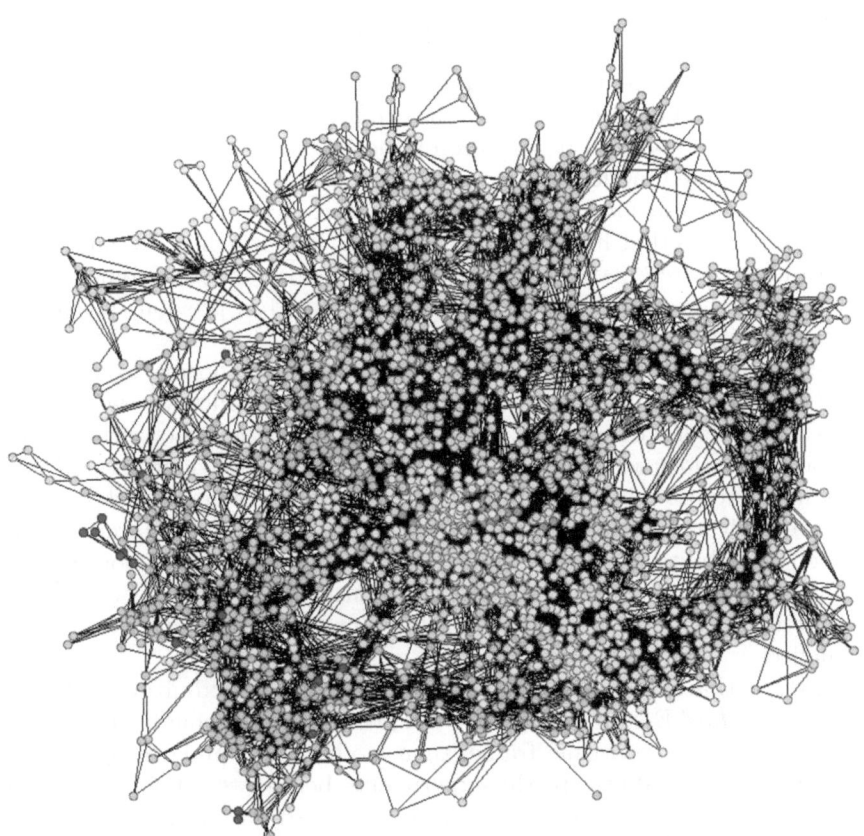

Fig. 2. CTN Segmentation. The giant component of a subject cellular tower network, segmented into 15 major location clusters (represented by 15 colors) using the Q-modularity community structure method.

The simplest dynamical Bayesian network we can use for location prediction is a Markov chain, in which the location y_t depends only on the location at the previous time step, y_{t-1}. The maximum likelihood transition probabilities $p(y_t|y_{t-1})$ can easily be estimated. Given evidence that a user is in a particular location at time t, this allows us to calculate the τ-step-ahead prediction $p(y_{t+\tau}|y_t)$.

We note that patterns of movement in practice are dependent on the time of day and the day of week. Subjects typically exhibit different dynamics on weekday mornings than on Saturday evenings, for example. Figure 3(a) shows an extended model where the probability of being in a location is also dependent on the hour of day h_t and the day of week d_t. In the experiments below, we code h_t to take on the values "morning", "afternoon", "evening" and "night", and code d_t to take on the values "weekday" or "weekend". After learning maximum likelihood parameters we can calculate the predicted density $p(y_{t+\tau}|y_t, d_{t+1:t+\tau}, h_{t+1:t+\tau})$ for new observations from the same user.

2.5   X-Factors for Abnormality Modeling

While there is strong structure in human behavior, there are also regular deviations from the standard routines. We incorporate an additional latent variable into our model to quantify the variation in behavior previously unaccounted for in the fully observed models above.

The model we use for this is shown in Figure 3(b). Here we factorize the location variable so that it depends on a hidden "abnormality" variable a_t. The model can now switch between "normal" and "abnormal" behaviour depending on whether a_t is 0 or 1 respectively, as demonstrated in previous physiological condition monitoring work [18].

We expect abnormal dynamics to be related to the normal dynamics but with a broader distribution. When estimating these dynamics, we therefore want to keep relevant structure in the dynamics (e.g. transitions between physically neighboring locations are still more likely), while allowing wider possibilities including non-zero probability of transitions not seen in the training data. We can achieve this effect by tying the parameters between the normal and abnormal transition probabilities such that $p(y_t|y_{t-1}, d_t, h_t, a_t = 1)$ are a smoothed version of $p(y_t|y_{t-1}, d_t, h_t, a_t = 0)$. To smooth the transition matrices for every combination of d_t and h_t we add a small constant ξ to each entry in the matrix and renormalize.

Learning of this model can be done with expectation-maximization (EM). We perform a standard E-step to calculate the probability of being in the normal or abnormal regime at each time frame, then modify the standard M-step to use the parameter tying above. In the experiments below, we set $\xi = .1$ by hand, though in principle this parameter can also be learnt using EM. Increasing ξ effectively specifies that a sequence has to depart further from normal dynamics in order to be considered "abnormal".

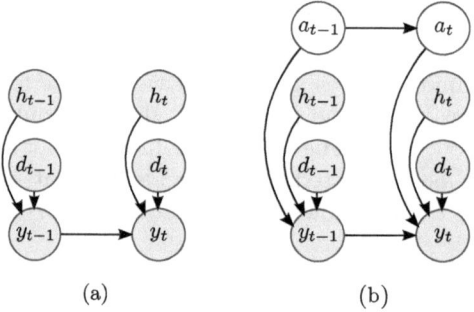

(a)   (b)

Fig. 3. Two DBN models used for location prediction. Shaded nodes are observed and unshaded nodes are latent; y_t denotes location, d_t denotes day of week, h_t denotes hour of day, and a_t denotes abnormal behaviour (all at time t). Panel (a) shows a fully observed model as a contextual Markov chain (CMC), and panel (b) shows the X-factor model, where location is additionally conditioned on the latent abnormality variable.

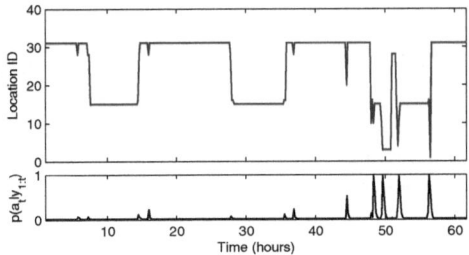

Fig. 4. Inferred points of abnormality using the X-Factor model. Each weekday the subject moves consistently between home (location 31) and work (location 15), but on the third day makes some extra, unusual journeys. The locations in this example were given by the Group Threshold segmentation method.

3   Results and Discussion

3.1   Segmentation Validation

We have shown how data collected from installed Bluetooth beacons can be used to create a known cluster of towers associated with each subject's home. We used this known cluster to validate each segmentation algorithm, selecting twenty locations for the Ncuts technique. Table 1 categorizes the community detecition algorithms by how well they detected the "home" towers as defined by the Bluetooth beacons, C_{BT}. The home cluster of towers generated by the Threshold Groups technique incorporated C_{BT} for every subject, $P(C_{BT} \subset C_H) = 100\%$, while this was the case for the Q-Modularity technique only 86% of the time. However, the other important statistic is the ratio of the number of the Bluetooth home towers, $N_{C_{BT}}$, to the number of towers in the inferred home cluster, N_{C_H}. This ratio describes how many additional towers were included in the inferred home location; for example, the Q-Modularity home cluster has a ratio of .18, indicating that its home cluster contains approximately five times as many towers as needed. Despite averaging the most number of clusters, the Ncuts home cluster has a ratio of .0061, implying that a few large clusters tend to dominate these segmentations.

3.2   Dwell and Movement Prediction

The three DBNs described above were trained on sequences of transitions between the locations that were inferred by each segmentation method. To compensate for the bias towards self-transitioning (at virtually every instance, the most likely event will be that the subject does not change locations), we compare the models success only on instances when a subject is about to transition between inferred locations. The DBNs are tasked with predicting the location where the subject is about to move. Table2 lists these prediction accuracies for the three segmentation methods and the two full-observed Markov models. While

Table 1. Segmentation Validation via Bluetooth Beacons. μ_{N_C} is the average number of clusters generated by each segmentation method. $P(C_{BT} \subset C_H)$ represents the probability that the set cellular towers associated the Bluetooth beacon at the subject's home, C_{BT}, is fully contained in a single cluster, C_H. The last column corresponds to the ratio of the actual number of home towers, $N_{C_{BT}}$ to the number of home towers inferred by the different segmentation methods, N_{C_H}. A small number corresponds to incorporating a large number of towers within the home cluster.

method	μ_{N_C} (σ)	$P(C_{BT} \subset C_H)$	$\frac{N_{C_{BT}}}{N_{C_H}}$
Ncuts	20 (0)	.93	.0061
Q-Modularity	13.3 (11.7)	.86	.18
Threshold Groups	6.8 (13.7)	1.0	.045

Table 2. Transition Accuracy and Dwell Errors. For every instance a subject moves between two clusters of towers, the DBN can be used to predict the subsequent cluster. The different accuracies between the segmentation methods are due to not only how well the clustering techniques performed at identifying the true salient locations, but also to the number and size of the clusters (described in Table 1). Given these high accuracies, the inclusion of the temporal observations in the Contextual Markov Chain (CMC) does not appear to provide significant improvement to the standard Markov chain (MC).

method	MC Transition Prediction	CMC Transition Prediction	MC Dwell Error (minutes)	CMC Dwell Error (minutes)
Ncuts	.932	.933	79.1	78.9
Q-Modularity	.953	.954	91.0	75.7
Threshold Groups	.992	.992	89.2	84.1

the X-factor model provides additional information about the regularity of a particular behavior, its accuracy is identical to the contextualized Markov model and was not included in the table. It is of interest that the highest accuracies did not come from the segmentation methods that provided the largest cluster sizes (Ncuts), but rather the smallest number of clusters (Threshold Groups). However, a direct comparison between these accuracies is not possible due to the differences in the dimensionality of the state spaces. A model with fewer inferred locations (N_C) should be expected to do better because it has less potential for a wrong prediction. In the extreme, a model with a single state will always be correct, yet obviously adds little value. Therefore, while the Threshold Groups segmentation method, with an average of 6.8 inferred salient locations ($\sigma = 13.7$), generated accuracies of over 99%, future work in predicting location dwell times may provide more conclusive information about the dominance of one particular segmentation method over the others. Given the extremely high accuracies using an unconditioned Markov model, incorporating information about the time of day and day of the week unsurprisingly adds little additional value.

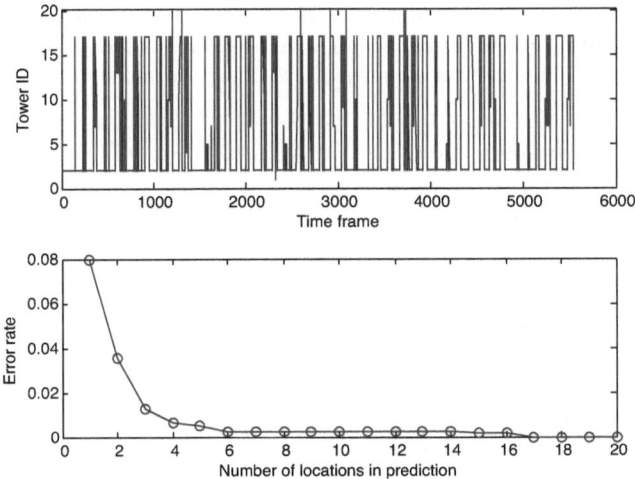

Fig. 5. A sequence of transitions between towers (top) and the average error rates for predicted transitions (bottom). The X-factor model was tested on approximately one month of movement segmented using Ncuts into 20 locations. While the top inferred location is 92% correct for this set of data, the subsequent location is in the top four locations over 99% of the time.

3.3 Entropic Individuals

By calculating the standard Shannon information entropy metric of the learned transition probabilities of the X-factor model, $H = -\sum p \times log_2(p)$, we are able to quantify the amount of behavioral regularity of each subject. The means and variances of this entropy metric are segmented across demographics in Table 3. Of particular note is the high entropy variance, indicating that there are individuals across all demographics whose behavioral patterns are seemingly unstructured. This finding runs contrary to previous research conducted on university students and staff which suggested behavioral entropy is correlated with demographics [3].

3.4 Future Work

This paper has provided the groundwork for the design of increasingly sophisticated models based on data from mobile phones that incorporate contextual and temporal variables and can use demographic priors for bootstrapping. For example, if the discovered Bluetooth devices can be clustered based on co-presence, it may be possible to classify particular Bluetooth phones as family, colleagues, and friends, incorporating the proximity of these individuals as observational variables. Additionally, the phones in this study also sample the ambient audio environment periodically to detect the subjects' media consumption, information that should also make for an intriguing additional observed variable in the DBN. Lastly,

Table 3. Demographic Entropy. The entropy of the conditional probability table from the X-factor model using the Group Threshold method was averaged across demographics. The results show extremely high variance, with entropic individuals in virtually every demographic as well as subjects with significant structure in their daily behavior.

demographic (N)	$\mu_{entropy}$ $(\sigma \times 10^2)$
Age:	
under 35 (107)	30.1 (4.2)
35 and over (108)	28.0 (4.2)
Gender:	
Male (136)	28.3 (4.4)
Female (79)	30.3 (3.8)
Income:	
over $60,000 (73)	34.2 (4.3)
$60,000 and under (140)	26.4 (4.0)
Education:	
College Grad (79)	31.2 (4.3)
No College Degree (125)	27.7 (4.1)

we would like to explore the potential of using demographic bootstrapping to aid in efficient model parameterization as introduced in similar models [12].

We have demonstrated the potential to repurpose algorithms developed originally to quantify community structure within graphs to identify salient locations within a cellular tower network. We have validated these unsupervised clustering algorithms on a known cluster of towers using the Bluetooth beacon installed in each of our randomly sampled subjects' homes. The resultant set of inferred clusters of towers correspond to salient locations and are incorporated as states in our DBN models. We introduced the X-Factor model to detect behaviors that deviate from a given routine by incorporating an additional latent variable corresponding a normal / abnormal switch. By calculating the entropy of the transition matrix from this model we were able to quantify the amount of structure in the daily routines of different demographics. It is our hope that these analytical methodologies will provide a framework for future studies of this rich behavioral data, currently being generated by the majority of humans today.

References

1. González, M.C., Hidalgo, C.A., Barabási, A.L.: Understanding individual human mobility patterns. Nature 453(7196), 779–782 (2008)
2. Onnela, J., Saramaki, J., Hyvonen, J., Szabo, G., Lazer, D., Kaski, K., Kertesz, J., Barabasi, A.L.: Structure and tie strengths in mobile communication networks. Proceedings of the National Academy of Sciences 104, 7332–7336 (2007)
3. Eagle, N., Pentland, A.: Reality mining: sensing complex social systems. Personal and Ubiquitous Computing 10, 255–268 (2006)

4. Chen, M., Sohn, T., Chmelev, D., Haehnel, D., Hightower, J., Hughes, J., LaMarca, A., Potter, F., Smith, I., Varshavsky, A.: Practical metropolitan-scale positioning for gsm phones. In: Dourish, P., Friday, A. (eds.) UbiComp 2006. LNCS, vol. 4206, pp. 225–242. Springer, Heidelberg (2006)
5. LaMarca, A., Chawathe, Y., Consolvo, S., Hightower, J.: Place lab: Device positioning using radio beacons in the wild. In: Gellersen, H.-W., Want, R., Schmidt, A. (eds.) Pervasive 2005. LNCS, vol. 3468, pp. 116–133. Springer, Heidelberg (2005)
6. Davis, M., King, S., Good, N., Sarvas, R.: From context to content: leveraging context to infer media metadata. In: Proceedings of the 12th annual ACM international conference on Multimedia, New York, NY, USA, October 10-16 (2004)
7. Sohn, T., Varshavsky, A., LaMarca, A., Chen, M.: Mobility detection using everyday gsm traces. In: Dourish, P., Friday, A. (eds.) UbiComp 2006. LNCS, vol. 4206, pp. 212–224. Springer, Heidelberg (2006)
8. Laasonen, K., Raento, M., Toivonen, H.: Adaptive on-device location recognition. In: Ferscha, A., Mattern, F. (eds.) Pervasive 2004. LNCS, vol. 3001, pp. 287–304. Springer, Heidelberg (2004)
9. Hightower, J., Consolvo, S., LaMarca, A., Smith, I.: Learning and recognizing the places we go. In: Beigl, M., Intille, S.S., Rekimoto, J., Tokuda, H. (eds.) UbiComp 2005. LNCS, vol. 3660, pp. 159–176. Springer, Heidelberg (2005)
10. Otsason, V., Varshavsky, A., LaMarca, A., de Lara, E.: Accurate gsm indoor localization. In: Beigl, M., Intille, S.S., Rekimoto, J., Tokuda, H. (eds.) UbiComp 2005. LNCS, vol. 3660, pp. 141–158. Springer, Heidelberg (2005)
11. Ashbrook, D., Starner, T.: Using gps to learn significant locations and predict movement across multiple users. Personal and Ubiquitous Computing 7, 275–286 (2003)
12. Liao, L., Patterson, D., Fox, D., Kautz, H.: Learning and inferring transportation routines. In: Proceedings of the Nineteenth National Conference on Artificial Intelligence, pp. 348–353 (January 2004)
13. Shi, J., Malik, J.: Normalized cuts and image segmentation. IEEE Transactions on Pattern Analysis and Machine Learning 22(8), 888–905 (2000)
14. Newman, M.: Modularity and community structure in networks. Proceedings of the National Academy of Sciences (January 2006)
15. Newman, M., Girvan, M.: Finding and evaluating community structure in networks. Physical Review E 69 (January 2004)
16. Clauset, A., Newman, M.E.J., Moore, C.: Finding community structure in very large networks. Physical Review E 70(6) (December 2004)
17. Schwaighofer, A., Grigoras, M., Tresp, V., Hoffmann, C.: Gpps: A gaussian process positioning system for cellular networks. Advances in Neural Information Processing Systems 16 (January 2004)
18. Quinn, J., Williams, C., McIntosh, N.: Factorial switching linear dynamical systems applied to physiological condition monitoring. IEEE Transactions on Pattern Analysis and Machine Intelligence (January 2008)

"It's Just Easier with the Phone" – A Diary Study of Internet Access from Cell Phones

Stina Nylander[1], Terés Lundquist[2], Andreas Brännström[3], and Bo Karlson[4]

[1] Swedish Institute of Computer Science, Box 1263, 16429 Kista, Sweden
[2] Luleå University of Technology, 971 87 Luleå, Sweden
[3] Umeå University, 901 87 Umeå, Sweden
[4] Squace AB, Döbelnsgatan 48, 113 52 Stockholm, Sweden
stny@sics.se, teres.lundquist@gmail.com, c02abm@cs.umu.se,
bo.karlson@squace.com

Abstract. We conducted a diary study of how 19 experienced users accessed the Internet from cell phones. Our data show that participants often chose the cell phone to access the Internet even though they had access to a computer, and the most common location for Internet access being the home. Reasons for choosing the phone over the computer were speed, convenience and a desire to use the phone for fun. Additionally, the phone is kept close and is always on which makes it convenient to use. The traditional motivation for mobile services "finding out something about where you are" only accounts for 15% of the user activity.

1 Introduction

Today, it is becoming more and more common to access Internet services from cell phones. The phone increasingly allows users to access online news, email, and other services in places and situations which were never possible using a traditional desktop computer.

However, the cell phone has several properties that make Internet access different than on an ordinary computer. So far, this has mostly been discussed in terms of restrictions: the small screen size and constraints on typing do limit the usability of the phone as an Internet access device. But there are other less tangible differences. The cell phone is not only mobile, it is also personal and communication-oriented in a way that computers are not. All of these aspects – the interaction abilities, the mobility, and the role of the cell phone as a personal communication device – will affect the usage patterns that emerge for the Internet on the cell phone. To shed light on how this might develop, we conducted a diary study of Internet usage on the cell phone to identify the emerging usage practices.

2 Related Work

Actual Internet access from cell phones has been little studied so far, but Lee et al. [7] have studied use contexts for the mobile Internet and their relation to types of services.

H. Tokuda et al. (Eds.): Pervasive 2009, LNCS 5538, pp. 354–371, 2009.

However, they did not investigate the influence of users' motivations on their mobile Internet use. We believe that it is important to investigate the motivation when examining user behavior. We have also chosen to analyze the locations for mobile Internet use in a deeper way than Lee et al.. For example, Lee et al. only analyzed indoor and private locations; they did not look at the home separately. In our material the home turned out to be an important location. Additionally, they did not look at to what extent their participants had access to a computer at the time of mobile Internet access. Our participants often chose the phone even though they had access to a computer.

Sohn et al. [12] have also investigated people's information needs when mobile, and how they met those needs. They found that people that had access to mobile Internet used it to meet many of their information needs, and those who did not have access to mobile Internet believed that it would have solved many of their information needs. Our material adds to this picture by showing that mobile Internet access also is used to meet information needs in many situations that are not mobile.

3   Method

The data for this study was collected using two methods. First, participants kept a diary of their Internet access from their cell phones for seven days and then were interviewed about their Internet habits in general and the study week in particular.

The diary method was chosen as a relatively unintrusive method to capture mobile Internet usage over extended periods of time during the course of everyday life. It has been used frequently to study usage of cell phone technology, see e.g. [4, 5], other mobile technologies [11], and mobile behavior [12]. The potential drawback of the method is that data from self-reporting is not always accurate and complete. However, an advantage is that diary entries act as triggers for reflection, generating rich narratives grounded in real life events which can be unpacked in the follow-up interview. We chose not to complement our diary data with log data from participants' phones since our main interest was in these rich narratives and in what they revealed about user motivation and context of use, little of which could be captured by automated logging. In addition, the geographic spread of our participants would have made it cumbersome to install such software on their phones.

A pre-printed paper diary that had several examples completed was sent to the participants by mail. When participants could be expected to have received this paper diary we called them to confirm that they had received it and to clarify any problems. At that point, a time for the follow-up interview was scheduled. Participants were provided with preprinted, stamped envelopes to return the log to the authors.

The participants were asked to log every session of Internet access from a cell phone during seven consecutive days. That is a fairly short time to monitor user behavior but we did not want to burden our participants unnecessarily and risk that they would under-report their Internet use. Since they were frequent users we believe that we gathered sufficient data from the seven day period. The paper diary contained fields for time and duration, web page or application, location, surrounding distraction, concurrent activity, purpose, why the phone was chosen and not a computer, and if any problems occurred during the session, (see Figure 1 for an example). For each day of the trial, they were also asked to give examples of web pages that they used on the computer and not on the cell phone.

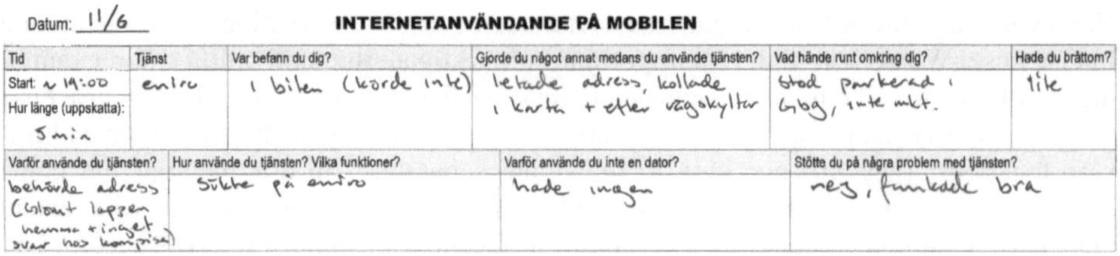

356   S. Nylander et al.

Fig. 1. The diary was designed in Swedish. The entry fields were named as follows. Upper row from the left: Time, Estimated duration, Service, Where were you?, Did you do anything else while using the service?, What was going on around you?, Were you in a hurry?. Lower row from the right: Why did you use the service? How did you use the service? What functionality?, Why did you not use a computer?, Did you encounter any problems with the service?.

The follow-up interviews were semi-structured and conducted shortly after the seven days of data collection. Questions were asked about the information users had provided in the diaries, together with more general questions about their Internet habits. Since the participants were recruited over the Web, the geographic spread made it necessary to conduct phone interviews. The interviews were conducted in Swedish by two of the authors they were approximately 30 minutes long.

The analysis of both the diary and the interview material was guided by grounded theory [13] and consisted of categorising the raw data accordingly. We looked for repeating themes as well as conflicting ones and did not start out with fixed categories.

3.1 The Participants

Participants were recruited through a small web survey that was published on web forums with mobile technology themes, as well as blogs with many visitors. This survey was also distributed to friends and colleagues of the authors, in addition to a large mailing list dedicated to mobile web surfing. Based on this initial response, we were able to recruit participants for the diary study. The participants selected for the diary study all stated in the survey that they accessed the Internet daily from their cell phone. The main reason for limiting the diary study to people who already were using the Internet on the phone was that they had already developed a set of usage practices. That way, we avoided novice and learning effects in the study and were also assured of gathering sufficient occurrences of Internet usage to draw conclusions from the data.

Nineteen participants were recruited for the study, seven women and twelve men with an average age of 30 years (max 55, min 20, mean 28). See Table 1 for basic data about the participants.

Two of the participants could only report the make of their phone but not the model. Of the 19 participants, seven had phones with a touch screen, and of these, two participants had hacked iPhones (the iPhone was not yet released in Sweden at the time of the study). The majority of the participants were working, but two were students and one was unemployed. Five participants reported that their Internet access from the phone had a connection to their work.

Table 1. Basic data about study participants

Id	Age	Gender	Work	Mobile phone	Net speed
P1	55	M	Electricity Engineer	Nokia N95	3G
P2	35	M	Truck driver	SE P1i	3G
P3	41	M	Consultant	SE W910i	3G
P4	32	M	Teacher	SE K800	3G
P5	20	M	Unemployed	Nokia N95	3G
P6	31	M	Interaction designer	Nokia N73	3G
P7	37	M	Management	SE T650i	3G
P8	22	M	Student	HTC Touch Cruise	Turbo 3G
P9	21	F	Nurse	SE	3G
P10	24	F	Student	iPhone	3G
P11	27	F	Sales	SE K800i	3G
P12	45	F	Telecom Research	SE K800	3G
P13	28	M	Project Managemnt	SE P1, SE W660i	3G
P14	28	F	Interaction designer	SE W960i	3G
P15	20	M	Industrial mainten.	HTC Cruise	3G
P16	29	F	Secretary	SE W850i	3G
P17	26	M	Student	SE W850i	3G
P18	26	F	Web Designer	iPhone	2G
P19	27	M	Dev. engineer	SE	3G
n=19	30,2	12M, 7F			

Our participants were in general fairly technologically savvy. They had high-end cell phones and many of them had tried, and succeeded to, install applications on their phones such as Opera's Mini browser or Gmail's Java application. Two participants even had iPhones that were not released in Sweden at the time of the study but needed to be hacked to function with a Swedish carrier. Using the cycle of market adoption (described in [10]), we would categorize them as early majority since they, in the interviews, described their cell phone Internet access with a utility perspective rather than mainly for the enjoyment of the technology itself. They liked what they could do with Internet access from the phone, rather than finding pleasure in mastering the technology.

Anne and Tom are two typical though very different examples of our participants. Anne is 45 years old, works in telecom research and her employer pays for her cell phone. She has a SE K800i phone and surfed 15 times from her phone during the seven days of the study. She mostly used her phone for Internet access during early mornings, and late evenings, and she primarily accessed news pages and the pages of her children's sports teams. Tom is 35 years old, works as a truck driver and pays for his cell phone himself. He has a SE P1i phone and surfed 12 times during the study. He mostly used his phone during the day to access news pages, map pages (to find customers' addresses), and his Internet bank.

Bob was not a typical participant, surfing an amazing 40 times during the study. He is 55 years old, unemployed and pays for his cell phone himself. He has a Nokia N95 which he used solely for web surfing (he has another cell phone for calling). He used his phone primarily in the morning accessing mainly web email and news sites.

4 Results

4.1 Overview of the Data

In all, 260 occurrences of Internet access from cell phones were recorded in the diaries, with an average of 13.6 per participant (max 40, min 4). Each participant reported on average 1.9 occurrences per day, (max 9, min 0). Eighty-four different Web sites were visited, and five online Java applications were used.

The average duration for a session was 12.6 minutes (max 180 min, min 0.5 min). Fifteen sessions were reported to last for 60 minutes or longer.

In 3.8% of the occurrences, participants reported that the service they tried to access did not work or that they gave up waiting for it to load. We have included these sessions in our analysis of user motivation, access to computer, and location since they do not depend on the success of the session. Most of the unsuccessful sessions did not take place in the vicinity of a computer and therefore did not allow participants to attempt to access the Internet via other convenient methods.

Table 2. Categories of services used during the study with no overlap between categories

Type	
News	27.69%
Mail	21.15%
Info site	15.77%
Travel/contact info	14.62%
Blog/forum	7.69%
Transactions	5.38%
Media	4.62%
Chat	2.69%
Download	0.38%
Sum:	100.00%

Table 2 gives an overview of the kinds of services used by all participants over the course of the study. The most frequently accessed web page in our diary data was Aftonbladet.se, a Swedish newspaper, accounting for 10% of the entries in the diaries. Gmail was the most frequently used Java application, accounting for 1% of the entries. In total, news was the most frequently used service category on the cell phone, followed by email.

4.2 Services Accessed from Both Cell Phones and Computers

In the interviews we asked participants which services they used both on cell phones and computers. News, email, and travel information were the three service groups that were mentioned most often as being used on both devices, each by nine participants.

Nine participants reported in the interview that they read news both on computers and cell phones, and eight of them read news from the cell phone during the study.

Travel information and other contact information was a service category that was also used both from computers and cell phones by nine participants. The most common service referred to was eniro.se, an online phone book/yellow pages service that also provides maps. Five of the nine participants that said they used travel or contact information retrieval services from both phone and computer used such a service from the phone during the study.

Email was the last of the three top services that were used both on cell phone and computer, also being mentioned by nine participants. All but one of them used email from cell phone during the study.

Seven participants reported using blogs or forum services from both computers and cell phones. Five of them had reported accessing blogs or forums from the cell phone during the study.

Online banking was reported by four participants as a service they used both from computer and cell phone. All four of them used their Internet bank from the cell phone during the study.

Several of our participants reported that all the online services that they used on their cell phone, they also used on the computer.

"Everything I use on the phone, I use it on the computer too." (P17)

One of them described accessing common services from both cell phone and computer, while services he only used occasionally were accessed from the computer. He also checked new services out on the computer before using them on the cell phone.

"I use all the regular services both on the phone and the computer, but services I only use once in a while or more by impulse [I use] on the computer." (P2)

In some cases, participants reported that they used a service on both computer and cell phone but different service functionality. Six participants report that they only read email on the phone and saved the writing part for the computer. Two participants report the same behavior on blogs (i.e. they only comment from the computer) and Facebook.

"Well, email and calendar I only read on the phone. On the computer I write too." (P13)

Differences between how participants use the services on the computer and on the cell phone can also originate from the fact that services provide different functionality on the devices. Our participants reported this for example on the betting site svenskaspel.se.

"On the mobile page of Svenska Spel there are fewer kinds of bets you can place, less to choose from." (P4)

Restricted functionality on cell phones are sometimes due to lack of support for security procedures. Our participants reported that they could not pay bills through their online bank from the phone because the security procedures were not compatible.

"Since, on the phone, you can't log in with the little box [that generates a secure code] so you can't pay." (P16)

The set of services our participants reported using on the cell phone was a subset of the services they use on the computer. Less functionality was used on the phone, in some cases because services offered restricted functionality on the phone, in other cases because interacting with the service on the phone was considered too cumbersome.

4.3 Services only Used from One Device

Participants only logged their habits of Internet access from cell phone over seven days, additionally providing examples of services they had used from the computer. This of course does not present a full picture of their Internet behavior but we would still like to note several details about the services that were only used from one device during the study.

Examples of services that were reported as only having been used from the computer during the study period were ikea.se (selling home furnishing products both online and in stores around the world) and hemnet.se (advertising real estate available in Sweden). These two sites share a heavy reliance on pictures to carry important information; nobody would be interested in buying a product for the home that they could not figure out how it looked from the picture, not to talk about a house. For these types of tasks, the larger screen of the computer is very important both to convey picture information and to facilitate collaboration.

"If I am looking for a place to stay or a new car where it's necessary with larger pictures it's better to use the computer. I avoid that stuff on the phone." (P7)

In addition, our examples of services that are only used on the computer concerns matters that often include a large amount of comparison and thought before a purchase is made. Usually, some time and effort is put into the purchase of a house or furniture and the process of doing that might be better suited both to the technical capabilities of a computer and to the home environment. It is easier to switch between different sites on the computer and longer sessions on a small device outdoors or in transit would be cumbersome.

"I feel like it's easier to evaluate different alternatives [on the computer]" (P13)

Two services were reported as only used from cell phone in this study, BBC World RSS and Jaiku. The two services were used by different participants, P11 and P6 respectively. BBC World is a news service and Jaiku is a microblogging service which allows users to update their status and follow their friends' status from the cell phone as well as a web page. What these services have in common is that the information they contain is frequently updated, which seem to be a characteristic of many of the services that were accessed frequently from cell phone in this study. The cell phone can satisfy users' need for constant updates, or simply sate their curiosity. It is highly possible that something new has arrived since they last checked, even if that check took place just minutes ago.

In the interviews, we asked participants if there was anything concerning Internet access or Web surfing that worked better on the cell phone than on a regular computer. Many participants could give examples of things that worked better on the phone, but no one preferred the phone over the computer in general. Two participants reported that it was quicker to access their email on the phone than on the computer, a fact that they greatly appreciated.

"You get to the email really fast with a single button press instead of going to the computer and then go to the Yahoo home page." (P10)

Two participants mentioned location-based services as being more meaningful on the phone than on the computer, and one of them appreciated the Google Maps functionality that shows your position on the map. That functionality is only available on the phone.

"Google Maps know where I am on the phone." (P6)

One participant also reported that the cell phone version of many web pages is not affected when the desktop version has trouble or is down.

"Sometimes the desktop version is down but the mobile version is up." (P3)

Our purpose has not been to provide a full comparison between participants Internet use on computers and cell phones but our material suggests that visual and explorative services tend not to migrate from the computer to the phone, while frequently updated information and/or communication services do. Finally, several participants stated that the phone is not better than the computer, but more convenient.

4.4 User Motivation

Much previous research on (and commercial development of) mobile services has focused on specifically mobile situations such as travelling or walking in a city [1], as well as providing information connected to such situations, information that is often tightly connected to a user's location [6, 8]. In our data, we certainly found that kind of use, such as cases where our participants were in a location and needed to know something specific about it (for example exactly from where their bus would leave). This sort of use was most common in outdoors situations (25% of the outdoors occasions) but less frequent in the data as a whole (15%).

However, for the majority of cases where the Internet was used on the phone, users were not mobile and did not search for information that had to do with their situation. For example, the most common motivation for Internet access was reading news (see Table 3), and this was mostly done at home.

Table 3. Motivations reported for Internet access from cell phone

Purpose	
Reading news	20.00%
Passing time	19.23%
Checking email	16.54%
Situated info search	15.77%
General info search	15.00%
Transactions	5.77%
Other	5.77%
Troubleshooting	2.31%
Sum	100.00%

In fact the top three purposes or motivations were, *Reading news* (20%), *Passing time* (19%), and *Checking email* (17%), none of which were connected to mobility or the current situation (see Table 3 for details).

The category *Situated information search* (15%) contained occurrences where participants used mobile Internet access to find information about their current situation or activity. Examples were checking the location of the bus they needed to catch, checking if it was necessary to run to catch the right bus, or finding out exactly where a meeting would take place. Most of these occurrences took place under time pressure.

"To see if I had to run." (P17)

Participants reported that the services they used on the phone for their situated information searches were also used on the computer. However, they often had a different purpose when they used maps or directory services on the computer. While the use from the phone was situated, the use from the computer was more connected to planning and finding information that would be needed for future travel.

"On the phone it's more 'I wanna go from here where I am right now' while on the computer it's more 'tomorrow I need to go to...' " (P13)

The category *General information* search (15%) contains occurrences where participants searched for information not specifically connected to their current location or activity. Examples include checking next week's practice schedule for their kids' teams, checking tomorrow's weather, or finding information about a vacation destination.

"I checked my son's practice schedule for this week" (P12)

The category *Transactions* (6%) contains those occurrences that concerned financial matters. We did not have any cases of participants using their cell phones to purchase something over the Internet but we did have examples of placing bets, booking tickets, moving money between accounts, and checking account balances.

"Moved money from one account to another." (P16)

We also had a number of occurrences that concerned troubleshooting the stationary Internet. Those occurrences took place under special circumstances, all of them by a participant whose stationary Internet connection at home was malfunctioning during the study. He used his cell phone to access the web page of the ISP to find out if there was a general problem or if the problem was with his connection. Even if this does not seem to be a representative use of the mobile Internet we note that since most users have different ISPs for computer and phone, the phone can actually be used to make the computer work.

"Troubleshoot the Internet... again!" (P18)

As shown in 4, in the 3% of the occurrences when participants chose a cell phone over a computer, they simply stated that they picked the phone because they wanted to use it. A closer inspection of those occurrences shows that participants sometimes found it more fun to use the phone for Internet access.

"It's quicker and more fun with the phone" (P10)

The motives for accessing the Internet from cell phones are thus connected to users' specific location and activity in only 15% of our material. The remaining 85% had other motivations.

4.5 Access to Computer

As described above, only a small part of our data concerns the use of location based services or otherwise situation based Internet use. To further investigate this, we looked at participants' access to computers. Certainly, many instances of Internet access from cell phones took place because participants did not have access to a computer. However, in 51% of the occurrences where they used phones to access the Internet, they also had ready access to a computer. Fifteen of our participants reported in their diaries that they had used a cell phone to access the Internet even though they had a computer available.

The most common reasons stated for choosing the phone over the computer were speed and convenience (24%). This explanation was particularly common in cases where mobile Internet access took place at home, where all participants had access to a computer (with the exception of one participant with malfunctioning broadband connection during the study, see Table 4).

Table 4. Reasons stated for choosing to access the Internet from a cell phone. No overlap between categories.

Reason for cell phone	
No available computer	49.23%
Convenience	23.85%
Laziness	10.77%
Other	5.38%
Internet not working	5.38%
Wanted to use phone	2.69%
Restrictions at work	1.54%
Computer occupied	1.15%
Sum	100.00%

Participants found it quicker and easier to perform certain tasks on the phone since they had the phone readily available, and often had the page they wanted to access bookmarked. Due to those factors, they needed little time and few key presses to get what they wanted. In many cases participants reported that if they would have used a computer they would have had to take it out of a bag and start it up, which was considered too cumbersome and slow.

"The cell phone was more convenient since I was in the kitchen." (P17)

A special case of convenience in our data was laziness (11%), which participants mainly stated as a reason for choosing the phone over the computer when they were at home. Our participants reported that they used the cell phone to access the Internet because they did not have the energy to get up from the TV couch, or because they did not feel like getting out of bed.

"Didn't have the energy to go downstairs to the computer." (P12)

The phone also fit better with other concurrent activities such as household chores, brushing teeth, feeding children, and watching TV.

In 3% of the occurrences, participants stated that they chose the phone over the computer because they specifically wanted to use the phone.

"Wanted to see how the blog looked in the phone." (P1)

In some of the occurrences this had to do with data stored in the phone, such as wanting to send a picture taken with the phone's camera or storing a shopping list in the phone. Preferring to create data or manipulate data on the device where it is stored is well in line with the findings from Dearman & Pierce [2] where users reported that managing information over multiple devices and transferring data between devices was a frequent source of trouble.

"I wanted the shopping list to be stored in the phone, my memory is so bad." (P16)

Interestingly though, some participants were not aware of the fact that they sometimes did use their phone for Internet access even though they had access to a computer. Four participants reported in the interviews that they never used a cell phone to access the Internet if they had access to a computer, but two of these still reported in the diaries that they did. One of them (P16) accessed a recipe site from the phone at home to create a shopping list and bring it to the grocery store. Alternately, P19 used a phone to read news twice very early in the morning and did not want to use the computer because that would have awoken other family members. Two participants reported in the interviews that they would not use a phone if they had a computer unless the circumstances were special

"Well, I guess it could happen at work where everyone can see my screen." (P14)

Both participants that stated that they would only choose the phone over the computer under special circumstances did so during the study. One of them (P17) had five occasions in the diary where the phone was chosen over a computer due to convenience and speed. Those occasions took place at home or in transit. P14 had two occasions in the diary, one where the phone was chosen for privacy, and the other because the information was stored in the cell phone.

As described above, our participants often chose their cell phone to access the Internet even though they had access to a networked computer. This suggests that cell phones offer advantages, such as always being close at hand and being quick to connect to the Internet. These advantages seem to outweigh their low bandwidth and tiny screen. In our material, this seems to be especially important in private situations such as the home which will be discussed below (see Places for Internet access from cell phone). We also believe that participants found that the cell phone integrates better with other activities that they are engaged in, an observation which will also be discussed later (see Activities combined with the cell phone Internet access).

4.6   Places for Internet Access from Cell Phone

Our participants recorded the location of their mobile Internet access in their diaries. An analysis of the locations shed further light on the Internet use that was not location-based or otherwise situated. The classification of the locations is shown in Table 5.

Table 5. Locations for Internet access from cell phone, no overlap between categories

Location	
Home	30.65%
Outdoors	23.37%
In transit	22.61%
Indoors	15.71%
Work	7.66%
Sum	100.00%

Much to our surprise, *At home* turned out to be the most frequent location for mobile Internet access (31%) even though all participants had a computer with Internet connection in their home. This suggests that, for our participants, the cell phone has its own role as a device for Internet access; it is not only used in situations where it is impossible to get computer access. The mobile world extends well into the home. This is in line with the results of O'Hara et al. [11] who found it common for their participants to watch video on mobile devices at home. However, they did not report how common the video consumption in the home was compared to other locations. The most common purposes for cell phone Internet access in the home were *Reading news* and *Checking email*. Interesting to note is that *Passing time* was a less common purpose at home (11%) than in the material as a whole (see Table 3 for motivation details).

A total of 23% of mobile Internet access during the study took place *Outdoors*. This category differs from the other location categories in that the most common purpose for mobile Internet use was *Situated information search*. This was due to situations such as looking for a restaurant while walking in a city and walking (or running) to bus stops or train stations being classified as outdoors situations. We also note that *Passing time* and *Checking email* was not as common as in other locations. It is worth mentioning that this study was conducted in Sweden in May, when the weather is fairly warm. We can speculate that if the study had been conducted during the winter, participants would probably have used their cell phones less outdoors due to the cold.

Not surprisingly, *In transit* was a common situation for mobile Internet access (23%). Our participants used the Internet from their cell phones in buses, subways, trains, tramways, taxis and cars (even while driving). The most common purposes for in transit situations were *Reading news* (27%), *Checking email* (22%), and *Passing time* (22%). Both types of information search were less common in transit than as a whole (8%) which is interesting. Our participants did not seem to search very much for information that pertained to with their traveling or their destination while in transit.

The *Indoors* category (19%) contains occurrences of mobile Internet access that took place in indoors locations that was not the participants' home or workplace, e.g. stores, cafes, train stations, and at friends' houses. *Checking email* and *Passing time* were the main purposes for the *Indoors* mobile Internet access, while *News reading* was not as frequent as in the other locations.

Our participants rarely used their phones for Internet access at *work* during the study; only 8% of cases (20 occurrences) were reported in the diaries and they mostly concerned *Checking email*. In total, only five participants reported accessing the Internet from cell phone at work. We had participants with no computer access at work, for example a truck driver and a nurse, and they used their phones for Internet access. In the interviews, six participants reported that their cell phone Internet access had connections to their work, but only two of them used Internet from their phone at work during the study. One reason for this might be that the phone is a tool for them to stay connected to work while they are not physically at work. Those who did not state that their cell phone Internet access had anything to do with work but used the phone to access Internet at work during the study reported work restrictions on web surfing (two participants) and that the phone integrated better with their work tasks and did not cause interruptions in the work (one participant) as motivation. Other reasons for the low frequency of mobile Internet access at work are the fact that people have little time during working hours, that many of our participants spend their working time in front of a computer, and that mobile Internet use mainly related to leisure activities.

4.7 Activities Combined with the Cell Phone Internet Access

Another aspect to examine is the extent to which participants reported being engaged in other activities while accessing the Internet from cell phones. This is an important part of the use context that helps explaining the non-mobile use found in our data.

A large portion of the reported occurrences of mobile Internet access is not combined with any other activities. In 24% of the occurrences, participants reported doing nothing else, and in 13% of the occurrences they reported doing very passive things such as resting or enjoying the sun (see Table 6). This is well in line with the top reasons for choosing the phone even if there was a computer available, *Convenience* and *Laziness* (see the section on Access to Computer).

Table 6. Activities that users reported they were engaged in during their Internet access

Type	
Doing nothing	24.47%
Home activities	18.09%
Relaxing	13.48%
Consuming media	9.22%
Walking	9.22%
Travel	7.45%
Sending SMS or calling	4.26%
Socializing	4.26%
Other	4.26%
Work	2.84%
Shopping	1.77%
Driving	0.71%
Sum	100.00%

Attending to various *Home activities* accounted for 18% of the concurrent activities. Participants reported accessing the Internet from their phones while doing laundry, eating, brushing their teeth, and even using the toilet. For example 6% of the occurrences took place while participants had breakfast, a time when they used the phone to find information they needed to start the day such as bus schedules or weather information for deciding what to wear.

"Had breakfast and wanted to know when the bus or subway was leaving." (P8)

We believe that the cell phone gives users local mobility [9] which makes it possible to combine moving around the home and attending to various chores with Internet access, as also has been shown in [3]. Users are not tied to a specific place but can go about their business and move freely in the home while having the Internet with them.

A total of 9% of the occurrences took place while participants were *Consuming media* such as watching TV or movies, reading books or listening to music, both live and on television. Many of these occurrences took place in the home, except for the live events (a live concert and a soccer game).

"Watching a movie on the TV." (P4)

Walking was reported as a concurrent activity with mobile Internet access in 9% of cases. These occurrences took place mostly outdoors and were described by participants as, for example "walking back to work from lunch" or "just walking and enjoying the sun". In those cases we cannot be entirely sure if participants were actually walking when they accessed the Internet or if they were walking but stopped to interact with the phone. In the interviews, several participants reported that they found it difficult to interact with the phone while walking, and comments about that were also made in the diaries.

"[I] had to stop to use it [the phone]". (P17)

Traveling by bus, train, tram or car was a common concurrent activity (8%), not surprisingly since *In transit* was one of the most common locations for cell phone Internet access. We only had two instances where participants admitted that they accessed the Internet while driving but we suspect that in some cases participants reported being in a car doing nothing else even though they were driving. The participants that reported driving while accessing the Internet did seem to feel bad about it. There is an ongoing debate in Sweden about the connection between talking on a cell phone while driving and accidents, and maybe participants did not want to admit that they did that.

"Driving the truck :-(" (P2)

Our participants also accessed the Internet from their cell phones while *Socializing*. They reported talking to other people or being in social situations with friends of family in 4% of the occurrences. We even had three cases of talking on the phone and surfing at the same time.

"Celebrated a friend's birthday. Drinking beer." (P8)

As described above, participants reported little use of phone Internet access at work. They also rarely reported work as a concurrent activity to the Internet access which means that they did not use the phone for Internet while they were working outside their workplace either. We only have one occurrence of accessing the Internet from the phone while working at home. This might be explained by the fact that most of our participants spent their working day in front of a networked computer and did not need the phone for Internet access. Another factor could be that the mobile Internet mostly concerned leisure activities and thus took place out of working hours or in breaks in work.

5 Discussion

Our data suggest that Internet access from cell phones certainly fits the traditional interpretation of "anytime, anywhere" usage, i.e. the cell phone is used to access the Internet in situations where users have no computer access such as outdoors and otherwise on the move and need quick information about their location. We found that 46% of the occurrences in our diary material took place *Outdoors* (23%) or *In transit* (23%). However, only 16% of the total number of occurrences was *Situated information search* reported as the purpose of the Internet access. This means that Internet access from cell phones certainly covers situated information search in traditional mobile situations, but it also covers much more. The most common place for Internet access from cell phone in our material was the *Home* (31%), which suggests that the mobile world does not only embrace users in transit and otherwise on the move, but also extends into the home. In other words, it is almost as likely to be used on the couch in front of the TV as it is on the move. In 84% of occurrences, participants also reported purposes for their Internet access other than situated information search. We believe that this opens up the design space for mobile services to include much more than the location-based service paradigm that has dominated the discussion on mobile services for the past few years.

One of the main explanations for why our participants used their phones to such a high degree for Internet access even though they had access to computers is clearly described in the diaries. Repeatedly, participants report that they chose the phone even though they had a computer because it was more convenient with the phone. Usually, participants had the phone in a pocket or otherwise close by so they could pick it up and access the Internet without having to drop what they were doing or leave the place they were to go to a computer. They also reported that the phone was always on and quick to connect compared to a computer that needed to start up or that was stored in a bag. Quick access and convenience made our participants choose a cell phone over a computer in 51% of our diary occurrences, and thus seem to have compensated for the hardware limitations of the phones such as screen size, key pad restrictions and network speed. As one of the participants put it

"Nothing [to do with Internet access] is better with the phone, but more convenient." (P15)

The cell phones also allowed participants local mobility, i.e. moving around for example in the home, making it possible for them to integrate Internet access with other activities. This was particularly obvious for the occasions that took place in the home, where users often combined Internet access with chores such as doing laundry or cooking, social activities such as watching TV with other family members, or brushing their teeth. All these activities either require certain mobility within the home, or occur in places where it is uncommon to have a computer. From this perspective, it would be interesting to know how many of our participants had laptop computers and wireless Internet connectivity in their homes. That would allow us to determine if the cell phone was the only device at home that allowed them local mobility. Unfortunately, we do not have that information about our participants. However, the cell phone does allow more local mobility than, for example, the laptop computer, since it does not need a flat surface to place it on to interact with it.

As described in the findings, the average Internet access session from a cell phone during this study was 12.6 minutes, and many of the sessions were as short as 30 seconds. However, we found it interesting that even though many sessions were very short, participants only reported being in a hurry in 20% of the occurrences. We believe that this is connected to what has previously been described as the main advantages of the cell phone when it comes to Internet access: *speed and convenience*. Our participants did not use Internet on their phones for short time spans because they were in a hurry, but because the phone works very well to quickly check email or read a few news headlines. If there is a minute or two to spend surfing instead of waiting or simply being bored, the phone does is ideal.

Participants described differences between their Internet access behavior on the cell phone and on the computer. They reported that they were not "sucked into" the phone in the same manner as with the computer. They reported that when they sat down in front of the computer to do something quick on the Internet the session could easily extend to an hour without them noticing it. This hardly ever happened when they accessed the Internet from the cell phone. They described it sometimes as "it is much easier just clicking around on the computer than on the phone". We believe that the main reasons for these differences are that the computer has much more visual power to capture users' attention and also that the computer is used in environments that in general are calmer and less invasive that the environments where the phone is used. We believe that this is true also for the use in the home, where it is common that a desktop computer is placed in a study or in a bedroom while our data shows that the phone is used in front of the TV with the rest of the family, in the kitchen and in other more distracting situations.

In this study, it seems that what the most popular and most frequently used services from a cell phone have in common is that they are updated continuously, such as email and news sites. The capacity of these services to trigger users' curiosity and (perceived) need for constant awareness seem to combine very well with the capabilities of the cell phone.

6 Conclusion

We have reported on a diary study of Internet access from cell phones with a primary finding that 51% of the reported occurrences took place in locations where participants had access to a computer but still chose a cell phone. Additionally, the most frequent location for mobile Internet access was the home. Even though these results need to be validated in further studies, this suggests that, for frequent users such as our participants, the mobile world extends well into the home and that the cell phone has its own role as a device for accessing the Internet. For them, the phone is not a mere backup solution for when there is no computer available, but a tool that often provides quicker and more convenient service than a computer. Moreover, the local mobility provided by the phone allows users to integrate their Internet access with other activities such as home chores or social activities.

Our participants also stated the typical mobile example "finding out something about the location you are in" as a reason for accessing the Internet from a cell phone, but those situations only constitute 15% of our material. The remaining 85% suggests that there is room for a wide range of on-line services for cell phones that are not specifically designed for mobile situations or connected to users' immediate activities.

Acknowledgements

We would like to thank our participants for taking the time and effort to complete the diary study, and Squace AB that supported the study in many ways. This research was funded by the Swedish Governmental Agency for Innovation (Vinnova).

References

1. Bellotti, V., Begole, B., et al.: Activity-based serendipitous recommendations with the Magitti mobile leisure guide. In: Proceedings of CHI, Florence, Italy, pp. 1157–1166 (2008)
2. Dearman, D., Pierce, J.: It's on my other Computer: Computing with Multiple Devices. In: Proceedings of CHI, Florence, Italy, pp. 767–776 (2008)
3. Dobashi, S.: The Gendered Use of Keitai in Domestic Contexts. In: Ito, M., Okabe, D., Matsuda, M. (eds.) Personal, Portable, Pedestrian. Mobile Phones in Japanese Life, pp. 219–236 (2005)
4. Grinter, R.E., Eldridge, M.: y do tngrs luv 2 txt msg? In: Proceedings of European Conference on Computer-Supported Cooperative Work (ECSCW), pp. 219–238 (2001)
5. Hinman, R., Spasojevic, M., et al.: They Call it Surfing for a reason: Identifying mobile Internet needs through PC deprivation. In: Proceedings of CHI, Florence, Italy, pp. 2195–2207 (2008)
6. LaMarca, A., Chawathe, Y., et al.: Place Lab: Device Positioning Using Radio Beacons in the Wild. In: Gellersen, H.-W., Want, R., Schmidt, A. (eds.) Pervasive 2005. LNCS, vol. 3468, pp. 116–133. Springer, Heidelberg (2005)
7. Lee, I., Kim, J., et al.: Use Context for the Mobile Internet: A Longitudinal Study Monitoring Actual Use of Mobile Internet Services. International Journal of Human-Computer Interaction 18(3), 269–292
8. Ludford, P.J., Frankowski, D., et al.: Because I Carry My Cell Phone Any-way: Functional Location-Based Reminder Applications. In: Proceedings of CHI 2006, Montréal, Canada, pp. 889–898. ACM Press, New York (2006)

9. Luff, P., Heath, C.: Mobility in Collaboration. In: Proceedings of CSCW, Seattle, WA, pp. 305–314 (1998)
10. Norman, D.A.: The Invisible Computer. MIT Press, Cambridge (1999)
11. O'Hara, K., Slayden Mitchell, A., et al.: Consuming Video on Mobile Devices. In: Proceedings of CHI, San Jose, CA, pp. 857–866 (2007)
12. Sohn, T., Li, K.A., et al.: A Diary Study of Mobile Information Needs. In: Proceedings of CHI, Florence, Italy, pp. 433–442 (2008)
13. Strauss, A., Glaser, B.: The Discovery of Grounded Theory. Strategies for Qualitative Research. Aldine (1967)

Does Context Matter ? - A Quantitative Evaluation in a Real World Maintenance Scenario

Kai Kunze[1], Florian Wagner[1], Ersun Kartal[2], Ernesto Morales Kluge[3],
and Paul Lukowicz[1]

[1] Embedded Systems Lab, University of Passau,
Innstr 43, 94032 Passau, Germany
www.wearable-computing.org, www.wearcomp.eu
first.lastname@uni-passau.de
[2] Carl Zeiss AG, Konzernfunktion Forschung und Technologie (KFT-TV)
73446 Oberkochen, Germany
[3] Bremer Institut für Produktion und Logistik, University of Bremen,
Hochschulring 20, 28359 Bremen, Germany
www.biba.uni-bremen.de
mer@biba.uni-bremen.de

Abstract. We describe a systematic, quantitative study of the benefits using context recognition (specifically task tracking) for a wearable maintenance assistance system. A key objective of the work is to do the evaluation in an environment that is as close as possible to a real world setting. To this end, we use actual maintenance tasks on a complex piece of machinery at an industrial site. Subjects for our study are active Zeiss technicians who have an average of 10 years job experience.

In a within subject Wizard of Oz study with the interaction modality as the independent variable we compare three interaction modalities: (1) paper based documentation (2) speech controlled head mounted display (HMD) documentation, and context assisted HMD documentation. The study shows that the paper documentation is 50% and the speech only controlled system 30% slower then context. The statistical significance of 99% and 95% respectively (one sided ANOVA test). We also present results of two questionnaires (custom design and standard NASA TLX) that show a clear majority of subjects considered context to be beneficial in one way or the other. At the same time, the questionnaires reveal a certain level of uneasiness with the new modality.

1 Introduction

Since early conceptual work on the use of context in pervasive systems (e.g. [5, 14, 11]) much research aims at implementing context and activity recognition systems. Interestingly, researchers devoted little work to a systematic, quantitative evaluation of the benefit such systems bring to diverse applications. This paper presents such a systematic, quantitative evaluation.

We focus on the domain of wearable maintenance systems. Many such systems are proposed and implemented since the early days of wearable computing (e.g. [1, 13, 15, 16]). These systems aim to provide maintenance personnel with access to complex

H. Tokuda et al. (Eds.): Pervasive 2009, LNCS 5538, pp. 372–389, 2009.

electronic information with as little interference as possible to the primary task at hand. Typically, they rely on head mounted displays (often with augmented reality), input modalities that minimize hand use (e.g. speech, special gloves) and interfaces that focus on reducing the cognitive load on the user.

It is widely believed that wearable maintenance systems can benefit from automatic work progress tracking. Main uses for such tracking are 'just in time' automatic delivery of information (seeing the manual page you need without having to explicitly demand it), error detection (e.g. "you forgot to fasten the last screw"), and warnings (e.g. " do not touch this surface"). In this paper we present a quantitative, statistically significant benefits evaluation of such functionality in a real industrial setting. The study involves 18 real technicians, most with over 10 years of job experience, doing 3 different, real maintenance task on a complex piece of industrial machinery. We compare three types of systems: (1) paper based documentation, (2) documentation displayed on a head mounted display controlled by a speech interface, and (3) context controlled documentation displayed on a head mounted display. Both speech recognition and context recognition are simulated using the Wizard of Oz technique to ensure perfect system performance and avoid system quality related artifacts. Key results are that the average time per task is around 50% longer when using paper based documentation than when using the context (level of confidence 99%) supported system. The speech controlled HMD system is a bit faster but still around 30% slower then the context controlled version (confidence level 95%). We also present and discuss the results of two questionnaires (NASA TLX and custom) assessing the subjective view of the participants.

2 Related Work and Paper Contributions

Following early conceptual work on the usefulness of context (e.g. [5, 14, 11]) there has been an extensive body of work on tracking a multitude of activity types from fitness, through furniture assembly to health care related issues. By contrast, only very few projects deal with the evaluation of the benefit that context recognition brings to different applications. Bristow et. al. demonstrate how context information speeds up access to environment related information from the Internet ([2]). A similar study related to the use of physical context for information retrieval was given by Rhodes ([10]). Smailagic et. al. study a mobile phone that provides the caller information about other persons context ([12]). They show that using such information to prompt the caller to speak slowly or make pauses reduces the risk of using the phone while driving. Some qualitative discussion of a context sensitive tourist guide application has been presented in [4]. In another qualitative study context sensitivity has been evaluated in a wearable nursing support system([6]).

Somewhat related to this paper is research evaluating wearable maintenance assistance systems in general without including the context issue. Examples include a wearable remote collaboration system evaluated on a bicycle repair task, an evaluation of wearable system for aircraft maintenance, studies using HMD technology for guided instructions in a medical setting and early work from in Mizell and Caudell that hints at the usefulness of HMDs in maintenance scenarios in a qualitative way [3, 8, 9, 12].

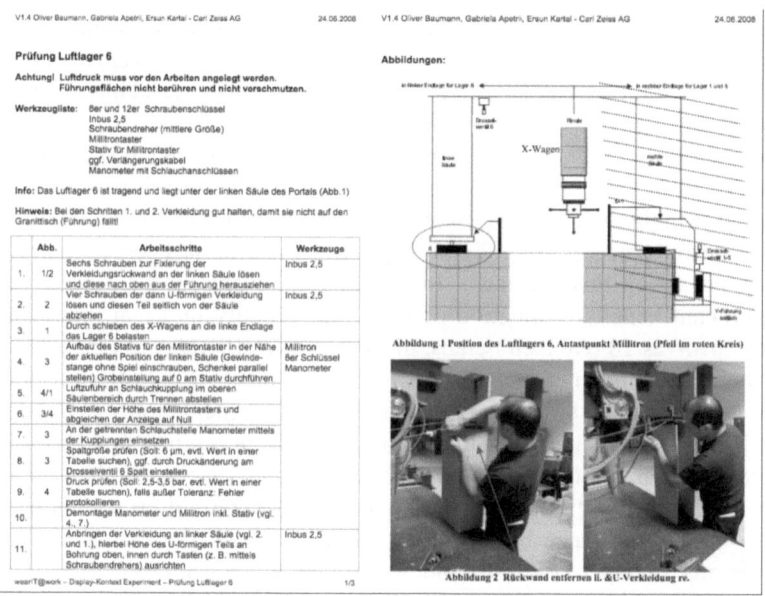

Fig. 1. A sample of the paper manual for the task 'checking bearing pressure'

Paper Contributions. The paper presents a quantitative, statistically relevant study showing the benefit of context recognition in wearable maintenance support systems. It does so in a real world, industrial environment with real, professional technicians and real maintenance task on a complex piece of machinery. This clearly goes beyond what has so far been investigated with respect to the usefulness of activity tracking in wearable maintenance systems. As sketched above, it also goes beyond much previous work on the benefits of context recognition in general.

We carefully describe out experimental design including a discussion of key considerations allowing other groups to learn from our experience. In addition to the quantitative results we describe a range of interesting qualitative observations. All results are discussed and put into perspective. We believe that this work constitutes an important piece of information for people designing context aware maintenance support systems as well as for more general context aware, assistive systems. It also provides a strong argument for continued research and development of such systems.

3   Experiments

Subsequently, we give an overview of the tasks, the selection process, the experimental design and setup.

3.1   Tasks

Task Selection Process. Task selection is a crucial step in our experimental design. We want to use a real maintenance task representative of the subjects' daily work, no artificial 'toy activities'. We want a complexity level that makes the use of some sort of documentation unavoidable. This also means that the specific task should be unfamiliar to our subjects (although of a general type to which they are used). In addition, the task needs to be not too short so that differences in performance can be resolved. On the other hand, the task can not be too long and too complex because the amount of technician time that we are allocated was limited. We also want the task to be doable by a professional without additional training. On the practical side, we need to find a machine that could be 'spared' for a couple of days and where a maintenance task could be performed repeatedly without fear of causing significant damage.

Finally, we need not one, but three tasks. We require each technician to use each of the three modalities (paper, HMD without context, HDM with context as described below). Yet, we want to avoid learning effect on the tasks.

The selection process involved several visits to the Zeiss facility, discussions with the responsible personnel and test runs of task candidates with a technician that was familiar with them. This was followed by test runs with ourselves and novice technicians.

Task Overview. Finally, we selected three maintenance tasks at a metrology system, more specifically at a Zeiss UMC 850 coordinate measuring machine (CMM) as shown in Figure 4. The specific machine is an older model taken back from a customer in an upgrade deal. This means that most technicians are not familiar with it and that it is 'not critical' in terms of any damage resulting from the experiments. The procedures that we selected can be summarized as follows:

1. Checking the bearing pressure on the left column. This task involved removing the casing, straining the air bearing, assembling the measurement apparatus, cutting the air supply, adjusting the militron, insertion of the manometer, measuring the gap size, measuring the pressure, disassembling the measuring apparatus, fixing the casing.
2. X-Motor installation with the following steps: attaching the belt pulley, attaching the oscillating element, fixing the belt at the engine shaft, attaching the basis plate, installing the motor, connecting the electronic cables.
3. Y-Gears installation with the following steps: affixing the gears, attaching the belt, threading the belt through the gears, attaching the belt, checking the friction clutch and the deflections towards the x and y-axis and adjusting the belt accordingly.

As seen from the descriptions, the tasks require being a trained technician to even understand them, not to speak of being able to perform them. They are by no means simple.

3.2 The Support Modalities

We investigate three support modalities: (1) traditional paper based documentation, (2) a speech controlled wearable support system with a head mounted display, and (3) a wearable support system with speech control and context aware support. Thus, we can determine how much of the improvement over paper based systems comes from the wearable system in general and how much is actually due to context.

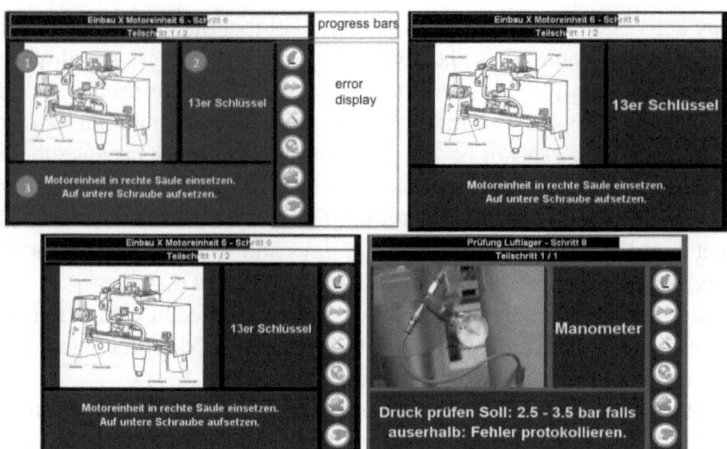

Fig. 2. The HMD UI as seen from the technician, first an overview, the second picture shows the UI in speech mode only, the last two with context recognition/error detection active

Depending on the modality, paper, speech or speech with context (we will refer to the later one as context for the rest of the paper), the technician has specific help to perform all three maintenance tasks.

To develop the UI and control application we used an iterative approach testing it during 2 test runs with 2 experienced technicians doing incremental improvements. Of course, these technicians have not participated in the later user study.

Paper Manual. As the official maintenance documentation manual contains over 800 pages and the information useful to the chosen maintenance tasks is spread throughout the manual, we decided to gather all relevant information and compile it into single compact document for each maintenance task. Our paper manual, a sample is depicted in Figure 1, contains general information on the top, a list of tools to use, the task steps in a table with the tools to use and references to pictures and pictures on the following pages. We evaluated our manual instructions with the 2 novice technicians, to be sure that they include all the necessary information to complete the three tasks.

Speech Controlled HMD GUI. The paper manuals are the basis for the instructions displayed in the HMD. We used only the information, pictures and text provided in the paper manual for the GUI instructions. No additional material/animation/video etc. is presented in the HMD GUI. The HMD user interface is depicted in Figure 2. The UI shows a task overview first, like a table of contents. Then, each step is displayed. If available, the technician sees a picture of the task at hand (1), the tools he needs to use (2) and a short description of what to do (3). On the top of the screen there are two progress bars, one for the overall task and one for the subtask as given in the table of contents. The technician has the following speech commands to navigate between task steps: next, previous, index (to display the table of contents), go to step no., zoom in and zoom out (for the images, one level of magnification only).

Fig. 3. The Wizard-Of-Oz control gui with preview enabled

The HMD UI is naturally constrained by the actual HMD we use. The colors we use, yellow on blue, provide the strong contrast ratios on the display and are save against ghost images, a problem we faced using black on white, for example.

The Context Aware HMD System. If context recognition is enabled an error bar is placed on the right side of the screen. The technician gets an alert if one of the following errors happen:

1. Touching the bearing surface. The bearing surface is not supposed to be touched as this can result in the need for recalibrate the machine which takes several hours.
2. Wrong task step. The technician missed an essential task previous to what he is currently working on.
3. Wrong tool. The technician is holding the wrong tool for the current task step.
4. Wrong position. The technician is not at the correct position relative to the machine to perform the task at hand.
5. Wrong part. The technician is operating/using at a wrong part of the machine.
6. Wrong tool usage. The technician uses the correct tool in a wrong way.

This error information is conveyed to the technician using the pictograms on the right. If a technician finishes a step successfully in context mode the UI switches to the next maintenance step.

The Wizard of Oz Control. We control the HMD display using a desktop application with the wizard of oz (WOZ) approach. This way, we avoid artifacts from the speech and context recognition system performance. This also saves us the effort of implementing the context recognizer which is highly non trivial.

The code for the HMD UI and the WOZ control application are written in Java. The complete implementation is open-sourced and published under http://sourceforge.net/projects/jwoz. The control application allows the wizard of oz to steer the UI the technician sees. He can display every task step and error information.

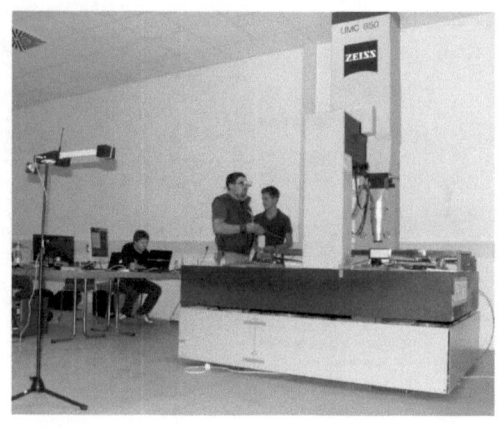

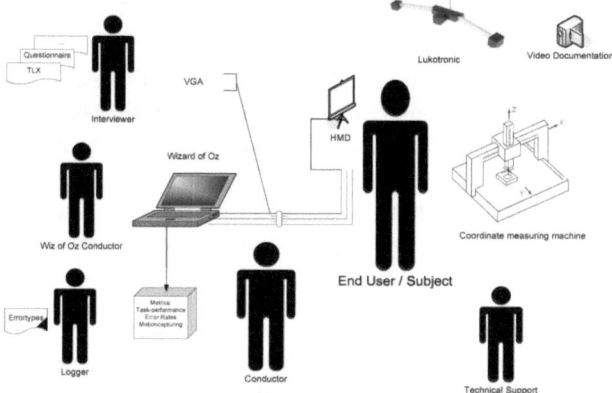

Fig. 4. The UMC 850 coordinate measuring machine and the experimental setup overview

The control gui is depicted in Figure 3. The wizard is able to load a given manual into the application and step with next and previous through the maintenance task steps (1). The task steps are displayed in a table view (2), highlighting the currently selected one. Below the table are radio buttons and checkboxes (3) for enabling context or speech mode, logging an unexpected error (an error that can not be put in one of the categories described above, in that case nothing is given as feedback, it is just written in the log-file), logging a technical problem and logging/displaying the error categories described on top. The Wizard of Oz is also able to display a preview window on the screen of the HMD screen. For the experiments we used a separate monitor. The control application is able to detect the monitor and convey the technician screen to it.

3.3   The Setup

For the experiments, we used the following setup shown in Figure 4. The test subject performed one of the three maintenance tasks at the machine using either the paper manual, the HMD with speech control or the HMD with speech control and context. A conductor introduces the test subject to the technology, he is the one contact person for the technician, if questions/technical problems etc. arise. He also signals to the wizard of oz the errors and completions of task steps. Over a separate monitor, the conductor can see the current UI the technician is looking at. The wizard of oz controls the UI over a desktop, which is directly connected over a VGA cable to the HMD and the monitor for the conductor. Another person is responsible for doing periodic consistency checks on the logs of the Wizard of Oz application, as well as running and maintaining the systems for ground truth information. We deployed two systems for ground truth, a simple video camera capturing the machine and the lukotronic active infrared marker system to capture the movements of the test subject (the information from this system was not evaluated in this paper). The logger also checks the error types helping the conductor/wizard of oz. A interviewer is located in a separate room interviewing the test subject using a custom questionnaire and the Nasa Task Load Index (TLX) after each experimental trial. In addition to this, at least one person is on stand-by for technical support. The test subject wears a vest on which the HMD, batteries and the cabeling is fixed together with 2 infrared markers for the Lukotronic systems. The subject also wears easy to strap on wrist bands with the remaining infrared marker (3 for each hand) for each trial.

3.4   The Studies

The Subjects. We selected a total of 18 subjects for participation – 16 males and 2 females aged between 17-56 years (mean 38.9 years). They were performed the tasks as part of their normal job. All 18 are professional maintenance operators from the Zeiss maintenance facility actively working in the maintenance field. Yet, they are not familiar with the CMM they have to perform the experiments on.

Study Design. The study uses a within subjects design with the interaction modality as the independent variable, meaning that all subjects will test every interaction modality in one of the three maintenance tasks.As there are three interaction modalities to test (and three different maintenance tasks to avoid bias caused by a learning effect), a Greco-Latin Square of the same order is used to distribute the 18 participants.

The LEGO Practice Round. To familiarize with the HMD, the speech commands and the error displays, we let each test subject perform several steps in building a simple LEGO Technic Forklift in a separate room. We picked the LEGO bricks assembly as it has nothing in common with the actual task and it is simple to explain the working of the UI and the error displays using this setup.

The Experimental Session. A test session consists of one practice round where the subject gets to practice each interaction modality, followed by the experimental rounds during which data is collected and interviews (questionnaires and TLX) are carried out for analysis. The practice round uses the same modalities of the experiment and an abstracted build task that allows subjects to get familiar with the modalities and that avoids fatigue. As explained above, the practice rounds are conducted using a LEGO

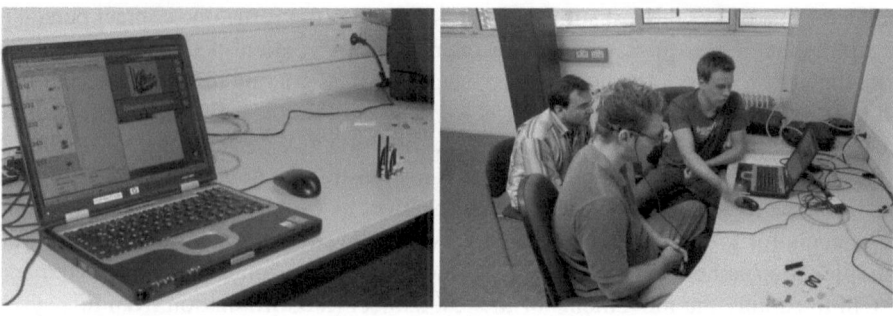

Fig. 5. The lego practice round

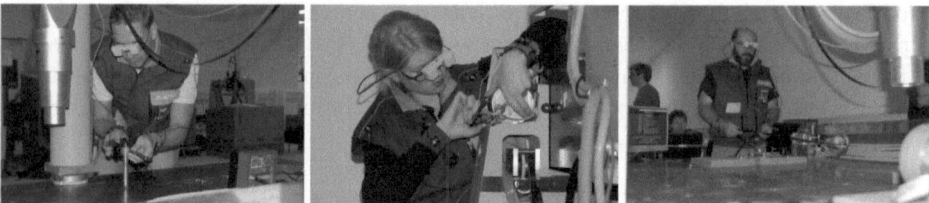

Fig. 6. Maintenance pictures from the experimental sessions. From left to right: Checking bearing pressure, installing the xmotor and ygears.

mockup. The total time required for a session is around 210 minutes. The three maintenance tasks are around 20-30 min each, however adding the time for the mockup and the interviews it takes substantially longer.

4   Quantitative Results

4.1   Objective Performance Metrics

Obviously, the two key objective performance metrics for a maintenance task are the time needed to perform the procedure and the number of mistakes.

Time. Figure 7 shows the average time per task for the three different modalities. The averages are taken over all workers and tasks. Since each worker performed each task only once and used each of the three modes only once we have 18 data points for each mode. These 18 instances are approximately equally distributed over the three tasks. Using paper documentation is on average around 50% (22.0 vs. 14.6 min) slower than context assisted HMD based documentation. The HMD without context is in between the two: 19.5 min, 30 % slower than context assisted case and about 15% faster than paper documentation. This is a significant difference, which however has to be seen in the context of high variation of individual times, in particular in case of paper manual.

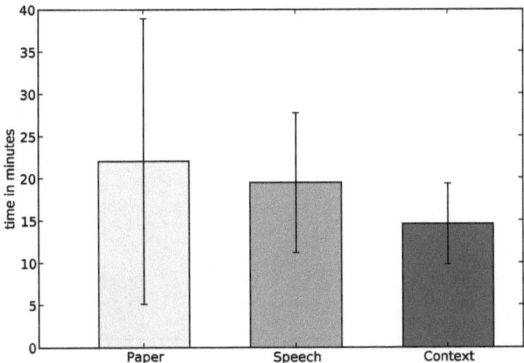

Fig. 7. The average time needed over all maintenance tasks split in modalities with standard deviation

As a consequence, when applying the one way ANOVA (F-Test)[1] to asses the statistical significance of the results we arrive at a confidence level of 99% (p-value: 0.01) for the comparison between context and paper,95% (p-value:0.04) between context and HMD, and only 80% (p-value:0.22) between HMD and paper.

In summary, we can say that the use of context improves the efficiency of our tasks in a relevant, statistically significant way. For HMD alone the results strongly point towards an improvement but the sample size and large variation between subjects mean that the results are not statistically significant (typically 95% is picked as the threshold for statistical significance).

4.2 Mistakes

The total number of mistakes made by all subjects in all tasks was 48 for paper, 31 for HMD without context and 29 for the context driven system. This clearly shows that the use of a wearable HMD based systems reduces the number of mistakes. It also indicates that the key factor in the error reduction is not context, but easy access to the information on the HMD.

Where context does make a significant difference is in the average time needed to recover from the mistakes, depicted in Figure 8. On average it has taken the workers about double as long to recover from a mistake when using paper then when using context. The use of HMD without context was only insignificantly faster then paper documentation. The statistical significance of the comparison between paper and context and HMD with and without context are both 99% respectively.

It is interesting to note that despite the huge relative difference in error recovery times, error recovery is not a relevant factor in the speedup in the overall average execution time. The error recovery speedup was in the range of 25sec, whereas the overall speedup is about 7min.

---

[1] The one-way ANalysis Of VAriance is used to test for differences among two or more independent groups and provides a likelihood to reject the null-hypothesis.

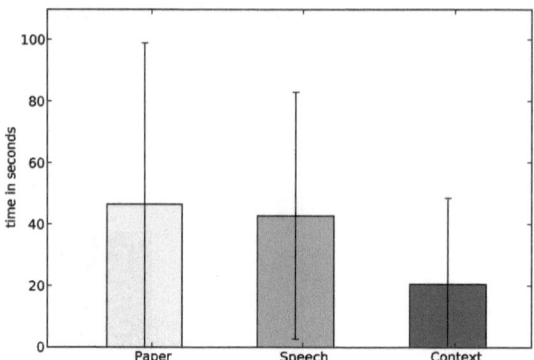

Fig. 8. The average time spent for resolving errors split in modalities with standard deviation

4.3   System Perception

We assess the subjective perception of the system with two questionnaires: one custom questionnaire designed for our specific study, and the standard NASA TLX questionnaire [7].

Custom Questionnaire. The custom questionnaire with a summary of the answers is shown in Figure 1. It contains 5 question groups. The first refers to the qualifications and background of the workers. Because of the small sample size correlating different backgrounds with the usefulness of context and HMD makes little sense, although it would be an interesting scientific question. Instead those questions are meant to provide an overview of the type of subjects that we were working with. It can be seen that (with the exception of two trainees) all consider themselves highly skilled in the repair of machines and mostly have 10+ years of experience on the job. Most, 15 out of 18, rate themselves as very good to average in terms of computer skills. The groups has a good age mix.

The second is the most relevant group of questions that reflects the perception of the context sensitive assistance system. The key questions are:

1. Overall impression of the system. On a scale from 1 (very good) to 6 (very bad) we have an average of 2,7 with 10 times 2, 5 times 3 , 2 times 4 and a single 6.
2. The favorite mode (paper, HMD with speech, HMD with context). Of the 18 participants 8 chose context, 6 chose HMD without context, 1 chose paper and 1 was undecided between context and paper.
3. The worst mode (paper, HMD with speech, HMD with context). Here 17 of the 18 participants named paper and one was undecided between HMD with and without context.
4. The improvement brought by context. On a scale from 1(very good) to 6(none) 12 participants picked 1, 3 picked 2 and 2 picked 3 while 1 participant failed to answer the question. This gives an average of 1.4. This might be see as an inconsistency with the question on favorite modality where only 8 subjects picked context. On the

Table 1. A summary of the questionaire filled out during the interviews with each participant

Questionaire	
Background	
How long are you employed in maintenance?	mean 14.81 years
How old are you?	mean 38.90 years
How much computer experience do you have?	mean 2.42 (Scale 1 to 6)
Are you satiesfied with your results?	mean 2.19 (Scale 1 to 6)
Perception of the context sensitive system	
How was your overall impression?	mean 2.72 (Scale 1 to 6)
Wich modality did you like best	P:1 HS:7 HC:10 counts
Which modality did you dislike most	P:16 HS+HC:1 counts
Wearing comfort	
How comfortable was the HMD?	mean 2.78 (Scale 1 to 6)
Would you wear the system for daily work?	yes:13 indifferent:4 no:1 counts
Do you felt relieved after taking of the HMD?	yes: 9 indifferent:1 no:8 counts
Do youl felt the system as obtrusive?	yes:6 indifferent: 2 no:10 counts
Do you had problems with the weight of the system?	yes: 4 no:14 counts
Would you have liked more help (inside the application)?	yes:13 indifferent:2 no:3 counts
How easy/difficult was it to put on the HMD?	mean 2.00 (Scale 1 to 6)
Was the screen big enough?	mean 2.11 (Scale 1 to 6)
How sharp/unsharp was the HMD image?	mean 2.83 (Scale 1 to 6)
How light/dark was the HMD image?	mean 1.89 (Scale 1 to 6)
Interface quality	
Did you need the offered resources and tools (system)?	yes:16 indifferent:2 no:0 counts
How comfortable was the navigation?	mean 2.11 (Scale 1 to 6)
How good/bad could you read the text?	mean 2.56 (Scale 1 to 6)
How much did you like the choice of colours and fonts?	mean 2.17 (Scale 1 to 6)
Motivation	
How motivated dou you felt during the experiment?	mean 2.00 (Scale 1 to 6)
How tensed dou you felt during the experiment?	mean 2.58 (Scale 1 to 6)

other hand, it is not unusual for people to understand that something might improve their work and still feel more comfortable with a different solution.

5. Willingness to use such a system in every day work. Again on a scale from 1 to 6 we get an average of 1.3 with 13 times 1, 4 times 2 and a single 3. Note that this question did not differentiate between a system with and without context.

In summary the above questions group indicates a positive subjective perception of the system.

The third questions group concentrates on the perception of the head mounted display. It was included out of two reasons, one due to interest of Zeiss (who manufactures the displays) and second, it should show whether there is reason to believed that the results were skewed by poor HMD quality. It can bee seen in Table 1 that most subjects rated the comfort and weight of the HMD to be average and were roughly evenly split between those say there were relieved to take the display off and those who say they were not. Thus, we conclude that there is no obvious indication for a HMD related skew.

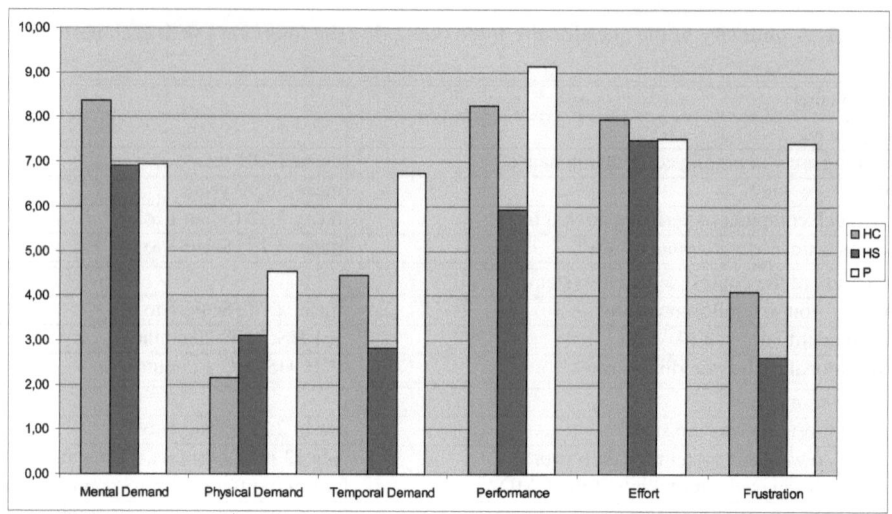

Fig. 9. Nasa TLX (see [7] for detailed description) results. Large values are "worse".

The fourth questions group dealt with the quality of the interface. Like the HMD related questions, it was meant to establish if there was reason to believe that the results were somehow skewed by the interface. Again the questionnaire shows no indication of such a skew. On all counts including navigation and readability the users rate the interfaces mostly between good and average.

The final two questions relate to the motivation and level of comfort during the experiments. On both issues the answers are mostly 2 and 3.

NASA TLX Questionnaire. From the background of the objective performance metrics and the custom questionnaire described above the results of the TLX questionnaire [7] shown in Figure 9 seem surprising at first. On every question with the exception physical effort the average score of the context based system is worse than that of a speech only HMD system (large score is worse for all items). In fact, in some cases (mental demand and effort) context even scores worse then paper. This is particularly surprising with respect to mental demand, since reducing mental demand is one of the key goals of context awareness.

However, when we examine the answers in more detail, a different picture emerges. Table 2 shows the breakdown of the answers by the number of subject who rate one modality to be better (lower score !), equal, or worse (higher score !) then the other on each of the criteria. It can be seen that with respect to mental load, effort and frustration more subjects consider context to be better then speech than the other way around (9 vs. 4, 7 vs. 4, and 15 vs. 3). The same is true for the comparison of context an paper. However, it seems that the comparison of paper with context is more polarizing. For mental load and effort there are 5 (4) subjects who consider both equal. For context and paper this numbers are 0 and 1.

It is interesting to note how clearly context wins against both speech only and paper with respect to frustration (15 vs. 3 against speech and 17 vs. 1 against paper).

Table 2. The number of people who have rated one modality better then the other on the different NASA TLX metrics

	Lower score is better so $A > B$ means A scores worse then B !					
	$HC > HS$	$HC = HS$	$HC < HS$	$HC > P$	$HC = P$	$HC < P$
Mental	4	5	9	7	0	11
Temporal	10	0	8	4	2	12
Performance	9	2	7	5	1	12
Effort	4	4	10	7	1	10
Frustration	3	0	15	1	0	17

Even in the breakdown in table 2 context "looses" with respect to two criteria. The majority (10 vs. 8) think they were faster with the speech only interface. In reality 11 subjects were actually faster with context and only 8 with speech. Here it is important to note that subjects performed different tasks with different modalities. At the same time, tasks were not equally long so that even if a subject was in principle more efficient with a given modality, he might have scored a better absolute time with a different modality. Therefore, these results must only be viewed as a trend over all users.

The second criterion where context "looses" is performance. Is is defined as the users subjective perception of how well he/she performed. The score here is clearly consistent with the skewed perception of the actual timing with different modalities.

In summary, the averaged TLX scores are a result of the subjects being against context are so by a large margin. Broken down by the number of people that score one modality better then another, the TLX results are reasonably well in line with the results of the custom questionnaire described in the previous section. The majority of the users see benefits from the context system (most clearly on the level of frustration). On the other hand, the subjective perception of the context systems seems to be worse then the objective data. To a degree, this is understandable as it is well known that people tend to perceive interaction modalities that are new and unusual for them in a more negative way. The key question, which we can not answer in this study, is whether there are other issues beyond context being new and strange. Thus, the question is if the negative bias will go away on its own, once people have worked with context aware systems long enough or whether there are some fundamental reasons why some people are uneasy about context controlled systems (e.g. they feel stressed by the system doing something on its own).

5   Qualitative Results and Observations

In this section we summarize some observations which are not backed by statistically relevant numbers but which we nonetheless consider noteworthy.

5.1   Most Proficient Subject

The overall fastest test subject managed the xmotor task in 9 min with the speech interface. With context he needed only 12 min for the ygears task. Of course, the two tasks are not really comparable. The subject did not do any mistakes, which is another indication of his skills.

It was interesting to see how this skilled subject used the system. We needed to relabel a lot of his data using video analysis, as he already skipped forward to the next steps to get an overview while working on earlier task steps.

This subject was also one of the few using the zoom function extensively. He used 15 times the picture zoom combined over all maintenance tasks, compared to an average mean of 3 zooms over all other participants.

5.2 Errors

Using the speech and context modalities not many bearing surface errors happened (an total of 3 compared to 9 in paper). We assume that although the paper has the information about the bearing surface in bold over the steps, it is often overlooked and not read carefully. Yet, in the speech and context mode the subjects are directly confronted with the text on screen and also often they asked the moderator again where the bearing surface is exactly located.

The errors in the context and speech are more deterministic than the ones during paper (happening at the same task steps), suggesting that an improved version of the system could fix this.

During paper and speech trials a total of 3 part errors for connecting the wrong electronic cables on the xmotor happened. This would have caused the motor to be destroyed when the machine is switched on. A context sensitive system detecting such errors can mitigate this problem.

5.3 HMD

One direct implication we saw, is that with HMD the technician focuses very much at the step displayed (and the last steps they saw). This is an important aspect for designing HMD systems. For example, one common error was in step 10 of checking the air pressure: The technician first employs two measuring devices and then needs to check the values for both of them. Most technicians built up both and measured first with the last one, which was the wrong one to use.

6 Conclusion

We have presented a systematic quantitative evaluation of the usefulness of context in a wearable maintenance assistance scenario. A key objective of the work is to perform the evaluation in an environment that is as close as possible to the setting in which such system would be deployed in a real world scenario. We have achieved this by picking real maintenance tasks on a complex piece of machinery, and using real technicians (who did not volunteer but did this as part of their normal job). We have carefully selected the procedures to be complex enough not to be doable without instructions but not too complex to be performed without extensive prior training. We also made sure that the procedures were long enough to resolve the effects we were looking for.

Despite the constraints of an industrial environment we managed to get enough subjects and runs to achieve statistical significance on the objective, quantitative metrics.

As an indication of the effort involved in the experiment consider the fact that initial discussion with Zeiss started more then a year before completing the experiment. This was followed by about 10 visits to the site to search for adequate machinery and tasks, evaluate manuals and interfaces with different technicians, and test the overall technical setup. The actual experiment took over a week during which we recorded and analyzed over 60 hours of data.

We believe that the experimental procedures described in this paper together with the scripts and software available from our www site are a significant additional contribution of the paper, from which other groups aiming at real life experiments can learn.

6.1   Key Conclusions

The most important conclusions from the study can be summarized as follows:

1. On average, the use of context information speeds up the procedures in a significant, statistically relevant way. Paper is 50% slower then context, speech control is 30% slower.
2. The amount of errors is already significantly reduced by the use of speech controlled HMD documentation. The addition of context brings a minimal additional improvement. However context reduces the time needed to correct the errors by an average of nearly 100% as compared to both speech and paper.
3. From our qualitative observations it seems that context is more useful for technicians who are less proficient. The technicians who were fastest with paper were the once who made no mistakes. The fastest technician achieved the result with the speech controlled system and used it in such a way (looking ahead of the task) that default context control (display manual for current task) would not have work. In fact, he was slower with context. An alternative conclusion could be that we need to rethink the way we use context.
4. A clear majority of subjects considered context to be beneficial in one or the other way. This was reflected in the respective questions of the custom questionnaires (12 participants very strongly and 3 strongly agreeing that context brought a benefit to their work) and in the answer counts (how many people though one modality was better then the other) in the TLX Frustration, Mental Load and Effort metrics.
5. However, the subjectively perceived advantages of context are less clear then the time measurement might suggest. Participants tend to subjectively underestimate their performance when using the context controlled system, which was reflected in the Temporal and Performance TLX metrics. Also, significantly less participants picked the context as their preferred modality that strongly agreed that context was beneficial (8 vs. 15). This suggests a degree of uneasiness towards the context control. At this stage, we can not say if this is the usual uneasiness towards the unknown or whether there are more fundamental issues behind it.
6. Interestingly, there was much less uneasiness towards the speech controlled HMD system. Context 'lost" mostly to speech, nearly never to paper. The only participant who picked paper as preferred modality wore special contact lenses and was unable to see clearly on the HMD.

7. When people felt that context was inferior, they tended to consider it significantly inferior. Similar applied to those who considered paper to be better. This is reflected in the average TLX scores, were context scored worse then speech controlled HMD.

6.2   Open Questions and Future Work

We are convinced that the study presented in this paper is a valid and relevant 'data point'. However, without doubt, it leaves a number open questions which need to be studied in the future. For one, it is unclear how much our interface design and the quality of the existing paper documentation influenced the results. We believe that our approach of taking the paper documentation 'as is' (except putting the relevant pieces in one document) and using it as 'blueprint' for the HMD interface is reasonable. However, it is conceivable that a more elaborate interface and/or documentation design might have lead to different results. Similar can be said about the way we use context and the choice of voice as 'non context' interface.

Another interesting question is how the results change when the subjects get used to the new modalities. Was the uneasiness towards context the usual result of the modality being new and strange, or was there something more fundamental? Finally, it is unclear if and how context interfaces can be adapted to help more proficient users, who seemed to benefit least in our study according to qualitative observations.

Acknowledgment

This work was funded by the European Commission through the WearIT@Work Project under grant 004216.

References

1. Boronowsky, M., Nicolai, T., Schlieder, C., Schmidt, A.: Winspect: A case study for wearable computing-supported inspection tasks. In: Fifth International Symposium on Wearable Computers (ISWC 2001), pp. 8–9 (2001)
2. Bristow, H., Baber, C., Cross, J., Wooley, S.: Evaluating contextual information for wearable computing. In: Proceedings of the Sixth International Symposium on Wearable Computers, pp. 179–185 (2002)
3. Caudell, T., Mizell, D.: Augmented reality: an application of heads-up display technology tomanual manufacturing processes. System Sciences (January 1992)
4. Cheverst, K., Davies, N., Mitchell, K., Friday, A., Efstratiou, C.: Developing a context-aware electronic tourist guide: some issues and experiences. In: Proceedings of the SIGCHI conference on Human factors in computing systems, pp. 17–24. ACM Press, New York (2000)
5. Dey, A., Abowd, G.: Towards a Better Understanding of Context and Context-Awareness. In: CHI 2000 Workshop on the What, Who, Where, When, and How of Context-Awareness (2000)
6. Drugge, M., Hallberg, J., Parnes, P., Synnes, K.: Wearable Systems in Nursing Home Care: Prototyping Experience. In: IEEE Pervasive Computing, pp. 86–91 (2006)
7. Hart, S., Staveland, L.: Development of NASA-TLX (Task Load Index): Results of empirical and theoretical research. Human Mental Workload 1, 139–183 (1988)

8. Nilsson, S., Johansson, B.: Acceptance of augmented reality instructions in a real work setting. In: CHI 2008: CHI 2008 extended abstracts on Human factors in computing systems (April 2008)
9. Ockerman, J., Pritchett, A.: Preliminary investigation of wearable computers for task guidancein aircraft inspection. In: Second International Symposium on Wearable Computers, 1998. Digest of Papers, pp. 33–40 (1998)
10. Rhodes, B., Innovations, R., Menlo Park, C.: Using physical context for just-in-time information retrieval. IEEE Transactions on Computers 52(8), 1011–1014 (2003)
11. Schmidt, A., Beigl, M., Gellersen, H.: There is more to context than location. Computers & Graphics 23(6), 893–901 (1999)
12. Smailagic, A., Siewiorek, D.: Application design for wearable and context-aware computers. IEEE Pervasive Computing 1(4), 20–29 (2002)
13. Smith, B., Bass, L., Siegel, J.: On site maintenance using a wearable computer system. In: Conference on Human Factors in Computing Systems, pp. 119–120. ACM Press, New York (1995)
14. Starner, T., Schiele, B., Pentland, A.: Visual contextual awareness in wearable computing. In: Proceeding of the Second Int. Symposium on Wearable Computing, Pittsburgh (October 1998)
15. Sunkpho, J., Garrett Jr., J., Smailagic, A., Siewiorek, D.: MIA: A Wearable Computer for Bridge Inspectors. In: Proceedings of the 2nd IEEE International Symposium on Wearable Computers, p. 160. IEEE Computer Society Press, Washington (1998)
16. Webster, A., Feiner, S., MacIntyre, B., Massie, W., Krueger, T.: Augmented reality in architectural construction, inspection and renovation. In: Proc. ASCE Third Congress on Computing in Civil Engineering, pp. 913–919 (1996)

On the Anonymity of Home/Work Location Pairs

Philippe Golle and Kurt Partridge

Palo Alto Research Center
{pgolle,kurt}@parc.com

Abstract. Many applications benefit from user location data, but location data raises privacy concerns. Anonymization can protect privacy, but identities can sometimes be inferred from supposedly anonymous data. This paper studies a new attack on the anonymity of location data. We show that if the approximate locations of an individual's home and workplace can both be deduced from a location trace, then the median size of the individual's anonymity set in the U.S. working population is $1, 21$ and 34,980, for locations known at the granularity of a census block, census track and county respectively. The location data of people who live and work in different regions can be re-identified even more easily. Our results show that the threat of re-identification for location data is much greater when the individual's home and work locations can *both* be deduced from the data. To preserve anonymity, we offer guidance for obfuscating location traces before they are disclosed.

1 Introduction

Location-based services offer valuable applications to mobile users. To receive these services, users must disclose their location to service providers. This raises privacy concerns [6]. Location records, when analyzed, can reveal sensitive facts about an individual, such as business connections, political affiliations or medical conditions. Misuse of location data can lead to damaged reputation, harassment, mugging, as well as attacks on an individual's home, friends or relatives.

Privacy policies and legislation address some of these concerns. But protection mechanisms rooted in policy or law are only effective when data collectors are honest and trusted. They offer no protection against a dishonest collector, or one whose data is compromised by malware, laptop theft or a weak password.

To minimize privacy concerns, the best practice is to collect the minimum amount of information needed. For location-based services, this *principle of minimal collection* typically means collecting anonymous or pseudonymous location data [2]. A restaurant recommendation service, for example, can give adequate recommendations based on locations reported anonymously, or under a pseudonym linked to a profile of dining preferences.

Anonymity is a useful, but imperfect tool for preserving location privacy. The problem is that ostensibly anonymous location data may be traced back to personally identifying information with the help of additional data sources.

H. Tokuda et al. (Eds.): Pervasive 2009, LNCS 5538, pp. 390–397, 2009.

Krumm [5] showed how to recover the home address (with median error below 60 meters) and identity (with success above 5%) of subjects who disclosed two weeks' worth of GPS data collected in their car. In Krumm's experiment, re-identification was made possible by joining GPS traces with a reverse geocoder and a Web-based white pages directory. Krumm also showed how to prevent re-identification attacks. Location obfuscation (via the addition of noise, rounding or cloaking) defeats re-identification, at the cost of some loss in location precision.

To be effective, anonymity requires understanding and countering the threat of re-identification. This paper contributes a fuller understanding of this threat for anonymous location traces. Krumm's work, while groundbreaking, considers only the threat of re-identification that comes from identifying a subject's home. We extend his analysis by considering also a subject's *place of work*. After the home, the workplace is arguably the second most easily identifiable location in a trace. Obfuscation techniques which prevent re-identification based on (approximate) home location alone may not be adequate if the subject's (approximate) work location is also known. In fact, we show that home and work locations, even at a coarse resolution, are often sufficient to uniquely identify a person.

We rely on data from the U.S. Census Bureau to estimate the threat of re-identification based on home and work locations. This data, collected for the Longitudinal Employer-Household Dynamics (LEHD) program [9], contains the home and work locations of 96% of American workers in the private sector ($103, 289, 243$ individuals). This very large dataset gives a precise indication across the U.S. working population of the threat of re-identification, and of the effectiveness of defenses such as location obfuscation (for comparison, Krumm's analysis was based on a study of 172 participants).

We adopt a strong definition of privacy based on the concept of an *anonymity set*. The anonymity set associated with a location trace is the set of people from whom this trace may have been collected, given all information known to the data collector (see section 2). A large anonymity set offers strong privacy. Conversely, a small anonymity set is cause for concern. A unique (or nearly unique) trace may not always be linkable to an identity, but it is prudent to make the conservative assumption that if a unique link exists between a trace and an identity, that link may be discovered. This definition of privacy is stronger than that described by Krumm in [5], which considers re-identification successful only when a link is positively established. Our privacy definition is identical to that used by Sweeney [7,8] in her well-known analysis of the uniqueness of simple demographic attributes in the U.S. population.

Our contributions. In summary, our contribution is threefold: 1) We study the threat of re-identification of anonymous location traces based on home and work locations; 2) We base our study on a very large dataset representative of the whole U.S. working population; 3) We adopt a strong and principled definition of privacy. Our analysis will help data collectors gain a better understanding of the sensitivity of location data, and the commensurate needs to obtain consent from users before collecting location data, to protect location data via obfuscation or access control, and to restrict the disclosure and publication of location data.

Organization. We review related work in the rest of this section. We discuss our model and assumptions in section 2. We present the statistical dataset that we used to estimate location privacy in section 3. We study the threat of re-identification for location traces, assuming approximate knowledge of home and work locations, in section 4. Finally, we conclude in section 5.

1.1   Related Work

Intentional degradation of location information quality, or obfuscation, is a well-known technique for preserving the anonymity, or pseudonymity, of location traces [2,3], but the question of *how much* obfuscation is required to preserve anonymity is often sidestepped. Our paper answers this question for location traces from which home and workplace locations can be deduced.

Krumm's analysis of inference attacks on location traces [5] is closest to our work. The main distinction between our work and Krumm's is that we also take into account workplace locations. Using a stronger definition of privacy, and a much larger data set obtained from the U.S. Census Bureau, we show that the threat of re-identification for anonymous location traces based on home and work locations is more severe than the threat reported in [5] for home location alone.

Hoh et al. [4] propose a time-to-confusion metric to characterize the degree of privacy of location traces. Their approach to anonymity consists of periodically withholding location information long enough for a location trace to be confused with sufficiently many others. This approach is well-suited to vehicular network applications. But it has two drawbacks: 1) it requires coordination from a centralized server that tells mobile devices how long to withhold location data, and 2) it is not applicable to pseudonymous location traces, since fragments of a pseudonymous trace can easily be linked. Pseudonyms are important to many location-based services (e.g. to personalize service or prove membership). In contrast to [4], our analysis of privacy applies to pseudonymous traces.

Our work is indirectly related to Sweeney's well-known study [7] of the uniqueness of simple demographic attributes in the U.S. population. Sweeney's analysis of census data showed that 87% of the U.S. population is uniquely identifiable given their full date of birth (year, month and day), sex, and the ZIP code where they live. We adopt Sweeney's strong definition of privacy [8], called *k-anonymity*, which is based on the concept of an anonymity set (see section 2).

2   Assumptions and Privacy Model

Anonymous location traces. The value of location data dissociated from identity was illustrated in the introduction: a location-based restaurant recommender can offer adequate recommendations based on location and a pseudonymous profile of dining preferences. Many other location-based services (e.g. friend-finding services) can similarly operate using a registered pseudonym only, without learning users' real identities. From a technical point of view, location traces can be anonymized or pseudonymized with help from a trusted network proxy. Mobile subscribers, for example, may trust their network provider to forward their location data anonymously to third party location-based service providers.

Threat of re-identification. We study the threat of *re-identification* for anonymous location traces. We focus specifically on the threat of re-identification under the assumption that the approximate home *and* work locations of the subject can be deduced from the trace (for example with a reverse geocoder). Approximate home and work locations may then be joined with employment directories, tax records or any other public or private dataset available to the adversary to map pairs of home and workplace locations to identities.

Model of privacy. For the sake of example, assume that a subject is the only person in the U.S. who lives in a certain region A and works in a certain region B. The subject's location trace is the only one with the home/workplace pair (A, B). It does not necessarily follow that the trace can be linked to the subject, as there may be no directory that links the pair (A, B) with the subject's identity. But since the datasets that an adversary may use to re-identify location traces are not known a-priori, it is best to make the most conservative assumptions about them. Accordingly, we assume that if a unique link exists, it will be discovered. Our measure of privacy is the set of all people associated with the pair (A, B), called the *anonymity set* [8] of the pair. The larger the anonymity set, the larger the crowd one is indistinguishable from, and consequently the better the privacy protection one enjoys. Enlarging the regions A and/or B (e.g. via location obfuscation) increases the size of the anonymity set, and thus the quality of privacy protection. The rest of this paper analyzes the size of the anonymity set of home/workplace location pairs for different region sizes, based on the census data described in the next section.

3   The LEHD Origin-Destination Dataset

The Longitudinal Employer-Household Dynamics (LEHD) program, run by the U.S. Census Bureau, compiles information about where people work and where they live, together with reports on their age, earnings and distribution across industries. LEHD includes all jobs covered by the reporting requirements of the states' unemployment insurance system. According to [1], "The prime exclusions are agriculture and some parts of the public sector, particularly federal, military, and postal works. Coverage varies across states and time, although on average, 96% of all private-sector jobs are covered."

LEHD lets us study the privacy implications of revealing (intentionally or indirectly) coarse-grained location information, such as the county or ZIP code where one lives and works. A person revealing this information allows her identity to be narrowed down to the set of people who live and work in the same geographic areas. The size of this *anonymity set*, which can be estimated with the LEHD dataset, is a good measure of the privacy loss that revealing coarse home and work locations entails.

The raw LEHD data is not publicly available for download, due to privacy concerns. But the Census Bureau releases privacy-preserving synthetic data derived from the raw data. Bayesian techniques are used "to synthesize workers' place of residence conditional on disclosable counts of workers by place of work, industry, age, and earnings categories." [1]. According to [1] (p. 6), "The key statistical property to preserve in the synthetic data is the joint distribution of workers across home and work areas." The synthetic data thus appears suitable for our privacy analysis of home/work location pairs. Three implicates (independent draws from the synthesizing algorithms) are available for download from [10]. To confirm the validity of our results, we repeat our analysis with all three implicates. We obtain nearly identical results, as described in section 4.

We focus on the "Origin/Destination" dataset, which reports where workers live and work at the granularity of census blocks. The most recent dataset available is from 2004. It includes information on workers from 42 states (the eight missing states are Arizona, Connecticut, Massachusetts, Nebraska, New Hampshire, New-York, Ohio and South Dakota). In total, the 2004 Origin/Destination dataset includes information on 103, 289, 243 individual workers.

The Origin/Destination dataset reports the home and workplace locations of workers along the following hierarchical geographic scale:

- **State.** The state is one of the 42 states included in the LEHD dataset.
- **County.** There are 2, 784 counties and county equivalents (boroughs, parishes) in the 42 states included in the LEHD dataset.
- **Census Tract.** Census tracts are county subdivisions defined by the U.S. Census Bureau. In terms of size and number of residents, they are roughly comparable to ZIP codes. The LEHD dataset contains 64, 881 tracts where workers live and 52, 852 tracts where they work. The mean number of workers living in a tract is 1, 592 and the median is 1, 479. The mean number of people *working* in a tract is 1, 954 and the median is 951 (it ranges from zero in completely residential tracts to 166, 680 in a tract of Los Angeles County).
- **Census block.** Blocks are the smallest area of census geography. They are subdivisions of census tracts, typically alongside streets. In urban areas, census blocks typically coincide with individual city blocks. In rural areas, blocks may cover many square kilometers.

4   Anonymity Set of Home/Work Location Pairs

In this section, we study the size of the anonymity set for workers who reveal where they live and where they work at various degrees of granularity (census block, census tract, or county). The knowledge of home and work locations at the census block granularity is information that could be learned from a lightly obfuscated location trace (with noise or rounding on the order of a city block or less). With more obfuscation (on the order of a kilometer or so), a location trace would reveal only the census tract where the person lives and works. Heavy obfuscation (on the order of tens of kilometers) may only allow for inference of the county or counties where a person lives and works.

Table 1. Size of the anonymity set for workers who reveal where they live and work

Location precision	Size of anonymity set			
	Median	10th percentile	5th percentile	bottom percentile
Census block	1	1	1	1
Census tract	21	3	1	1
County	34,980	446	92	6

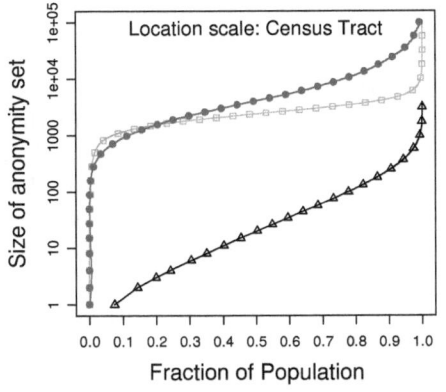 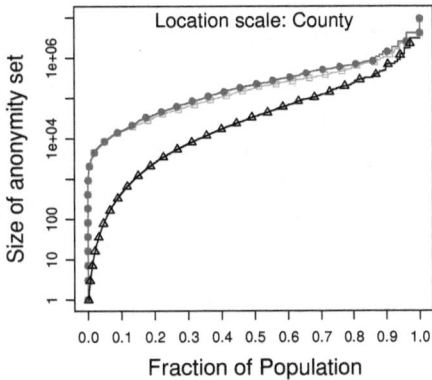

Fig. 1. Size of anonymity set under disclosure of work location (red circles), home location (green squares) or both (black triangles). Location granularity is either census tract (left graph) or county (right graph). Note the different scales on the Y-axes.

We compute the anonymity set for U.S. workers assuming knowledge of where they live and work at these three levels of granularity. Table 1 summarizes our findings. It shows the median size of the anonymity set, as well as the 10th, 5th and bottom percentiles of anonymity sets sorted by size. The table shows that revealing where one lives and works at the granularity of census blocks is uniquely identifying for a majority of the U.S. working population. Revealing where one lives and works at the relatively coarse level of census tracts is uniquely identifying for 5% of U.S. workers, and offers little privacy to the majority. Revealing the counties where one lives and works poses little threat to privacy.

The graphs in Fig. 1 give more detail on the size of the anonymity set of workers who reveal the location where they live, the location where they work, or both. The graphs show what fraction of the U.S. working population (on the X-axis) falls in an anonymity set of less than a given size (on the Y-axis) after revealing location information (where they live, where they work, or both) at different precisions (census tract granularity for the graph on the left; county granularity for the graph on the right).

The red circle curve plots the anonymity of workers who reveal only where they work. The green square curve plots the anonymity of workers who reveal only where they live. The black triangle curve plots the anonymity of workers who reveal both where they live and where they work. Both at the granularity of census tracts and at the granularity of counties, we observe that revealing both the locations where one lives and works is strikingly more identifying than revealing only one of them (note that the Y-axis follows a logarithmic scale). For example, disclosing both the census tracts where one lives and where one works places 24.2% of the working population in an anonymity set that contains 5 or fewer individuals, and 7.4% in a set of 2 or fewer individuals.

While these statistics are based on synthetic data, we obtained the same results in Fig. 1 with all three synthetic implicates of the 2004 LEHD dataset (the curves are so similar as to appear indistinguishable).

4.1 Factors Influencing Anonymity

Anonymity differs dramatically between individuals who live and work in the same region, and individuals who work in a different region from where they live. Fig. 2 plots these differences. It shows that workers who live and work in different locations (brown stars) have a much smaller anonymity set (i.e. are less anonymous) than those who live and work in the same location (blue diamonds). This holds true both for locations revealed at census tract (left graph) and county granularity (right graph).

The traces of workers who cross location boundaries to go to work are particularly vulnerable to re-identification attacks. Across the states included in the LEHD dataset, 94.1% of workers live and work in different *census tracts*. The percentages range from a high of 97.2% in California to a low of 80.3% in Wyoming. These numbers show that the extra anonymity afforded by living and working in the same census tract is uncommon. The fraction of workers who live and work in different *counties* is 43.5%. The percentages range from 63.8% in Virginia to 10.7% in Hawaii. Living and working in the same county is more frequent, and therefore had a bigger effect on the national average (see Fig. 2).

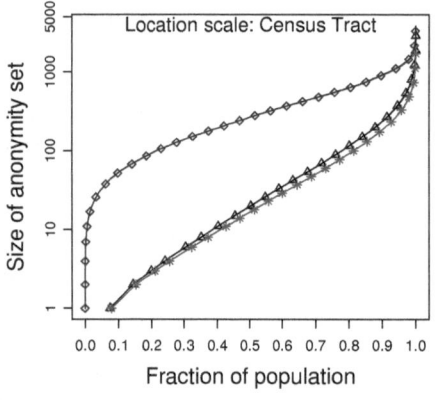 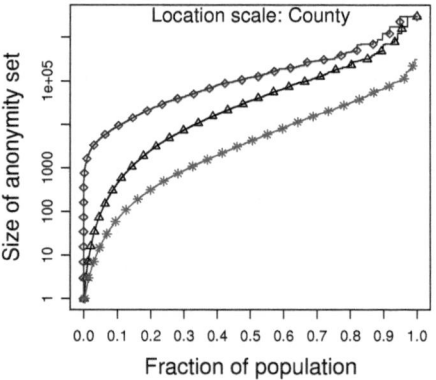

Fig. 2. Size of anonymity set under disclosure of home and work locations, for workers who live and work in the same location (blue diamonds) or in different locations (brown stars). For reference, black triangles show the anonymity set of all workers as in Fig. 1. Location granularity is census tract (left graph) or county (right graph). Note the different scales on the Y-axes.

5 Conclusion

We studied the threat of re-identification of anonymous location traces in the U.S. based on obfuscated home and work locations. We showed that this threat is substantially greater with the disclosure of *both* home and workplace locations then either one alone. This result is important, because the workplace is arguably the second most easily identifiable location in a subject's trace, after the home.

Obfuscation techniques which prevent re-identification based on home location alone [5] may not be adequate if the subject's work location is also known. When both home and work locations can be deduced, our results show that a considerable amount of location obfuscation (at the granularity of counties) is required to protect the anonymity of location traces. An alternate approach to privacy would be to maintain different "home" and "work" personas, in applications where these personas can be kept strictly separate and unlinkable.

Our study distinguishes itself from previous work in that it adopts a principled definition of privacy based on the concept of anonymity sets, and relies on a large dataset that is representative of the whole U.S. working population. While our analysis is based on U.S. data, we speculate that our results apply to other developed countries with broadly similar densities and commute patterns.

Acknowledgements

We thank the anonymous reviewers and our shepherds James Scott and Alastair Beresford for their valuable comments and suggestions.

References

1. Andersson, F., Freedman, M., Roemer, M., Vilhuber, L.: LEHD OnTheMap Technical documentation (February 21, 2008)
2. Beresford, A.R., Stajano, F.: Location privacy in pervasive computing. IEEE Pervasive Computing 2(1), 46–55 (2003)
3. Duckham, M., Kulik, L.: A formal model of obfuscation and negotiation for location privacy. In: Gellersen, H.-W., Want, R., Schmidt, A. (eds.) Pervasive 2005. LNCS, vol. 3468, pp. 152–170. Springer, Heidelberg (2005)
4. Hoh, B., Gruteser, M., Xiong, H., Alrabady, A.: Preserving Privacy in GPS Traces via Density-Aware Path Cloaking. In: Proc. of ACM Conference on Computer and Communications Security (CCS) (2007)
5. Krumm, J.: Inference Attacks on Location Tracks. In: LaMarca, A., Langheinrich, M., Truong, K.N. (eds.) Pervasive 2007. LNCS, vol. 4480, pp. 127–143. Springer, Heidelberg (2007)
6. Schilit, B., Hong, J., Gruteser, M.: Wireless Location Privacy Protection. Computer 36(12), 135–137 (2003)
7. Sweeney, L.: Uniqueness of Simple Demographics in the U.S. Population. Laboratory for International Data Privacy, Carnegie Mellon University (2000)
8. Sweeney, L.: K-anonymity: a Model for Protecting Privacy. International Journal on Uncertainty, Fuzziness and Knowledge-based Systems 10(5), 557–570 (2002)
9. U.S. Census Bureau. Longitudinal Employer-Household Dynamics, http://lehd.did.census.gov/led/
10. VirtualRDC OnTheMap Data, http://www.vrdc.cornell.edu/onthemap

Working Overtime: Patterns of Smartphone and PC Usage in the Day of an Information Worker

Amy K. Karlson[1], Brian R. Meyers[1], Andy Jacobs[1], Paul Johns[1], and Shaun K. Kane[2]

[1] Microsoft Research, One Microsoft Way, Redmond, WA 98052, USA
{karlson,brianme,andyj,pauljoh}@microsoft.com
[2] The Information School, DUB Group, University of Washington, Seattle, WA 98195, USA
skane@u.washington.edu

Abstract. Research has demonstrated that information workers often manage several different computing devices in an effort to balance convenience, mobility, input efficiency, and content readability throughout their day. The high portability of the mobile phone has made it an increasingly valuable member of this ecosystem of devices. To understand how future technologies might better support productivity tasks as people transition between devices, we examined the mobile phone and PC usage patterns of sixteen information workers across several weeks. Our data logs, together with follow-up interview feedback from four of the participants, confirm that the phone is highly leveraged for digital information needs beyond calls and SMS, but suggest that these users do not currently traverse the device boundary within a given task.

Keywords: Mobile information work, multiple devices, cross-device interfaces.

1 Introduction

Information workers typically rely on an ecosystem of desktop and portable computers that span both work and home to support a wide range of professional and personal tasks [1]. Smartphones are a relatively recent addition to this ecosystem, and with increasing connectivity and computing power, they are well-poised to support productivity tasks beyond the basic capabilities of voice calls and SMS. For example, it is becoming commonplace for smartphones to handle email, and increasingly these devices can be used for reviewing and even editing documents. Yet relatively little is known about how activities performed on the phone relate to productivity tasks that people perform at their more capable desktop and laptop computers.

This paper seeks a better understanding of tasks that span smartphones and other computing resources. Such knowledge is crucial for creating additive experiences for situations when devices of differing form factors and computing capabilities are used together. With such a range of affordances, it may be infeasible to simply replicate users' desktop computing experiences across all of their devices [2]. Instead, we envision systems that allow tasks to be seamlessly shared across devices, while offering interfaces that are appropriately tailored to the device and usage context. To better understand the opportunities and requirements for supporting the continuation of tasks across devices, we began by examining how information workers currently use their smartphones and personal computers.

H. Tokuda et al. (Eds.): Pervasive 2009, LNCS 5538, pp. 398–405, 2009.

The study of mobile work practices, and the challenges mobile users face in managing multiple devices, has been an active area of research for several years. In 2001, Perry et al. [3] studied the issues that mobile workers face in planning for and working during scheduled business trips, primarily aided by laptops, paper, and standard cell phones. Recently, Oulasvirta and Sumari [4] studied how members of a highly mobile IT workforce manage and select among a variety of co-located mobile devices (laptops, smartphones, and standard cell phones) in an effort to balance convenience, input efficiency, content readability and information accessibility throughout a typical workday. Dearman and Peirce [1] extended this investigation to a population of academics and industry workers, and documented how these users manage data and tasks across primarily their personal and work laptops and PCs.

Together these reports paint a picture of how the increasing power and portability of mobile devices can boost productivity, but can also lead to pragmatic challenges in management and resumption of tasks across devices. The role of smartphones in these scenarios, however, has received relatively little attention, particularly the extent to which phones can be leveraged to continue tasks started on the desktop, or start tasks that are then transferred back to more capable devices for completion.

To gain insight into the value of enabling the transfer of activities or tasks between devices of vastly differing capabilities, we examined both the temporal and data access patterns that people exhibit when using their smartphones and PCs over the course of a typical day. While prior studies have employed interviews and self-reporting [1,3,4], we deployed automated logging tools to track sixteen participants' usage of their primary work computer and smartphone across several weeks. The detailed usage characteristics we gathered can be used to complement the findings drawn from previous ethnographic methods. For example, Dearman and Pierce [1] found that users might benefit from sharing web browser state between users' devices; our data can help suggest how and when this state might be shared when one of the devices in question is a mobile phone.

2   Understanding Patterns of Multi-device Use

While information workers often have several computing devices at their disposal [4], we focused our investigation on users' interactions with their smartphone and primary work computer. Smartphones and work PCs are a particularly compelling duo because they differ considerably in their level of portability, support for traditional information work, and acceptability for extending use into non-work hours. To avoid the potential for bias associated with self-reporting as well as to gather detailed cross-device usage behavior, we conducted a logging-based field study.

2.1   Method

Our study consisted of a data collection phase where interaction activity was logged on each participant's smartphone and primary work computer. To gather descriptive complements to our raw data traces, as well as to help us understand opportunities for improving multi-device use, follow up interviews were conducted with four participants who demonstrated varying patterns of device usage (Table 1).

Apparatus. Two software systems were deployed to capture participants' computing activities. The PersonalVibe [5] logger runs on Windows machines and records both window-level events (e.g., titles, applications, active times, and durations) and the existence of keyboard and mouse activity. A smartphone version of the logger was developed for Windows Mobile devices to capture users' phone interactions such as application switches, calls, device locks, and web access.

Participants. Sixteen (12 male, 4 female) Windows Mobile smartphone users, with a median age of 34 (μ=33.9), were recruited through an internal mailing list at a technology company. Seven of the participants' phones had touchscreens. To ensure that we captured web access as one type of cross-device activity, we screened for regular (at least ten minutes per day) users of the mobile web. Four participants (3 male, 1 female) from a range of professional roles (publishing manager, software developer, product manager, and recruiter) and with a median age of 34.5 (μ=36.5), participated in a follow-up interview. Participants were compensated monetarily in accordance with the length of their participation.

Procedure. After providing consent to participate in the study, participants were instructed on how to remotely install the logging software on their work PC and personal mobile phone. Logging for each participant was scheduled for two weeks, but individuals' schedule constraints and interest levels resulted in participation that ranged from 5–30 (median=21) days, during August and September of 2008.

Follow-up interviews were semi-structured, took place in the participant's office or a nearby conference room, and were restricted to one hour. Participants were asked about their job role, percentage of time typically spent in meetings, the set of computers (phones, laptops, and desktops) they interacted with both at work and at home, the general types of tasks they performed on their phone, and whether their phone activities were primarily related to work, personal, or a combination. Next, participants were presented with a graphical representation of their phone and PC activity over the duration of the study (see Fig. 2) and were questioned about the actions performed on their devices for five key scenarios: "pre-work", "start of workday", "end of workday", "concurrent phone-PC use", and "interleaved phone-PC use." During each walkthrough, participants were asked to comment on how typical the behaviors we recorded were and what relationship, if any, existed between the activities they performed on each device. The interview concluded with questions designed to elicit ideas for improving the interoperability of such devices, such as what might be supported by sharing task context between devices.

3 Study Results

We structured our analysis of the log data into three stages. First, we looked only at the types of applications people ran on their devices. Next we examined only the temporal patterns of device access across each day. Last, we crossed these approaches by looking at the temporal patterns of two important activities, email and Web use.

3.1 Activity Types by Device

The desktop logger recorded an aggregate of 55,011 minutes of active PC use across 269 days, with a median of 17 days per participant. Fig. 1 shows the total number of application activations (in 1000s), by application type, that our loggers captured on each device type across all participants. Although the figures suggest the relative usage rates of different applications overall, the relationships varied by user and day of study. If we examine the usage duration of each application we find that email and web access dominated other activities on the desktop. Based on the actual duration of use, the corporate email program accounted for 33.8% of participant activity on the desktop and Internet Explorer accounted for 24.1% of the activity.

The phone logger recorded activity on 262 days with a median of 16.5 days per person. We found that email was by far the most common activity performed on the phone, accounting for 55.3% of the application activations over the study period. To allow for comparison with desktop behavior, we report phone-based email access as either corporate email (6319) or other/personal email (704). Calls were the next most frequent activity, at 10.8% of total activations. Web usage was predictably lower at 9.4%. SMS, typically considered a large percent of phone usage, accounted for only 6.4% of the activations. As an additional comparison, we found that users loaded a total of 34,155 web pages on their desktops versus 8,999 pages on their phones.

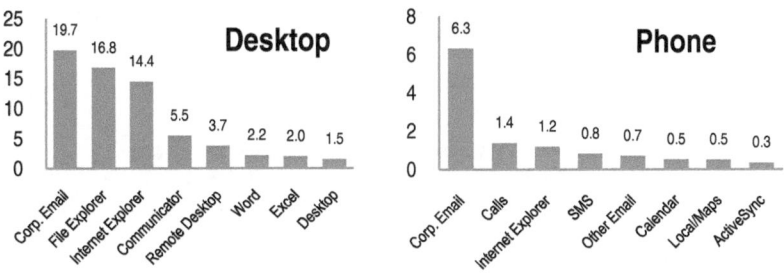

Fig. 1. Application use on the Desktop and the Phone (# of application activates in 1000s)

3.2 Interaction Patterns between and Across Devices

Next we examined the temporal patterns between participants' phone and PC usage. Fig. 2 depicts five consecutive workdays for three of our participants, and demonstrates the wide range of behaviors we observed in our logs. P9 used her phone for long periods at a time, nearly to the exclusion of her PC, and including hours typically reserved for sleeping. P10 used his PC extensively and sometimes into evening hours, but also interacted frequently with his phone for short bursts of time. Finally, P19 spent most of each day interacting with his desktop PC, accessing the phone relatively infrequently and generally between PC sessions.

Given the importance of email and the Web for productivity tasks and their high usage on both devices (Fig. 1) we decided to further investigate access patterns to these two applications. Fig. 3 illustrates the email and Web use for each device during one of P19's work days: 3a shows overall device usage, 3b shows the email and Web

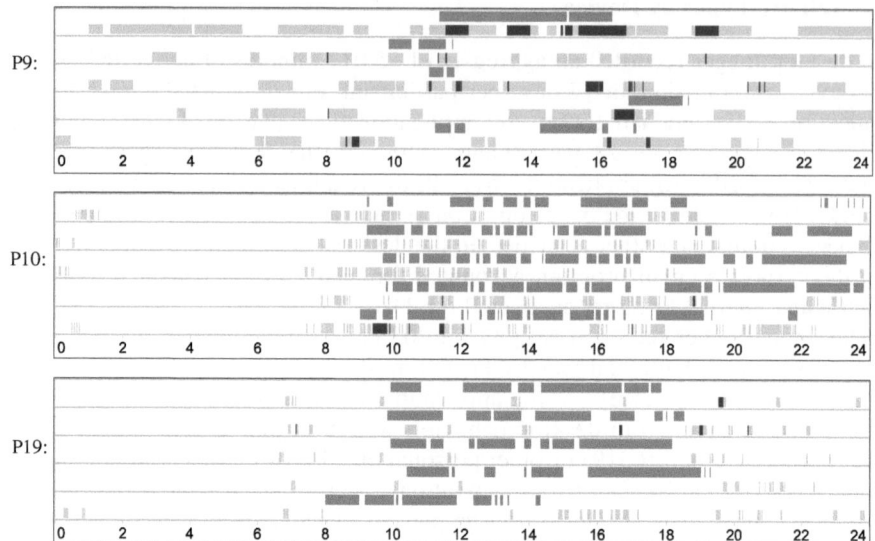

Fig. 2. Usage of the PC and phone during 5 consecutive work days for 3 different participants. For each day, the top line is desktop activity and the bottom line is phone activity. For the phone, light gray represents the phone being unlocked and black represents phone calls.

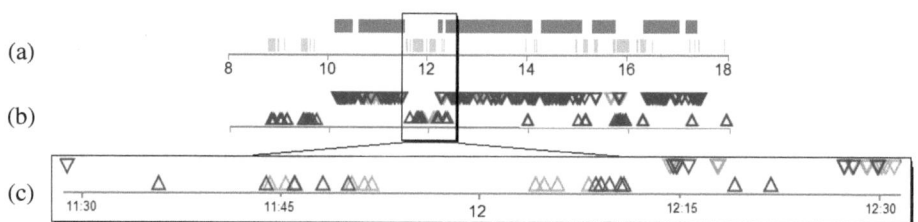

Fig. 3. (a) P19's PC and phone use during one workday; (b) email (dark gray) and web (light gray) access during the same period for each device; (c) a detailed view of P19's lunch hour

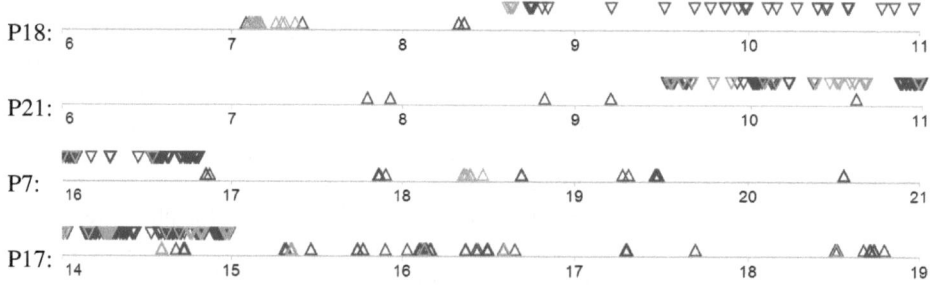

Fig. 4. Web and email use on both the PC (down pointing triangles) and phone (up pointing triangles) for four participants. Dark gray is email access and light gray is web access.

activity during that time period, and 3c shows the hour from 11:30am to 12:30pm at high detail. Our follow-up interview confirmed that Fig. 3c represents a lunch break, with the participant leaving the office just before 11:30, checking email on the phone at 11:35, again at 11:43, and then followed by a combination of web surfing and email triage until 11:53. Email and Web access resume on the phone at 12:04 and continue until P19 returns to his desk just before 12:15. This example demonstrates users' device transitions, and their persistent connection to email and the Web.

Fig. 4 illustrates common patterns of phone access we observed at opposite ends of the workday. P18 and P7 exemplify pre- and post-work activity, respectively, and in both cases we find that participants access work email outside of typical work hours. P18's morning includes frequent accesses to work email, starting shortly after waking up and continuing after the morning commute. At the end of the work day, P7 accesses email on the way out of the office, shortly after the commute home, and periodically throughout the evening. P21 and P17 demonstrate similar morning and evening patterns, even though their work hours are less typical than P18's and P7's.

These usage patterns were not relegated to these participants alone. Across all participants, on the 226 days in which both devices were used in the same day, the phone was used before the desktop 88.1% of the time. In addition, in 87.6% of those days, the phone was accessed after the last time the desktop was used. A systematic review of all participants' usage throughout the day led us to conclude that people do this not merely in extraordinary situations, but as part of their daily routines. Even people who do not use their phone heavily still generally touch it every morning, evening, and when they are away from their PC for extended periods.

3.3 Interview Findings

While the above data revealed a variety of compelling usage patterns, a logging study alone cannot provide insight into user intentions, needs, or desires which might then be used to inform the design of supportive technologies. For example, understanding the daily balance that participants maintain between work and personal use of the phone, together with feedback on how their current phone activities differ from their *preferred* activities, could suggest quite different directions for cross-device system design. We therefore conducted follow-up interviews to help us develop a more complete representation of multi-device use. In order to capture as wide a range of responses as might be expressed in our participant pool, we selected four participants who exhibited distinctive device usage patterns (Table 1), and served different roles within the company. Despite our efforts at diversity, we were surprised at the consistency of user responses, the most striking of which are summarized below.

Continuous Connectivity: During the interviews we discovered all four informants used their phones primarily or extensively for work purposes, and reported "always" keeping their phone with them. All informants expressed that the most valuable role of the phone was for accessing and responding to work and personal email, so much so that it was the first thing they each checked upon waking up. This confirms our observations from the logged data. Perhaps unsurprisingly, all gave work email higher priority than personal email, rarely accessing their phones to check personal email only. As P21 put it, "none of [my personal email] is really urgent. If somebody needs me [...] they'll text me." Having continuous phone-based access to work email was viewed very positively, helping users stay current (P19: "If I'm headed to work, then I usually want to know if I'm headed into any problems."), and on top of their inboxes (e.g., P7: "The phone's always with me and I frequently triage emails with it"; P4: "I'm a little OCD when it comes to my inbox. So, if I can resolve it on my phone, then I just do it").

Although we might expect the bleeding of work into personal time to cause stress, the continuous access to work email instead seemed to provide peace of mind, (e.g., P21: "I'll wake up in the middle of the night, check [work] email, if it's something I can answer right then, I'll go ahead and do it. It's more convenient to go ahead and do it, and then I can forget about it"; P7: "I triage emails in the evening for work. [...] If I can answer it quickly and I have the time, I might just punch it out"). These comments help explain our observations from the log traces: the phone is being used to help keep a close watch on email throughout the day because users prefer to be kept up-to-date.

Table 1. Profiles for interviewed participants

		Studied devices, *other owned devices*	Level of phone use: primary work purpose Sample 24 hour Timeline (as described in Fig. 2)
P7	M 35	**HTC Tilt, work PC**, *work laptop, home PC & laptop*	Heavy during day, tapers at night: email triage, calendar
P9	F 34	**Blackjack II, work laptop**	Heavy, 24 hr: email read/writing, doc review, calendar
P19	M 25	**HTC Tilt, work PC**, *home PC, home laptop*	Light but consistent across day: email awareness
P21	M 52	**HTC Touch, work laptop**, *home PC, home laptop*	Modest during day and night: email triage, calendar, news

Fast and Always On. The advantages that the participants felt the phone offered over their PC echoed those that have been reported previously (e.g., [1,4]). Besides the phone's high portability, informants often found it was faster and easier to check email and calendar on their phones, even while sitting near their PCs. Reasons included nonexistent boot-up, fast password entry, high density and at-a-glance information presentation, and discretion in meeting settings.

Phone as Primary Computer. For one informant (P19), the phone was clearly only used to fill in the gaps between access to a more capable system. The others instead seemed to prefer the ultra-portability their phone offered, and only fell back to desktops and laptops when absolutely necessary. In their own words: P7: "I've pretty much selected the most powerful one they make. I want this to be powerful enough to do all the tasks I want", P21: "If I have this [phone], I can go a long time without my laptop", and P9: "I'd just prefer to use my phone for everything."

Cross-Device Data, not Tasks. Because we consistently observed patterns of activity hand-off between phones and PCs in the data (e.g., Figs. 3 and 4), we asked users about the tasks they typically performed at these transition points, and ways we might create more seamless experiences during their shifts between devices. To our surprise, but as noted in previous research [1], participants rarely if ever engaged in an activity on one device that they were interested in continuing on another, and expressed very little interest in the type of "hand-off" capability we suggested. Of course, email and calendar were clearly being used extensively on both devices, but interviewees felt the current synchronization tools were sufficient for their needs. Rather than having their devices work in concert to stitch an activity together, the "seamlessness" participants desired was much more about data synchronization and universal data access. Participants were otherwise proficient and content with using the phone as an independent, auxiliary communication and information channel.

4 Summary

Support for the notion that people are using their phones extensively throughout the day was abundant, both from our logged data and from the interviews. Although this finding might seem obvious, we believe that the extent to which this happens in practice has not been fully appreciated. Far from "keeping work and personal life separate" we find that our participants were, and in fact *preferred* to be, continually connected to their work email. Whereas in the past this might be associated with a person obsessed with work, it now seems that people are engaging in this activity to maintain a sense of calm and control in their work lives during their personal time.

We observed that the phone is emerging as a primary computing device for some users, rather than as a peripheral to the PC. The amount of email and web activity recorded in the logs alone supports this, but in addition, each of the people we interviewed stated that they wanted their smartphone to be as powerful as a laptop, and in fact, frequently preferred it to the overhead of using their laptop.

Finally, the concept of task carryover between the phone and the PC was not widely embraced by our participants. Log analysis and interview feedback suggest that tasks were either completed on the phone or delayed until the PC was available. Of course with current technology, such a crossover is inherently difficult, and so we believe there is still value and opportunity for further development in this arena.

References

1. Dearman, D., Pierce, J.: "It's on my other computer!": Computing with Multiple Devices. In: Proc. CHI 2008, pp. 767–776. ACM Press, New York (2008)
2. Satyanarayanan, M., Zoxuch, M., Helfrich, C., O' Hallaron, D.: Toward seamless mobility on pervasive hardware. In: Pervasive and Mobile Computing, vol. 1, pp. 157–189. Elsevier, Amsterdam (2005)
3. Perry, M., O'Hara, K., Sellen, A., Harper, R., Brown, B.: Dealing with Mobility: Understanding access anytime, anywhere. In: ACM TOCHI, vol. 8(4), pp. 323–347. ACM Press, New York (2001)
4. Oulasvirta, A., Sumari, L.: Mobile kits and laptop trays: managing multiple devices in mobile information work. In: Proc. CHI 2007, pp. 1127–1136. ACM Press, New York (2007)
5. Brush, A.J.B., Meyers, B.R., Tan, D.S., Czerwinski, M.: Understanding memory triggers for task tracking. In: Proc CHI 2007, pp. 947–950. ACM Press, New York (2007)

Author Index

Abowd, Gregory D.   115
Åkesson, Karl-Petter   220
Alt, Florian   9

Beckwith, Richard   176
Ben Allouch, Somaya   77
Borriello, Gaetano   59
Brännström, Andreas   354
Brown, Lorna M.   133
Buzeck, Markus   1, 17

Chang, Keng-hao   151
Choudhury, Tanzeem   176
Clauset, Aaron   342
Clear, Adrian K.   327
Consolvo, Sunny   176

Dey, Anind K.   168
Dobson, Simon   327

Eagle, Nathan   342
Evers, Christoph   9
Exeler, Juliane   1, 17

Golle, Philippe   390
Grzeszczuk, Radek   59
Gupta, Manu   95

Harle, Robert   238
Hightower, Jeffrey   151, 176
Hile, Harlan   59
Holland, Thomas   327
Hornecker, Eva   42
Hudson, Scott E.   168

Ilic, Alexander   291
Intille, Stephen S.   95

Jacobs, Andy   398
Jay, Tim   1
Johns, Paul   398
Jonsson, Staffan   220

Kane, Shaun K.   398
Karlson, Amy K.   398
Karlson, Bo   354
Kartal, Ersun   372
Kern, Dagmar   42
Kientz, Julie A.   115
Klasnja, Predrag   176
Košecka, Jana   59
Konomi, Shin'ichi   202
Krüger, Antonio   1, 17
Krumm, John   25
Kunze, Kai   372
Kuznetsov, Stacey   168
Kveton, Branislav   151

Larson, Kent   95
Lehtonen, Mikko   291
Lindt, Irma   220
Liu, Alan   59
Ljungstrand, Peter   220
Lukowicz, Paul   372
Lundquist, Terés   354
Lyons, Kent   184

Marshall, Paul   42
Meyers, Brian R.   398
Michahelles, Florian   291
Molloy, Mike   133
Morales Kluge, Ernesto   372
Müller, Jörg   1, 17

Nixon, Paddy   327
Nylander, Stina   354

Ostojic, Daniel   291

Partridge, Kurt   390
Patel, Shwetak N.   256
Pering, Trevor   184
Peters, Oscar   77
Pirttikangas, Susanna   202

Quigley, Aaron   327
Quinn, John A.   342

Robertson, Thomas   256
Rogers, Yvonne   42
Rosario, Barbara   184

Schmidt, Albrecht   1, 9, 42
Schmidtke, Hedda R.   309
Scott, James   133
Seshan, Srinivasan   274
Sezaki, Kaoru   202
Shannon, Ross   327
Sheth, Anmol   274
Stuntebeck, Erich P.   256
Sud, Shivani   184
Suzuki, Ryohei   202

Thepvilojanapong, Niwat   202
Tobe, Yoshito   202

van Dijk, Jan A.G.M.   77
Vedantham, Ramakrishna   59

Waern, Annika   220
Wagner, Florian   372
Want, Roy   184
Wetherall, David   274
Wetzel, Richard   220
Wilmsmann, Dennis   1
Woo, Woontack   309
Woodman, Oliver   238